Mechanical Properties of Asphalt and Asphalt Mixtures

Mechanical Properties of Asphalt and Asphalt Mixtures

Guest Editors

Guoqiang Liu
Yanqiu Bi
Xiaokang Zhao

MDPI

Basel • Beijing • Wuhan • Barcelona • Belgrade • Novi Sad • Cluj • Manchester

Guest Editors

Guoqiang Liu
School of Mechanics and Civil Engineering
China University of Mining and Technology
Xuzhou
China

Yanqiu Bi
School of Civil Engineering
Chongqing Jiaotong University
Chongqing
China

Xiaokang Zhao
School of Highway
Chang'an University
Xi'an
China

Editorial Office
MDPI AG
Grosspeteranlage 5
4052 Basel, Switzerland

This is a reprint of the Special Issue, published open access by the journal *Buildings* (ISSN 2075-5309), freely accessible at: https://www.mdpi.com/journal/buildings/special_issues/53D8JJ17B2.

For citation purposes, cite each article independently as indicated on the article page online and as indicated below:

Lastname, A.A.; Lastname, B.B. Article Title. *Journal Name* **Year**, *Volume Number*, Page Range.

ISBN 978-3-7258-7945-8 (Hbk)
ISBN 978-3-7258-7946-5 (PDF)
https://doi.org/10.3390/books978-3-7258-7946-5

Contents

About the Editors

Guoqiang Liu

Guoqiang Liu is a distinguished researcher affiliated with the School of Mechanics and Civil Engineering at the China University of Mining and Technology in Xuzhou, China. His academic and professional interests are rooted in asphalt materials and advanced pavement engineering. Throughout his career, Dr. Liu has focused on exploring the fundamental mechanical behaviors of pavement infrastructure to enhance material durability and performance. His ongoing research initiatives continue to contribute to the optimization of road construction, management, and maintenance strategies.

Yanqiu Bi

Yanqiu Bi is an accomplished researcher based at the School of Civil Engineering, Chongqing Jiaotong University, in Chongqing, China. Her extensive research interests cover a wide array of road engineering topics, including asphalt pavement performance, durable pavement design, and the complex rheological characteristics of paving materials. In addition to traditional pavement engineering, Dr. Bi actively engages in exploring innovative fields such as smart road technologies, energy conservation, and vehicle exhaust purification systems. Her multidisciplinary approach is instrumental in the development of sustainable transportation infrastructure.

Xiaokang Zhao

Xiaokang Zhao serves as a dedicated researcher at the School of Highway within Chang'an University in Xi'an, China. His primary area of expertise lies in pavement materials and the detailed mechanical evaluation of asphalt pavements. Dr. Zhao's specific research interests include the study of fatigue damage, meso-scale simulation, and micro-scale characteristics utilizing the discrete element method. By investigating mesoscopic cracking and complex failure mechanisms, his work provides important insights into the structural integrity of pavements and thereby supports the design of more resilient highway systems.

Preface

This Reprint gathers twelve peer-reviewed articles originally published in the Special Issue "Mechanical Properties of Asphalt and Asphalt Mixtures" published by *Buildings*. The primary purpose of this collection is to showcase the most recent achievements in the research of the physical, mechanical, and rheological properties of asphalt materials. Since it is driven by the need to improve the durability and sustainability of highway infrastructure, this compilation is aimed towards researchers, pavement engineers, and materials scientists. The contents span from micro-scale characteristics and discrete element method simulations to the practical application of recycled asphalt and smart pavement technologies. We extend our sincere gratitude to all contributing authors for contributing their valuable work and to the rigorous peer reviewers who helped to ensure the high scientific quality of this Reprint.

Guoqiang Liu, Yanqiu Bi, and Xiaokang Zhao
Guest Editors

buildings

MDPI

Article

Study of Pavement Performance and Temperature Regulation Capacity of Asphalt Binders Modified with Dual-Phase-Change Materials

Donghui Huo [1], Debao Hou [1], Shuaixiang Zhang [2,3], Wei Gao [1], Changchun Yu [1], Ling Jia [1], Baolin Chang [1], Rui Zhang [2,3] and Meng Guo [2,3,*]

1 Hebei Jiaotou Road and Bridge Construction and Development Co., Ltd., Qinhuangdao 066004, China; qhdlqbgs@126.com (D.H.); m17733508639@163.com (D.H.); 13703356931@139.com (W.G.); 13933646818@139.com (C.Y.); m13933532824@163.com (L.J.); changbaolin2005@163.com (B.C.)

2 State Key Laboratory of Bridge Engineering Safety and Resilience, Beijing University of Technology, Beijing 100124, China; zhangshuaixiang@emails.bjut.edu.cn (S.Z.); zhangrui19971104@163.com (R.Z.)

3 The Key Laboratory of Urban Security and Disaster Engineering of Ministry of Education, Beijing University of Technology, Beijing 100124, China

* Correspondence: gm@bjut.edu.cn; Tel.: +86-18001205566

Abstract: Due to the temperature changes caused by seasonal changes and extreme weather, asphalt pavement suffers from rutting, cracking, and other damage. With commonly used pavement additives, the high-temperature performance and the low-temperature performance of asphalt pavement show opposite trends, with related research endeavoring to find a balance between the two. In this study, a dual-phase-change material (DPCM) with both high- and low-temperature effects was prepared. The chemical stability and rheological properties of modified asphalt binders were characterized using Fourier transform infrared spectroscopy (FTIR) and a dynamic shear rheometer (DSR). Temperature control tests of the DPCM-modified asphalt binders were carried out with an indoor simulation device. The results show that the DPCMs could improve the rutting resistance of the asphalt binders at a high temperature, but the fatigue performance of the modified asphalt binder with different DPCM contents was reduced. The FTIR results showed that no chemical reaction occurred in the mixing of the asphalt binder and the DPCM. In the indoor simulation temperature control test, the 40% DPCM-content-modified asphalt binder reduced the high-temperature extreme value by 4.2 °C and increased the low-temperature extreme value by 2.5 °C, showing a good temperature control effect and practical application value.

Keywords: phase-change materials; rheology; modified asphalt; rutting resistance

1. Introduction

Asphalt binder is a viscoelastic material with high-temperature sensitivity, and it is negatively affected by temperature changes in practical engineering applications. Asphalt pavement is prone to rutting when subjected to vehicle loads in high-temperature environments [1,2]. In winter, it cracks easily in low-temperature environments [3,4]. Meanwhile, cyclic temperature stress also hardens asphalt binder and accelerates the aging of asphalt pavement, thus reducing its service life [5].

In recent years, researchers have developed many effective technologies to alleviate the damage to asphalt pavement caused by temperature changes. These include heat-reflective technology [6], water retention and cooling technology [7], thermal resistance technology [8], and energy conversion technology [9,10]. However, these temperature control technologies can only achieve either high- or low-temperature control; they cannot achieve both a delay in the road temperature rise in high-temperature environments and a delay in the road temperature drop in low-temperature environments. Obviously, this limited temperature regulation is not applicable to monsoon climate areas with distinct seasons and a large temperature range. As an energy storage, temperature regulation technology that has been widely used in recent years, phase-change materials (PCMs) have the potential to delay the temperature rise of road surfaces in high-temperature environments and delay the temperature drop of road surfaces in low-temperature environments [11–13].

At present, there are many types of phase-change materials, and, according to their different phase-change forms, PCMs can be divided into four categories: solid–solid, solid–liquid, solid–gas, and liquid–gas [14]. Among these, researchers consider solid–liquid organic phase-change materials to be the most effective heat storage materials because of their large phase-change enthalpy values and small volume changes before and after phase change [15,16].

Paraffin compounds are one of the most popular PCMs for asphalt pavement. Ryms et al. analyzed the temperature variation rule of asphalt mixtures containing paraffin in the aggregate. They found that the temperature of an asphalt mixture made of aggregate containing 20% paraffin was 7~8 K lower than that of a common asphalt mixture [17]. However, it has been found that the thermal conductivity of paraffin compounds is low and that the temperature regulation efficiency is limited. Adding materials with a high thermal conductivity can improve the temperature regulation efficiency of paraffin compounds [18]. Using a differential scanning calorimeter (DSC), Chen et al. measured the melting temperature of lauric acid, finding that it ranged from 43 to 44 °C and that the latent heat of phase transition was about 160 kJ/kg, showing that it could be used for the temperature control of asphalt pavement [19]. Through scanning electron microscopy (SEM) and DSC tests, Kong et al. found that lauric acid has good compatibility with asphalt binders. A lauric acid-modified asphalt binder showed good heat absorption and release performance [20]. Polyethylene glycol (PEG) is an organic polymer whose phase-transition temperature, enthalpy of melting, and enthalpy of crystallization depend on its molecular weight. At present, PEGs with molecular weights of 2000 [21] and 4000 [22] are widely used to reduce the rutting of asphalt pavement at high temperatures, while PEG with a molecular weight of 400 is used to alleviate icing and snow on asphalt pavement in winter [23].

Many researchers have investigated the phase transition mechanism of PCMs. Studies have shown that most organic PCMs control the temperature by melting and by absorbing heat during temperature rise [24]. In the cooling stage, Anupam et al. believe that the solidification process of PCMs mainly includes three stages: sensible heat release, latent heat release, and sensible heat release. The solidification process of PCMs begins with the formation of initial crystals or initial nuclei, which is called nucleation [14]. Kuznik's research found that some PCMs undergo the phenomenon of overcooling. Due to the low nucleation rate, PCMs remain in the liquid state below the melting temperature, which is not conducive to the PCM's role in regulating temperature within a certain range [25]. Fan et al. found that paraffin and alcohol can effectively promote the formation of PCM crystals, thereby reducing the overcooling of PCMs [26].

In conclusion, solid–liquid organic PCMs, such as paraffin wax, fatty acids, and PEG, can be applied to asphalt pavement. However, the low thermal conductivity of solid–liquid organic PCMs in the application process affects the thermal response speed in asphalt pavement, thus affecting its temperature regulation efficiency.

The objectives of this study were to (1) prepare a dual-phase-change material with both high- and low-temperature regulation abilities and to test the pavement performance of modified asphalt, and to (2) clarify the optimal dosage of dual-phase-change materials through an indoor simulation temperature control test to verify their practical application value.

2. Materials and Methods

2.1. Experimental Materials

2.1.1. Asphalt Binder

A modified asphalt binder specific for an ultra-thin surface was selected and tested according to the American Society of Testing Materials (ASTM) standard. The results are shown in Table 1.

Table 1. Basic performance index of asphalt binder.

Index	Unit	Result	Specification Requirement	Standard
Penetration (25 °C)	0.1 mm	57	≥50	ASTM D5 [27]
Softening point	°C	84	≥65	ASTM D36 [28]
Rotary viscosity (135 °C)	Pa·s	2.725	≤3.0	ASTM D4402 [29]
Ductility (5 °C)	cm	27	≥20	ASTM D113 [30]

2.1.2. Dual-Phase-Change Material (DPCM)

Beijing has a typical temperate continental monsoon climate. Its monthly temperature trend throughout the year is shown in Figure 1. It can be seen in Figure 1 that the temperature in Beijing is below 0 °C for five months in a year. Similarly, there are five months in which the temperature is above 25 °C. In such months, the daytime temperature of the road surface can reach more than 45 °C [31]. In order to solve the problems of the low thermal conductivity and low thermal storage efficiency of PEG, in this study, externally expanded graphite (EG) was chosen as a carrier matrix to adsorb PEG with different molecular weights. A PEG/EG composite phase-change material was prepared, which encapsulated PEG and improved its thermal storage efficiency. A PEG400/EG composite phase-change material was prepared for paving at 0 °C. A PEG1500/EG composite phase-change material was used for pavements at 45 °C and above. Finally, the two composite phase-change materials were mixed at a ratio of 1:1 to obtain a dual-phase-change material, as shown in Figure 2. The dual-phase-change material was a black granular solid with good dispersion, and its apparent density was 1.07 g/cm^3.

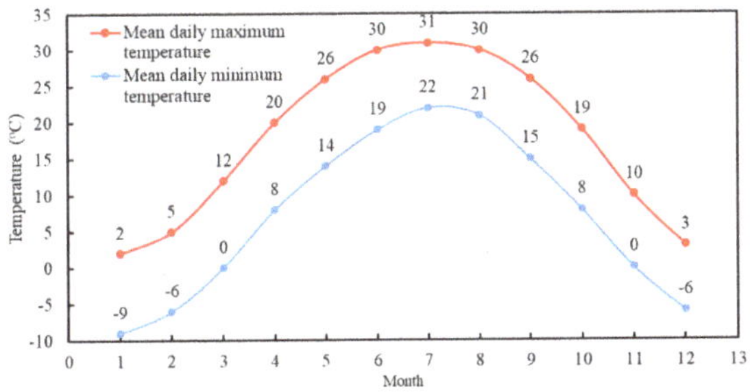

Figure 1. Monthly temperature trend of Beijing throughout the year.

Figure 2. Dual-phase-change material.

2.1.3. DPCM-Modified Asphalt

The DPCMs were directly mixed with asphalt binders according to the mass ratios of 10%, 20%, 30%, and 40% to prepare DPCM-modified asphalt binders. The specific processes were as follows: The asphalt binder was heated in an oven at 165 °C to reach the flowing state and then transferred to a constant temperature oil bath at 170 °C. The DPCMs were slowly poured into the asphalt binders during the oil bath and stirred with a glass rod at a speed of 60 r/min. After that, a high-torque electric mixer was used to stir at a speed of 500 r/min for 20 min. Finally, a glass rod was used to manually stir at a speed of about 60 r/min for 10 min to remove the bubbles in the DPCM-modified asphalt binders.

2.2. Test Methods

2.2.1. Temperature Scanning Test

A DHR-1 instrument produced by TA Company in the USA was used to conduct a temperature scanning test on the various asphalt binders modified with dual-phase-change materials. Parallel plates with a diameter of 25 mm and a thickness of 1000 μm were selected to test the rutting factor. Plates with a diameter of 8 mm and a thickness of 2000 μm were selected to test the fatigue factor. Plates with a diameter of 4 mm and a thickness of 2200 μm were selected to test the low-temperature performance. The test parameters were as follows:

- Temperature range: 30~70 °C (high-temperature performance), 10~40 °C (fatigue performance), and −20~10 °C (low-temperature performance).
- Strain: 0.01% (for −20~−5 °C) and 1% (for −5~70 °C).
- Scanning frequency: 10 rad/s.

The complex modulus frequency curve and phase angle frequency curve at each temperature were adapted according to the time–temperature superposition principle (TTSP).

2.2.2. Scanning Electron Microscope (SEM) Test

The microstructures of the EG and PEG/EG composite PCMs were studied via scanning electron microscopy (SEM). The electron microscope model was SU9000. The test parameters were as follows:

- Acceleration voltage: 15,000 V.
- Magnification: 1500 times.

2.2.3. Fourier Transform Infrared Spectrometer (FTIR)

A spectrum II infrared spectrometer produced by PE Company of the United States was selected for the test. The spectrum had a resolution of 0.5 cm^{-1} and a test range of 4000–500 cm^{-1} [32]. A heated spatula was dipped into the asphalt binder sample and then used to evenly apply the sample to a diamond crystal for scanning. The result was analyzed using OMNIC.

2.2.4. Temperature Detection and Recording Device

A PT100 platinum resistance sensor and temperature recorder (as shown in Figure 3) were selected to record and analyze the temperature control effect of the dual-phase-change materials on the asphalt binders.

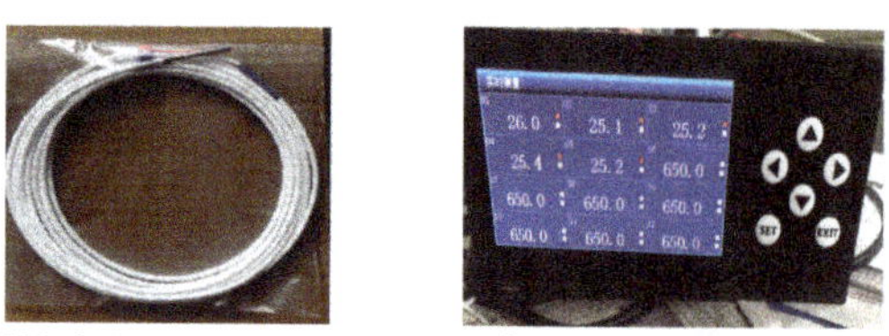

(a) PT100 platinum resistance sensor **(b)** Temperature recorder

Figure 3. Temperature detection and recording device for temperature regulation test.

- Recording frequency: once per minute.
- Temperature deviation: ±0.1 °C.

2.2.5. Indoor Simulated Temperature Regulation Test

An indoor simulated temperature regulation test was carried out by using an environment chamber. The steps were as follows: (1) 300 mL of the DPCM-modified asphalt binders with different DPCM contents was divided into 5 beakers, and then the beakers were heated to reach a flowing state. The metal probe of the PT100 platinum resistance sensor was inserted into the asphalt binder of each beaker and kept in the middle position. (2) All the beakers were put into the environmental chamber. The initial temperature of the high-temperature regulation test was 25 °C, and it was 20 °C in the low-temperature regulation test. (3) In the high-temperature regulation test, the DPCM-modified asphalt binders with a stable temperature were heated from 25 °C to 65 °C. When one sample reached 65 °C, all samples were cooled from 65 °C to 25 °C. In the low-temperature regulation test, the DPCM-modified asphalt binders with a stable temperature were cooled from 20 °C to −20 °C. When one sample reached −20 °C, all samples were heated from −20 °C to 20 °C. The temperature changes in the DPCM-modified asphalt binders were recorded per minute using a temperature recorder. The main process of the indoor simulation temperature regulation test is shown in Figures 4 and 5.

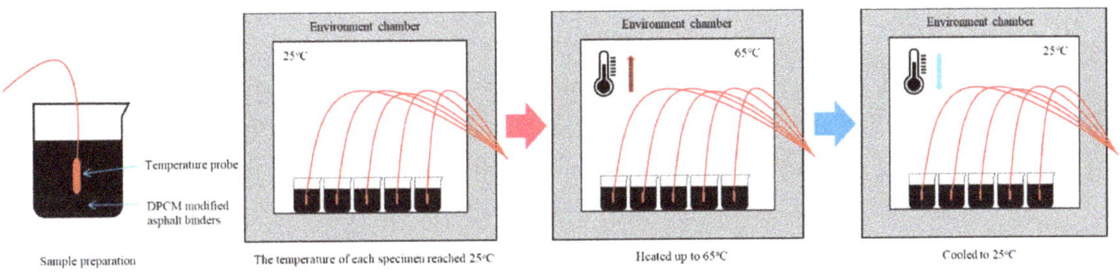

Figure 4. Flowchart of indoor simulated high-temperature regulation test.

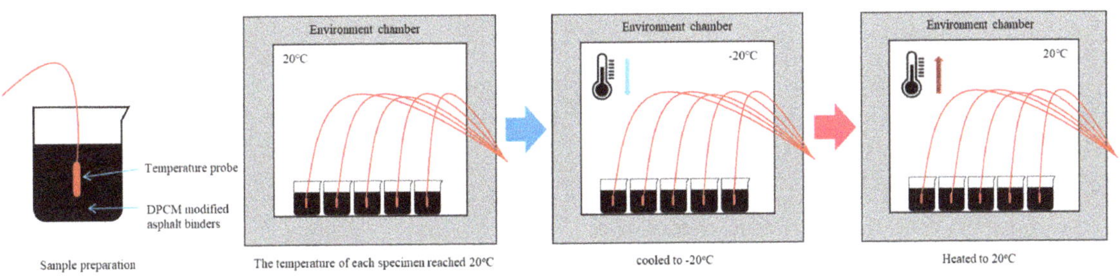

Figure 5. Flowchart of indoor simulated low-temperature regulation test.

3. Results and Discussion

3.1. Microstructures of EG and PEG

Figure 6 shows that, during the melt blending and vacuum adsorption, a large amount of liquid PEG entered the pore structure of EG under the actions of capillary force and vacuum negative pressure. EG was connected closely to a network of structures, which were conducive to improving the thermal conductivity efficiency of PEG/EG, thus improving the temperature regulation effect of the DPCM.

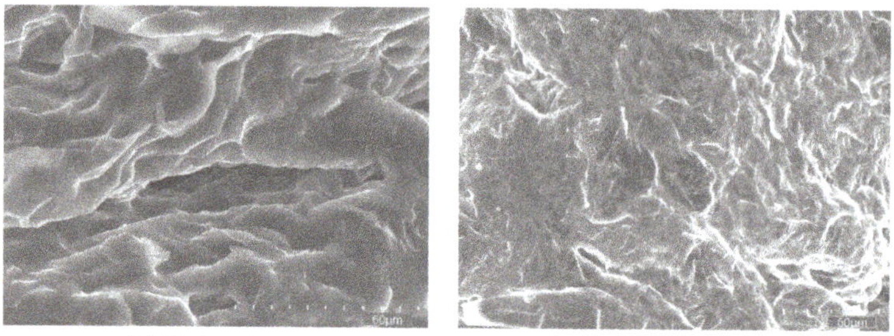

(**a**) PEG400/EG (**b**) PEG1500/EG

Figure 6. Microstructures of EG and PEG.

3.2. Rheological Properties of DPCM-Modified Asphalt Binders

DSR was used to study the influence of the DPCM content on the viscoelastic properties of the asphalt binders. The road performance of the modified asphalt binders at different temperatures was evaluated.

3.2.1. High-Temperature Performance of DPCM-Modified Asphalt Binders

$$\text{Rutting factor} = G^*/\sin\delta$$

$$\text{Fatigue factor} = G^* \times \sin\delta$$

In the Superpave specification, the rutting factor can effectively characterize the high-temperature stability of asphalt binders. An asphalt binder with a larger rutting factor has a better rutting resistance.

The fatigue factor represents the energy lost by asphalt samples under repeated loading. An asphalt binder with a smaller fatigue factor has a better fatigue resistance under repeated loading.

The changes in the complex modulus, phase angle, and rutting factor of the DPCM-modified asphalt binders with different DPCM contents in the temperature range of 30~70 °C were studied using a temperature scanning test. The test results are shown in Figures 7 and 8.

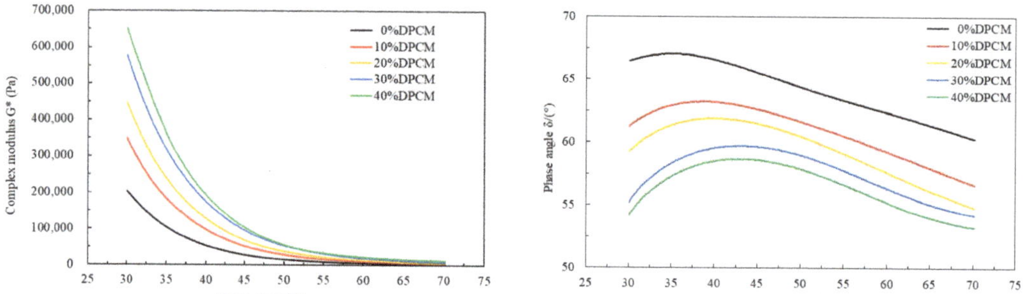

(**a**) Complex modulus (**b**) Phase angle

Figure 7. Effect of DPCM content on complex modulus and phase angle of DPCM-modified asphalt binders at high temperatures.

As shown in Figures 7 and 8, in the temperature range of 30~70 °C, the difference in the complex modulus of each DPCM-modified asphalt binder gradually decreased with the increase in the temperature. Meanwhile, the phase angle first increased and then decreased with the increase in the temperature, which was mainly related to the viscoelastic properties of the original asphalt [33]. At the same temperature, the complex modulus of the modified asphalt increased, and the phase angle decreased with the increase in the DPCM content, which led to an increase in the rutting factor. The physical property of the DPCM was similar to that of the mineral powder, which improved the viscosity performance of the modified asphalt binders. This is one of the reasons for the improvement of the rutting resistance of the DPCM-modified asphalt. In addition, no effect of the phase change on the rheological properties of the asphalt binders was found during the heating process, which was due to the greater function of the DPCM than the mineral powder in the asphalt binders. Compared with the function of the mineral powder, the effect of the phase transformation on the rheological properties at high temperatures was too slight to be observed.

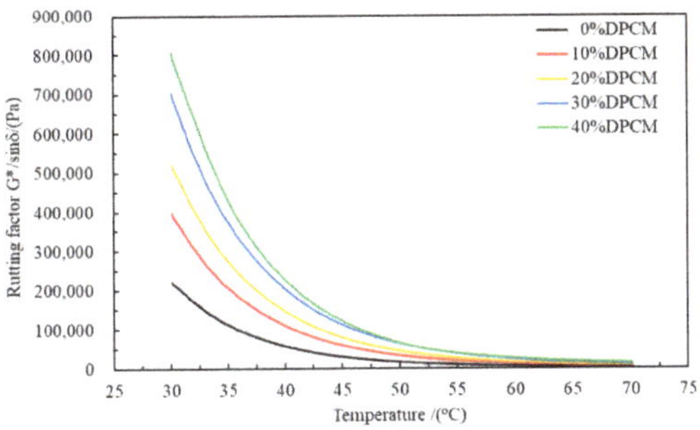

Figure 8. Effect of DPCM content on rutting factor of DPCM-modified asphalt binders at high temperatures.

3.2.2. Fatigue Properties of DPCM-Modified Asphalt Binders

The changes in the complex modulus, phase angle, and fatigue factor of the DPCM-modified asphalt binders with different DPCM contents in the temperature range of 10~40 °C were studied using a temperature scanning test. The results are shown in Figures 9 and 10.

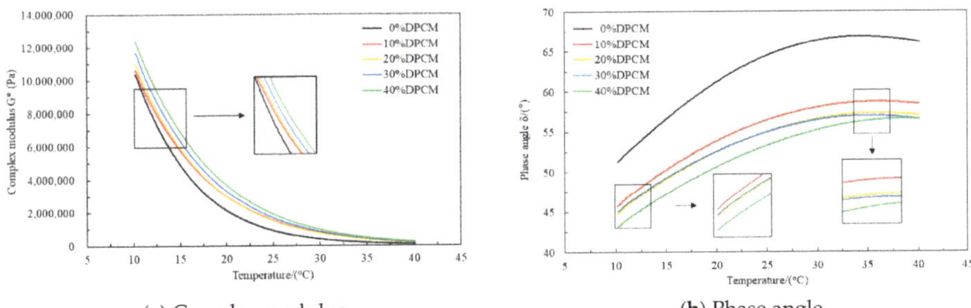

(**a**) Complex modulus (**b**) Phase angle

Figure 9. Effect of DPCM content on complex modulus and phase angle of DPCM-modified asphalt binders at medium temperatures.

As shown in Figures 9 and 10, in the temperature range of 10~40 °C, with the increase in the DPCM content, the complex modulus of the modified asphalt binders increased. The phase angle of the asphalt binder with the DPCM was significantly lower than that of the virgin asphalt binder at each temperature. This is due to the physical morphology of the DPCM being similar to that of the mineral powder. The powder–binder ratio increased with the content of the DPCM. The DPCM itself does not have viscosity, so the viscous component of the asphalt mastic decreased, and the elastic component increased. The fatigue factor was the product of G* and sinδ, which increased with the content of the DPCM. This means that the asphalt mastic is more inclined to crack under repeated loading, and the fatigue life became worse. Therefore, the content of DPCM in asphalt binders should not be too high.

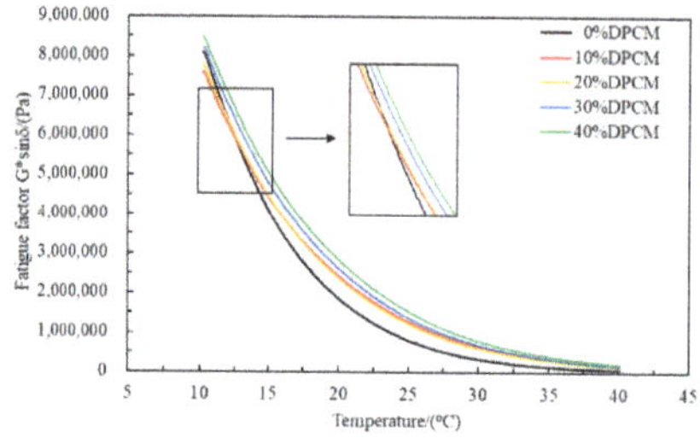

Figure 10. Effect of DPCM content on fatigue factor of DPCM-modified asphalt binders.

3.2.3. Low-Temperature Properties of DPCM-Modified Asphalt Binders

The changes in the complex modulus and phase angle of the DPCM-modified asphalt binders with different DPCM contents in the temperature range of −20~10 °C were studied using a temperature scanning test. The result is shown in Figure 11.

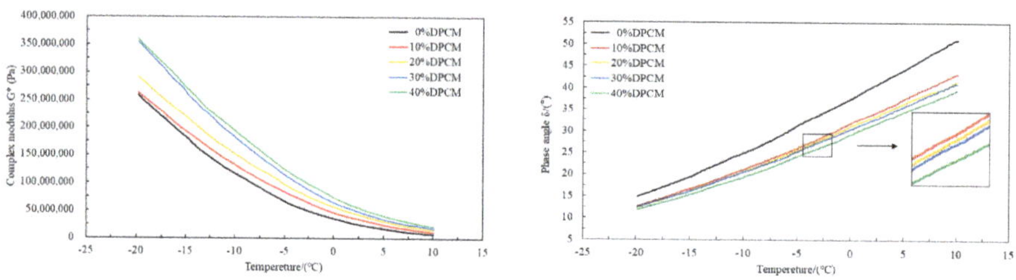

(**a**) Complex modulus

(**b**) Phase angle

Figure 11. Effect of DPCM content on complex modulus and phase angle of DPCM-modified asphalt binders at low temperatures.

As shown in Figure 11, in the temperature range of −20~10 °C, the complex modulus of the DPCM asphalt increased, and the phase angle decreased with the increase in the content of the DPCM. This shows that the influence of the amount of DPCM on the complex modulus and phase angle in this temperature range is consistent with that in the high-temperature and medium-temperature ranges. At low temperatures, the DPCM still mainly assumed the role of mineral powder, and its thermodynamic properties had little influence on the rheological properties of the asphalt binders.

Sui et al. proposed a test method using DSR to test the low-temperature performance of asphalt binders. There was a strong linear correlation between the test data and BBR test data [34]. On the basis of the existing data, the m values and stiffness at −6 °C, −12 °C, and −18 °C were calculated. The results are shown in Figure 12.

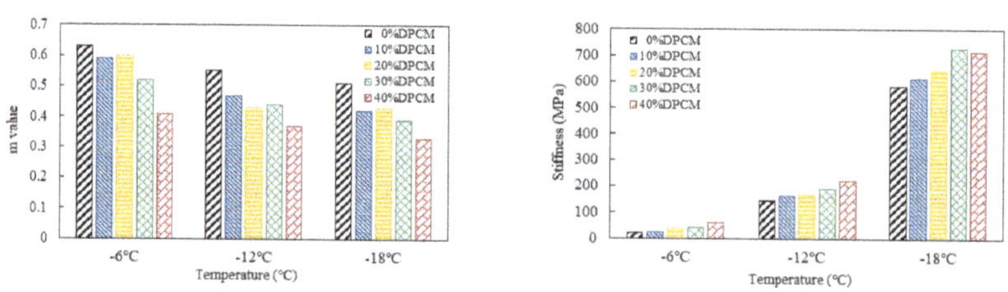

(**a**) m value

(**b**) Stiffness

Figure 12. The m values and stiffness obtained via temperature scanning test.

The increase in the m value means that the tensile stress in the material reduced at low temperatures, and the possibility of cracking also decreased. As can be seen in Figure 12a, with the increase in the DPCM content, the m value of the asphalt binder decreased at all temperatures, making it more prone to cracking. The DPCM improved the stiffness modulus of the asphalt to different degrees under three different low-temperature conditions. It is worth noting that the stiffness modulus of all samples was already greater than 300 MPa at −18 °C, indicating that these samples are more prone to cracking when serving at low temperatures of −18 °C.

The change in the m value and stiffness modulus proves that, in the case of continuous extremely low temperatures, when the DPCM no longer implemented phase transformation, the asphalt binders tended to crack at low temperatures due to the excessive DPCM content.

3.3. Chemical Stability of DPCM-Modified Asphalt Binders

By comparing the infrared characteristic peaks of the DPCM-modified asphalt binders with those of the original asphalt binders in the FTIR test, the potential chemical reaction in the blending process of the DPCM and asphalt was studied. The test result is shown in Figure 12.

There were obvious infrared characteristic peaks at 3411 cm^{-1}, 2888 cm^{-1}, and 1111 cm^{-1}, corresponding to the stretching vibration peaks of -OH, $-CH_2$, and C-O, respectively [35,36]. In the infrared spectrum of the asphalt binders, 2920 cm^{-1} and 2851 cm^{-1} corresponded to the C-H stretching vibration peaks in alkanes and cycloalkanes. Furthermore, 1454 cm^{-1} and 1375 cm^{-1} corresponded to the stretching vibration peaks in the C-H plane of C-CH3 and $-CH_2$. It can be clearly seen in Figure 13 that the infrared characteristic peaks of the DPCM-modified asphalt binders included the characteristic peaks of the DPCM and the original asphalt binders. There were no new structural characteristic peaks, indicating that there was no chemical reaction in the DPCM modification process. The modification of the DPCM was a physical mixing process.

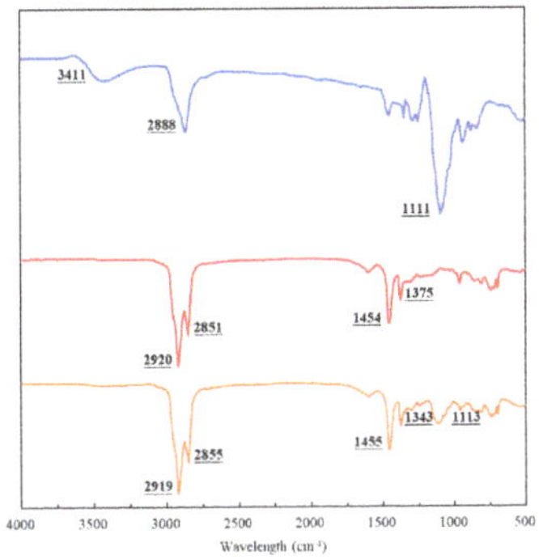

Figure 13. Infrared spectrum of DPCM-modified asphalt binders.

3.4. Temperature Regulation Performance of DPCM-Modified Asphalt Binders

The temperature regulation effect of the DPCM-modified asphalt binders with different DPCM contents was studied by controlling the sample temperature in the environment chamber. The indoor simulated temperature regulation tests of the DPCM-modified asphalt binders were carried out in temperature ranges of 25~65 °C and −20~20 °C.

3.4.1. Indoor Simulated High-Temperature Regulation Test

An indoor simulated high-temperature regulation test was carried out on the modified asphalt binders. The temperature range was 25~65 °C. The maximum temperature and heating rate of the DPCM-modified asphalt binders during the heating process were selected to evaluate the effect of the DPCM content on the temperature regulation effect. The heating rate of the DPCM-modified asphalt binders is expressed by Formula (1):

$$Q_1 = \frac{T_1 - T_0}{\Delta t} \quad (1)$$

In the formula, Q_1 is the heating rate of the sample during the heating process, in unit °C/min or °C/h; T_1 is the final temperature of the sample during the heating process, in unit °C; T_0 is the initial temperature of the sample during the heating process, in unit °C; and Δt is the time required for the sample to heat up, in unit min or h.

The cooling rate of the DPCM-modified asphalt binders is expressed by Formula (2):

$$Q_2 = \frac{T_2 - T_3}{\Delta t} \quad (2)$$

In the formula, Q_2 is the cooling rate of the sample during the cooling process, in unit °C/min or °C/h; T_2 is the final temperature of the sample during the cooling process, in unit °C; T_3 is the initial temperature of the sample during the cooling process, in unit °C; and Δt is the time required for the sample to cool down, in unit min or h.

Based on the above evaluation index of the temperature regulation effect, the influences of the different contents of the DPCM on the temperature regulation effect in the modified asphalt binders at high temperatures are shown in Table 2 and Figure 14.

Table 2. Influence of the content of DPCMs on the temperature regulation effect of DPCM-modified asphalt binders at high temperatures.

Content (%)	High-Temperature Extreme (°C)	Heating Rate Q_1 (°C/min)	Cooling Rate Q_2 (°C/min)
0	64.7	1.588	0.167
10	63.3	1.532	0.158
20	61.7	1.468	0.151
30	61.1	1.444	0.150
40	60.5	1.420	0.146

As can be seen in Table 2 and Figure 14, with the increase in the DPCM content, the high-temperature extreme value in the heating and cooling stages decreased. Meanwhile, the heating rate also decreased significantly. During the 25 min heating process, with the addition of the DPCM, the temperature rise rate decreased from 1.588 °C/min in the original asphalt binders to 1.420 °C/min in the asphalt binders with 40% DPCM. The temperature rise rate decreased by 10.6%. The temperature of the asphalt without the addition of DPCM rose to the highest value of 64.7 °C during heating, while the extreme value of the high temperature decreased with the increase in the DPCM content. The cooling range of the asphalt binders with 40% DPCM content was the highest, up to 4.2 °C. In the cooling stage, the cooling rate of the modified asphalt binders also decreased with the increase in the DPCM content. The cooling rate of the asphalt binders with 40% DPCM was 12.6% lower than that of the asphalt binders without DPCM. In the heating process, the DPCM melted and absorbed heat through the phase change when the external temperature rose to about 40~50 °C. The DPCMs stored part of heat, thus delaying the temperature rise rate of the asphalt binders and reducing their high-temperature extreme value. In the cooling process, when the temperature reached a certain range, the DPCM solidified and released the stored heat through the reverse phase change, thus delaying the cooling rate of the asphalt binders. With the increase in the DPCM content, the temperature regulation effects at 25~65 °C gradually improved. The effects of peak cutting and valley filling displayed in Figure 14 were more intuitive and obvious.

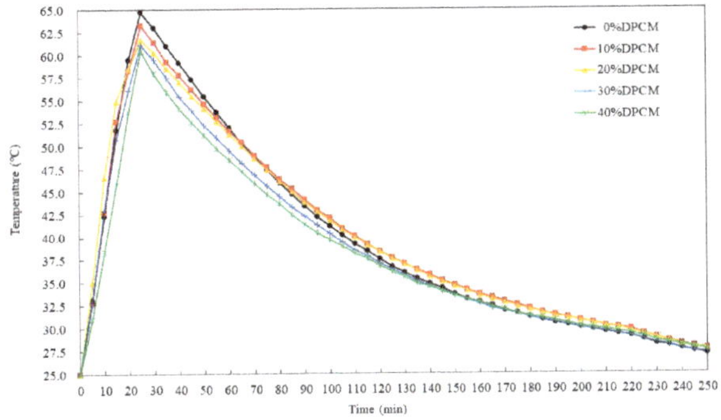

Figure 14. Influence of the content of DPCMs on the temperature regulating effect of DPCM-modified asphalt binders at high temperatures.

3.4.2. Indoor Simulated Low-Temperature Regulation Test

The temperature range was −20~20 °C. In the same way, the minimum temperature and cooling rate of each DPCM-modified asphalt binder were selected to evaluate the influence of the DPCM content on the temperature regulation effect. The heating and cooling rates were still calculated according to Formulas (1) and (2). The results of the low-temperature regulation effect test are shown in Table 3 and Figure 15.

Table 3. Influence of the content of DPCMs on the temperature regulation effect of DPCM-modified asphalt binders at low temperatures.

Content (%)	Low-Temperature Extreme (°C)	Cooling Rate Q_2 (°C/min)	Heating Rate Q_1 (°C/min)
0	−19.9	0.363	0.304
10	−19.5	0.359	0.288
20	−18.7	0.352	0.282
30	−17.9	0.345	0.279
40	−17.4	0.340	0.274

As can be seen in the table and figure above, with the increase in the content of DPCMs, the low-temperature extreme value of the DPCM-modified asphalt binders gradually increased during the rising and cooling process from −20 °C to 20 °C. The cooling rate also significantly decreased. In the cooling process, when the temperature of the DPCM-modified asphalt binders was reduced to −19.9 °C, the cooling amplitudes of the DPCM-modified asphalt binders with 10%, 20%, 30%, and 40% DPCM contents were 0.4 °C, 1.2 °C, 2.0 °C, and 2.5 °C, respectively. The cooling rate decreased from 0.363 °C/min in the original asphalt binders to 0.340 °C/min in the modified asphalt binders with 40% DPCM. The cooling rate was decreased by 6.3%. In the heating process, the heating rate of the DPCM-modified asphalt binders also decreased with the increase in the DPCM content. Obviously, in the cooling process, the DPCM in the modified asphalt binders released the stored heat through endothermic and exothermic phase changes when the external temperature reduced to about −10~0 °C, thus delaying the cooling rate of the asphalt binders and increasing their low-temperature extreme value. In the heating stage, the DPCM melted and absorbed heat through phase change when the temperature reached a certain range. It stored part of the heat for the next exothermic phase-change process. With the increase in the DPCM content, the temperature regulation effects of the DPCM-modified asphalt binders at −20~20 °C also gradually improved.

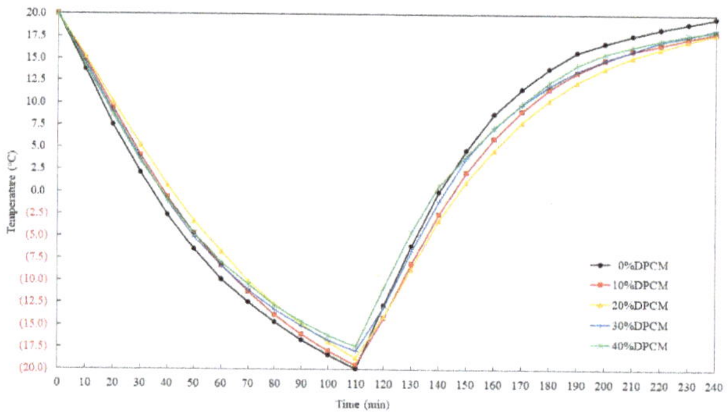

Figure 15. Influence of the content of DPCMs on the temperature regulation effect of DPCM-modified asphalt binders at low temperatures.

4. Conclusions

PEG400/EG and PEG1500/EG composite phase-change materials were used as the raw materials of dual-phase-change materials. DPCM-modified asphalt binders were prepared. The thermal storage properties, rheological properties, chemical stability, and temperature regulation properties of the DPCM-modified asphalt binders were studied. The main conclusions are as follows:

(1) The effect of the DPCM on the rheological properties of the asphalt mastic is due to the increase in the powder–binder ratio. In the research on the pavement performance of the modified asphalt binders based on rheological parameters, the influence of the DPCMs on the rheological properties of the asphalt binders was mostly similar to influence of the mineral powder. The addition of the DPCMs increased the high-temperature rutting resistance of the asphalt binders, but a high DPCM content showed a negative influence on the fatigue performance of the modified asphalt binders.

(2) PEG could be adsorbed by the pores of EG to form a stable microstructure. In the SEM test, continuous and stable network structures of PEG400/EG and PEG1500/EG were observed, which were conducive to improving the thermal conductivity efficiency and temperature regulation effect.

(3) The modification of asphalt by DPCM is only a physical mixing process. In the FTIR test, the results showed that the infrared characteristic peaks of the DPCM-modified asphalt binders included the characteristic peaks of the DPCM and the original asphalt binders. No new structural characteristic peaks appeared, indicating that there was no chemical reaction in the DPCM modification process. The chemical properties of both the DPCM and the original asphalt binders were not changed.

(4) Increasing the content of DPCM can improve the temperature change rate and temperature extremum of DPCM-modified asphalt. In the indoor simulated temperature regulation test, the DPCM-modified asphalt binders showed a good temperature regulation effect in the two temperature domains of −20~20 °C and 25~65 °C. Their temperature regulation effect increased with the increase in the DPCM content. The modified asphalt binders with a 40% DPCM content reduced the high-temperature extremum by 4.2 °C during the heating process. The asphalt heating rate was reduced by 10.6%. In the cooling process, the low-temperature extreme value of the asphalt was increased by 2.5 °C, while the cooling rate of the asphalt was reduced by 6.3%.

(5) The use of these DPCM-modified asphalt binders in pavement in monsoon climate areas will reduce the occurrence of pavement defects and prolong its service life.

In future studies, the effects of DPCMs in asphalt mixtures should be further confirmed, including but not limited to the cycle stability, durability, and heat storage capacity. In addition, more application scenarios suitable for DPCMs can be explored, such as bridge paving and airport roads, to improve the durability of road surfaces in key areas.

Author Contributions: Conceptualization, C.Y. and M.G.; methodology, R.Z.; software, D.H. (Debao Hou) and R.Z.; validation, S.Z. and D.H. (Donghui Huo); formal analysis, B.C. and S.Z.; investigation, S.Z.; resources, B.C. and M.G.; data curation, L.J. and R.Z.; writing—original draft preparation, W.G., D.H. (Donghui Huo) and S.Z.; writing—review and editing, D.H. (Donghui Huo), D.H. (Debao Hou) and S.Z.; visualization, W.G. and M.G.; supervision, C.Y., L.J. and M.G.; funding acquisition, C.Y. and M.G. All authors have read and agreed to the published version of the manuscript.

Funding: The authors would like to acknowledge the financial support from the National Key R&D Program of China (2022YFE0137300) and the National Natural Science Foundation of China (52078018).

Data Availability Statement: The data presented in this study are conditionally available upon request to the corresponding author.

Conflicts of Interest: The authors declare no conflict of interest.

References

1. Yang, S.; Lee, L. Characterizing the chemical and rheological properties of severely aged reclaimed asphalt pavement materials with high recycling rate. *Constr. Build. Mater.* **2016**, *111*, 139–146. [CrossRef]
2. Jiang, X.; Titi, H.; Ma, Y.; Polaczyk, P.; Zhang, M.; Gabrielson, J.; Bai, Y.; Huang, B. Evaluating the performance of inverted pavement structure using the accelerated pavement test (APT). *Constr. Build. Mater.* **2022**, *346*, 128489. [CrossRef]
3. Guo, M.; Liang, M.; Sreeram, A.; Bhasin, A.; Luo, D. Characterisation of rejuvenation of various modified asphalt binders based on simplified chromatographic techniques. *Int. J. Pavement Eng.* **2021**, *23*, 4333–4343. [CrossRef]
4. Liu, J.; Zhao, S.; Li, L.; Li, P.; Saboundjian, S. Low temperature cracking analysis of asphalt binders and mixtures. *Cold Reg. Sci. Technol.* **2017**, *141*, 78–85. [CrossRef]
5. Guo, M.; Liang, M.; Jiao, Y.; Zhao, W.; Duan, Y.; Liu, H. A Review of Phase Change Materials in Asphalt Binder and Asphalt Mixture. *Constr. Build. Mater.* **2020**, *258*, 119565. [CrossRef]
6. Chen, Z.; Ma, D. Effectiveness of Solar Heat Reflective Coatings in Reducing Asphalt Concrete Temperature. *J. Test. Eval.* **2012**, *40*, 740–745. [CrossRef]
7. Starke, P.; Goebel, P.; Coldewey, W. Urban evaporation rates for water-permeable pavements. *Water Sci. Technol.* **2010**, *62*, 1161–1169. [CrossRef]
8. Du, Y.; Dai, M.; Deng, H.; Deng, D.; Wei, T.; Li, W. Evaluation of thermal and anti-rutting behaviors of thermal resistance asphalt pavement with glass microsphere. *Constr. Build. Mater.* **2020**, *263*, 120609. [CrossRef]
9. Duarte, F.; Ferreira, A.; Fael, P. Road Pavement Energy-Harvesting Device to Convert Vehicles' Mechanical Energy into Electrical Energy. *J. Energy Eng.* **2018**, *144*, 04018003. [CrossRef]
10. Yu, H.; Ma, T.; Wang, D.; Wang, Z.; Lv, S.; Zhu, X.; Liu, P.; Li, F.; Xiao, Y.; Zhang, J.; et al. Review on China's pavement engineering reseach·2020. *China J. Highw. Transp.* **2020**, *33*, 1–66.
11. Šavija, B.; Schlangen, E. Use of Phase Change Materials (Pcms) to Mitigate Early Age Thermal Cracking in Concrete: Theoretical Considerations. *Constr. Build. Mater.* **2016**, *126*, 332–344. [CrossRef]
12. Souayfane, F.; Fardoun, F.; Biwole, P. Phase Change Materials (Pcm) for Cooling Applications in Buildings: A Review. *Energy Build.* **2016**, *129*, 396–431. [CrossRef]
13. Ren, S.; Li, J.; Zhang, B.; Huang, K.; Bai, Y. Preparation of a Composite Phase Change Material with High Thermal Storage Capacity Using Modified Expanded Graphite as the Matrix. *Diam. Relat. Mater.* **2022**, *121*, 108736. [CrossRef]
14. Anupam, B.; Sahoo, U.; Rath, P. Phase change materials for pavement applications: A review. *Constr. Build. Mater.* **2020**, *247*, 118553. [CrossRef]
15. Li, F.; Zhou, S.; Chen, S.; Yang, J.; Zhu, X.; Du, Y.; Yang, Z. Low-temperature organic phase change material microcapsules for asphalt pavement: Preparation, characterisation and application. *J. Microencapsul.* **2018**, *35*, 635–642. [CrossRef]
16. She, Z.; Wei, Z.; Young, B.A.; Falzone, G.; Neithalath, N.; Sant, G.; Pilon, L. Examining the effects of microencapsulated phase change materials on early-age temperature evolutions in realistic pavement geometries. *Cem. Concr. Compos.* **2019**, *103*, 149–159. [CrossRef]
17. Ryms, M.; Lewandowski, W.M.; Klugmann-Radziemska, E.; Denda, H.; Wcisło, P. The Use of Lightweight Aggregate Saturated with Pcm as a Temperature Stabilizing Material for Road Surfaces. *Appl. Therm. Eng.* **2015**, *81*, 313–324. [CrossRef]
18. Fallahi, A.; Guldentops, G.; Tao, M.; Granados-Focil, S.; Van Dessel, S. Review on Solid-Solid Phase Change Materials for Thermal Energy Storage: Molecular Structure and Thermal Properties. *Appl. Therm. Eng.* **2017**, *127*, 1427–1441. [CrossRef]
19. Chen, M.; Zheng, S.; Wu, S.; Xu, G. Melting Intercalation Method to Prepare Lauric Acid/Organophilic Montmorillonite Shape-Stabilized Phase Change Material. *J. Wuhan Univ. Technol. (Mater. Sci. Ed.)* **2010**, *4*, 674–677. [CrossRef]
20. Kong, W.; Liu, Z.; Yang, Y.; Zhou, C.; Lei, J. Preparation and Characterizations of Asphalt/Lauric Acid Blends Phase Change Materials for Potential Building Materials. *Constr. Build. Mater.* **2017**, *152*, 568–575. [CrossRef]

21. He, L.; Tang, B.; Zhu, H.; Cao, X. Phase Change Heat Storage Properties of Peg/Sio2 Shape-Stabilized Phase Change Materials in Asphalt. *Funct. Pavement Des.* **2016**, *7*, 1069–1075.
22. Chen, J.; Li, J.; Wang, H.; Huang, W.; Sun, W.; Xu, T. Preparation and Effectiveness of Composite Phase Change Material for Performance Improvement of Open Graded Friction Course. *J. Clean. Prod.* **2019**, *214*, 259–269. [CrossRef]
23. Min, Z.; Wang, Q.; Xie, Y.; Xie, J.; Zhang, B. Influence of Polyethylene Glycol (PEG) Chain on the Performance of Epoxy Asphalt Binder and Mixture. *Constr. Build. Mater.* **2021**, *272*, 121614. [CrossRef]
24. Ma, B.; Si, W.; Ren, J.; Wang, H.-N.; Liu, F.-W.; Li, J. Exploration of Road Temperature-Adjustment Material in Asphalt Mixture. *Road Mater. Pavement Des.* **2014**, *15*, 659–673. [CrossRef]
25. Kuznik, F.; David, D.; Johannes, K.; Roux, J.-J. A Review on Phase Change Materials Integrated in Building Walls. *Renew. Sustain. Energy Rev.* **2011**, *15*, 379–391. [CrossRef]
26. Fan, Y.; Zhang, X.; Wang, X.; Li, J.; Zhu, Q. Super-Cooling Prevention of Microencapsulated Phase Change Material. *Thermochim. Acta* **2004**, *413*, 1–6. [CrossRef]
27. *ASTM D5*; Standard Test Method for Penetration of Bituminous Materials. ASTM International: West Conshohocken, PA, USA, 2013.
28. *ASTM D36*; Standard Test Method for Softening Point of Bitumen (Ring-and-Ball Apparatus). ASTM International: West Conshohocken, PA, USA, 2012.
29. *ASTM D4402*; Standard Test Method for Viscosity Determination of Asphalt at Elevated Temperatures Using a Rotational Viscometer. ASTM International: West Conshohocken, PA, USA, 2015.
30. *ASTM D113*; Standard Test Method for Ductility of Asphalt Materials. ASTM International: West Conshohocken, PA, USA, 2017.
31. Guo, M.; Yin, X.; Du, X.; Tan, Y. Effect of aging, temperature and relative humidity on adhesion between asphalt binder and mineral aggregate. *Constr. Build. Mater.* **2023**, *363*, 129775. [CrossRef]
32. Guo, M.; Zhang, S.; Du, X.; Tan, Y.; Luo, D. Study on effect of various rejuvenators on virgin asphalt binders suffering various aging conditions and the unified evaluation index. *Constr. Build. Mater.* **2023**, *397*, 132331. [CrossRef]
33. Wang, G.; Cao, D.; Wang, Z.; Qiu, W.; Zhang, S. Comparative Study on Rheological Properties of Asphalt Binder Modified with High Content Crumb Rubber. *J. Highw. Transp. Res.* **2022**, *39*, 7–14.
34. Sui, C.; Farrar, M.J.; Tuminello, W.H.; Turner, T.F. New technique for measuring low temperature properties of asphalt binders with small amounts of material. *Transp. Res. Rec. J. Transp. Res. Board* **2010**, *2179*, 23–28. [CrossRef]
35. Bai, G.; Fan, Q.; Song, X. Preparation and characterization of pavement materials with phase-change temperature modulation. *J. Therm. Anal. Calorim.* **2019**, *136*, 2327–2331. [CrossRef]
36. Jin, J.; Xiao, T.; Zheng, J.; Liu, R.; Qian, G.; Xie, J.; Wei, H.; Zhang, J.; Liu, H. Preparation and thermal properties of encapsulated ceramsite-supported phase change materials used in asphalt pavements. *Constr. Build. Mater.* **2018**, *190*, 235–245. [CrossRef]

buildings

MDPI

Article

Effect of Fractionation Process and Addition of Composite Crumb Rubber-Modified Asphalt on Road Performance Variability of Recycled Asphalt Mixtures with High Reclaimed Asphalt Pavement (RAP) Content

Wenwu Wei [1], Chao Ji [1], Honggang Song [1], Zhigang Li [2], Zhen Liu [2], Lijun Sun [2,*] and Zhou Zhou [2]

1 China Communications Construction Group Sixth Engineering Co., Ltd., Beijing 102600, China
2 School of Transportation, Southeast University, Nanjing 211189, China; 230208344@seu.edu.cn (Z.L.)
* Correspondence: lijunshun@seu.edu.cn

Abstract: The application of reclaimed asphalt pavement (RAP) can help reduce resource waste and environmental pollution in road construction. However, so far, only a small percentage of RAP materials can be used in road construction. The key obstacles to the application of a recycled asphalt mixture (RAM) with high RAP content are the variability of RAP materials and the difficulty of fully rejuvenating aged asphalt. However, there is still a lack of research on the effect of the variability of RAP materials and recycled asphalt on the quality control of a RAM. Therefore, this study investigates the effects of sieve pretreatment of RAP material using 4.75 mm sieve mesh and the use of composite crumb rubber-modified asphalt (CCRMA) to reclaim aged asphalt on the road performance and frame variability of reclaimed asphalt mixtures. Therefore, this study investigates the effects of the fractionation process of RAP material using 4.75 mm sieve mesh and the use of CCRMA to reclaim aged asphalt on the road performance of a RAM. The results show that the fractionation process can effectively reduce the mitigation of RAP agglomeration and reduce the variability of gradation, which in turn reduces the variability of road performance. The incorporation of CCRMA can effectively improve the high-temperature stability performance and low-temperature cracking resistance. The dynamic stability and the fracture energy of the CRAM (RAM prepared using CCRMA) were four and one and a half times as large as that of the NAM (RAM prepared using base asphalt), respectively. The fractionation process of RAP material and the utilization of CCRMA could help reduce the variability of the RAM while improving the road performance of the RAM.

Keywords: recycled asphalt mixture; variability; reliability design; fractionation process; road performance

1. Introduction

Utilizing recycled materials instead of virgin materials in the production of hot-mix asphalt mixtures helps to reduce construction costs, lessen the reliance on natural aggregates, and reduce greenhouse gas emissions [1–3]. When considering the performance of a recycled asphalt mixture (RAM), the reclaimed asphalt pavement (RAP) content in the actual projects is generally controlled within 30% ("RAP" is a mixture of old asphalt pavement that has been excavated, recycled, crushed, and screened, and "RAM" is a mixture made by remixing RAP with new asphalt material, new aggregate, etc. in a certain proportion) [4,5]. However, the continuous growth of RAP materials in recent years has caused unpredictable environmental pollution. The use of a RAM with low RAP content cannot solve the growing RAP problem [6,7]. Therefore, the RAM with a high content of RAP has received more and more attention in practical applications.

When the RAP content exceeds 25%, the road performance of the RAM gradually decreases as the RAP content increases [8–10]. Barros et al. [11] found that the cleavage expansion rate increased when the RAP content in the mixture increased. Zhu et al. [12] stated that mixtures with high RAP content suffered from weak cracking resistance. The variability of RAP and the low performance of recycled asphalt are the main reasons for limiting the maximum RAP blending. Al-Qadi et al. [13] noted that the lack of understanding of aggregate and binder properties and the uncertainty of RAP grades resulted in the RAP content remaining at a low level. To keep RAP variability within acceptable limits, some studies have pretreated RAP material by fragmentation and fractionation, etc. Previous research has shown that fractionation treatments provide better retention of rutting potential, improved fatigue performance (higher fracture energy), and better control of the volumetric properties of asphalt mixtures [14]. Zaumanis et al. [15] found that pretreating RAP by fragmentation and fractionation reduced the variability of RAP and increased the RAP content in the RAM. Pan et al. [16] proposed an optimized gradation design method after RAP fractionation and found that the fractionation process reduced the effect of RAP variability on the RAM. Most of the current research focuses on the effect of the RAP fractionation process on the road performance of the RAM [17]. However, there is a lack of research on the effect of the RAP fractionation process on the variability of RAM road performance.

Another major factor affecting the performance of a RAM with high RAP content is the method of reclaiming asphalt in the RAM. Various methods have been used to effectively rejuvenate aged asphalt, including the incorporation of rejuvenators and soft binders into RAP mixtures [18–20]. Many studies have been conducted on the utilization of recycling agents to regenerate asphalt mixtures [21–23]. Wang et al. [18] investigated the impact of the HRA-2 rejuvenator on a RAM containing high RAP content and found that the addition of the HRA-2 rejuvenator reduced the high-temperature performance of the RAM and improved its water stability and low-temperature crack resistance. Yousefi et al. [24] investigated the effect of two types of rejuvenators (one was aromatic extracts and one was triglycerides/fatty acids) on the cracking properties of asphalt mixtures containing 25% RAP and found that recycling agents improved the low-temperature cracking resistance of asphalt mixtures. In addition, the other method is to use a softer asphaltene blend and rejuvenate aged asphalt by adding virgin asphalt [25]. Recent research showed that composite crumb rubber-modified asphalt (CCRMA) was significantly more effective than virgin asphalt in blending aged asphalt containing high RAP content [26,27]. Therefore, CCRMA has great potential for enhancing the road performance of the RAM with high RAP content. In recent years, many studies have been conducted on the effect of CCRMA on the properties of aged asphalt. Liu et al. [27] found that the high-temperature and fatigue resistance performances of aged asphalt were improved by adding CCRMA. Chen et al. [28] concluded that CCRMA had a significant effect on the chemical, microscopic, and rheological properties of aged asphalt binders. However, compared to aged asphalt, the effect of CCRMA on the performance of the RAM has been less studied, especially on the performance of the RAM with high RAP content.

Therefore, the objectives of this study were (1) to evaluate the effect of the fractionation process on controlling the variability of the road properties for the RAM and (2) to evaluate the effect of the combined use of CCRMA as a blending asphalt and fractionation process on the road properties for the RAM and their variability.

2. Materials and Methods

2.1. Materials

2.1.1. RAP Characterization

RAP was obtained from Wuxi Road and Bridge Municipal Co., Ltd. (Xihu Middle Road, Xishan District, Wuxi City), as shown in Figure 1. The RAP material was fractionated using a 4.75 mm sieve and divided into two portions based on the aggregate size. The aggregate gradation for the two portions of the fractionated RAP material is displayed in Figure 2. The old asphalt content (by mass of mixture) in the RAP was 5.57%.

Figure 1. RAP material.

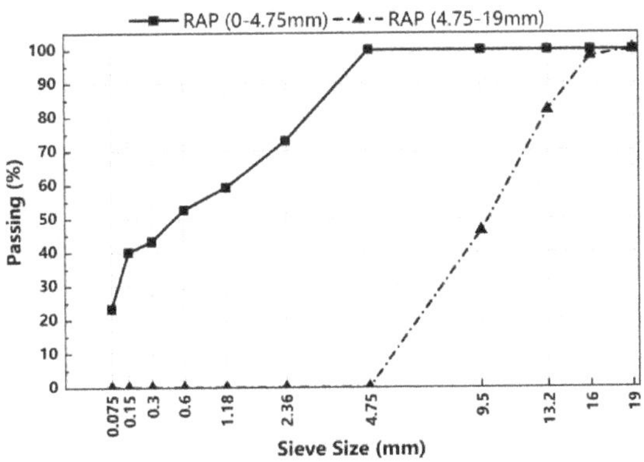

Figure 2. The aggregate gradation of RAP material.

2.1.2. Properties of Virgin Binders

The base asphalt was chosen as the virgin binder to prepare the RAM in this study. According to our previous tests [27], CCRMA can effectively improve the high-temperature performance and fatigue resistance of aged asphalt, so CCRMA was also chosen to prepare recycled asphalt mixtures. Table 1 lists the technical indicators of the asphalt, where G* is the complex modulus and δ is the phase angle.

Table 1. Technical indicators of base asphalt and CCRMA.

Technical Indicators		Base Asphalt	CCRMA	Test Method
Penetration (25 °C, 5 s, 100 g) (0.1 mm)		70	83	T0604-2011
Ductility (5 °C, 5 cm/min) (cm)		-	59	T0605-2011
Ductility (10 °C, 5 cm/min) (cm)		25	-	T0605-2011
Softening point (°C)		48.5	87	T0606-2011
G*/sinδ@64 °C	Virgin asphalt	1.24	-	T0628-2011
	RTFOT aged asphalt	3.01	-	T0628-2011
G*/sinδ@70 °C	Virgin asphalt	0.63	-	T0628-2011
	RTFOT aged asphalt	2.94	-	T0628-2011
G*/sinδ@94 °C	Virgin asphalt	-	1.35	T0628-2011
	RTFOT aged asphalt	-	2.26	T0628-2011
G*/sinδ@100 °C	Virgin asphalt	-	0.83	T0628-2011
	RTFOT aged asphalt	-	1.54	T0628-2011

2.1.3. Design and Preparation of RAP Mixtures

For the preparation of RAM, basalt was selected as coarse and fine aggregates, and limestone mineral powder with a particle size of less than 0.075 mm was selected as filler. Table 2 shows the technical specifications of the aggregate.

Table 2. Technical specifications of the aggregates.

Technical Specifications	Unit	Coarse Aggregate	Fine Aggregate	Fillers	Test Method
Apparent relative density	g/cm^3	2.78	2.66	2.61	T0328-2005
Crushing value	%	11.4	-	-	T0316-2005
Los Angeles abrasion loss	%	18.2	-	-	T0317-2005
Needle flake particle content	%	11.7	-	-	T0312-2005
Water absorption rate	%	1.3	-	-	T0307-2005
Firmness	%	10.4	-	-	T0314-2000
Sand equivalent	%	-	86	-	T0334-2005
Angularity	%	-	51	-	T0345-2005
Hydrophilic coefficient	-	-	-	0.4	T0353-2000
Plasticity coefficient	-	-	-	3.1	T0354-2000

The RAM was designed as a dense-graded mixture with a nominal maximum aggregate size of 13 mm and 50% RAP content. Referring to the grading control range of AC-13, three gradation curves were determined and named AC-13a, AC-13b, and AC-13c, as shown in Figure 3. The asphalt content was selected as 5.57%, which is the same as the old asphalt content in RAP. To investigate the effect of the fractionation process on the performance of the RAM, RAM with fractionated RAP materials (RAM-FR) and unfractionated RAP materials (RAM-UR) was prepared.

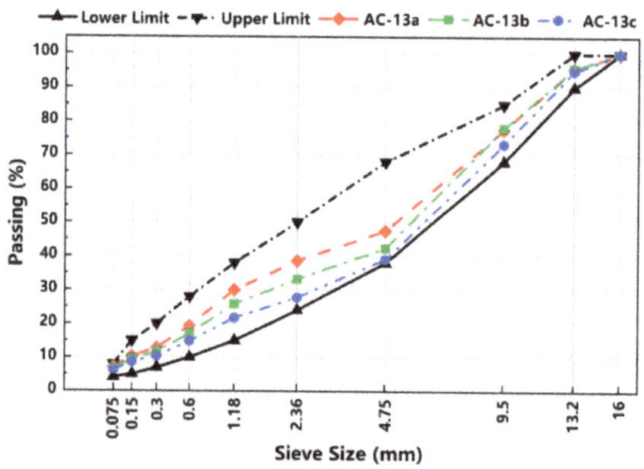

Figure 3. Gradation curves of RAM.

For the preparation of RAM, asphalt, new aggregates, and RAP material were first placed in ovens at different temperatures for insulation. Among them, the asphalt was kept at 150 °C for 2 h, the new aggregates were insulated at 180 °C for 4 h, and the RAP material was insulated at 120 °C for 2 h [29]. After insulation, the RAP material was mixed in a mixing pot at a temperature of 180 °C for 90 s. After adding the new aggregates to the RAP material, the mixture was mixed for 90 s to ensure a uniform mixture. Subsequently, the new asphalt (base asphalt) was added and mixed for 90 s to obtain a loose asphalt mixture. The asphalt mixture specimens of target heights were obtained using a Superpave gyratory compactor (SGC) by controlling the number of gyrations to 100 times and adjusting the quality of the asphalt mixture [30]. Finally, the specimens were placed at room temperature for 48 h for further testing.

2.2. *Experimental Design*

2.2.1. Rutting Test

The rutting test was used to evaluate the rutting resistance of RAM [31]. According to the JTG E20-2011 [12], the RAM was formed into plate specimens measuring 300 mm × 300 mm × 50 mm in size and repeatedly rolled on the same track with a solid rubber wheel with a wheel pressure of 0.7 MPa at a temperature of 60 °C to form rutting grooves. The rutting resistance of RAM was evaluated in terms of the rutting depth (*RD*) and the dynamic stability (*DS*) of the asphalt mixture specimens, as calculated by Equation (1):

$$DS = \frac{(t_2 - t_1) \times N}{d_2 - d_1} \times C_1 \times C_2 \tag{1}$$

where d_1 and d_2 correspond to the rut depths at t_1 and t_2, respectively; N is the crushing speed of the round-trip process of the test wheel; and C_1 and C_2 are the correction factors of the testing machine and specimen factor, respectively.

2.2.2. Indirect Tensile Test

The cylindrical specimens with a diameter of 100 mm and a height of 63.5 mm were prepared for the indirect tensile (IDT) test. Considering that the actual load acting on the pavement is constant and the deformation of the pavement grows with the growth of the constant load, the stress-controlled mode was used to perform the IDT test at 25 °C. To obtain the fatigue life of RAM, the average indirect tensile strengths of the different types of mixtures under dry conditions were first obtained from the IDT test, and then the stress ratio (0.25) for the fatigue test was determined [32,33]. Finally, the fatigue life was obtained based on the vertical permanent strain curve of the asphalt mixture, which was used to evaluate the medium-temperature fatigue performance of RAM.

2.2.3. Semi-Circular Bending Test

To investigate the low-temperature performance of RAM, a single-load semi-circular bending (SCB) test was performed at −12 °C using a 50 mm/min displacement loading rate [34,35]. Based on the obtained vertical load–displacement curves, the fracture energy (G_f) was calculated according to Equation (2):

$$G_f = \frac{W_f}{A_{lig}} \tag{2}$$

where W_f represents the fracture work, corresponding to the area below the vertical load–displacement curve, and A_{lig} represents the area of the fracture surface.

The flow chart of the performance-related tests for RAM is illustrated in Figure 4. The blue line portion represents the experimental information for the RAM prepared using RAM-FR and CCRMA.

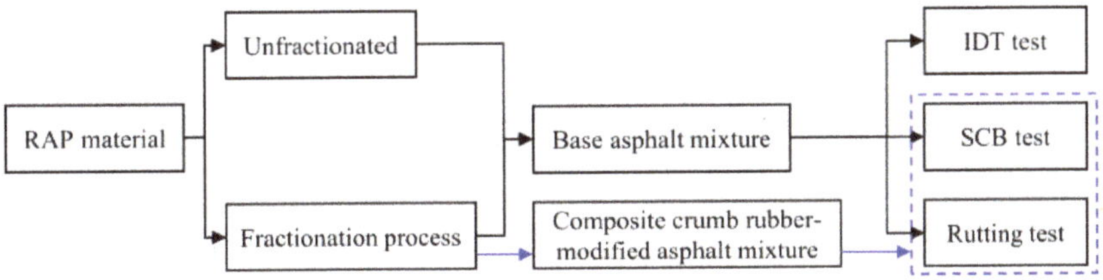

Figure 4. Laboratory experimental design flow chart.

3. Results

3.1. *Analysis of the Variability in Road Properties for Recycled Asphalt Mixtures with High RAP Content*

Figure 5 shows the results of the rutting test, IDT test, and SCB test. The data dispersion of the RAM-UR was significantly greater than that of the RAM-FR. The experimental results of the RAM were subjected to a normal distribution test to intuitively display the results, as listed in Tables 3–5. The results show that the performance indicators of the RAM obeyed a normal distribution at the 0.05 level, i.e., all data obeyed a normal distribution at the 95% confidence level. The normal distribution curves were fitted with different performance indicators and their frequency of occurrence as horizontal and vertical coordinates, as shown in Figure 6 [36,37]. The combination of Tables 3–5 and Figure 6 shows that for the RAM containing 50% RAP, the curve of the RAM-FR was significantly narrower than that of the RAM-UR, implying that the dispersion of road performance test results for the RAM-FR became lower. This indicates that the road performance variability of the RAM can be effectively controlled by the fractionation process of RAP material [38]. Compared with the curves of the RAM-UR, the curves of the RAM-FR showed a significant forward shift, indicating that the road performance of the RAM can be improved by controlling the variability. This is because the fractionation process can reduce the agglomeration phenomenon produced by the bonding action of the aged asphalt in the RAP, reduce the degree of RAM grade separation, and realize the fine fractionation of the RAP [39]. That is, the variability of RAM road performance is reduced by reducing the variability of RAP materials, which is consistent with the studies of Wang et al. [18]. and Feng et al. [17].

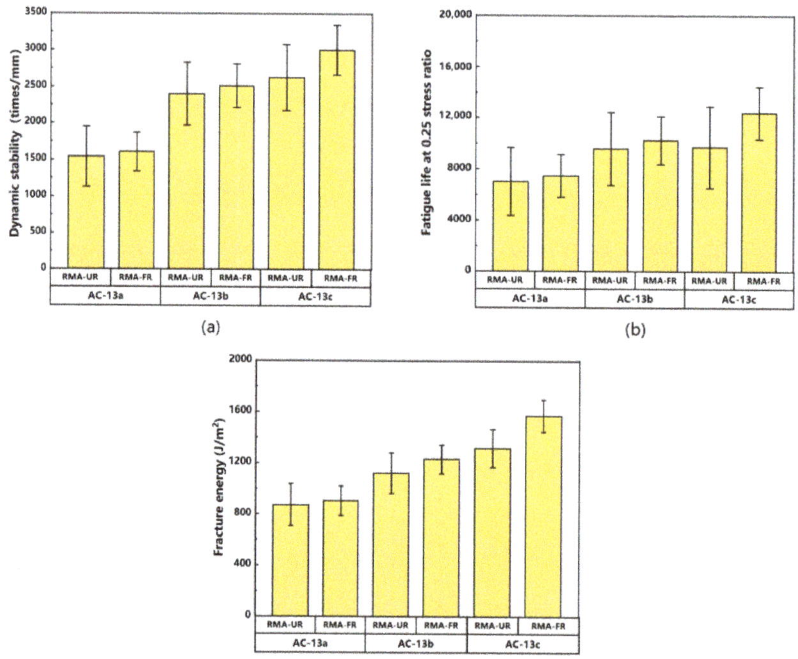

Figure 5. Test results on road performance of RAM-UR and RAM-FR: (**a**) dynamic stability; (**b**) fatigue life at 0.25 stress ratio; and (**c**) fracture energy.

Table 3. Parameters of the normal distribution curve on dynamic stability of RAM with various gradations.

Gradation	Material	Statistics	*p*-Value	μ	σ
AC-13a	RAM-UR	0.903	0.267	1539	410.61
AC-13a	RAM-FR	0.874	0.136	1606	265.69
AC-13b	RAM-UR	0.886	0.181	2400	431.19
AC-13b	RAM-FR	0.952	0.709	2512	298.37
AC-13c	RAM-UR	0.897	0.232	2626	452.15
AC-13c	RAM-FR	0.966	0.859	3003	339.27

Table 4. Parameters of the normal distribution curve on fatigue life of RAM with various gradations.

Gradation	Material	Statistics	p-Value	μ	σ
AC-13a	RAM-UR	0.947	0.679	7017	2646.00
AC-13a	RAM-FR	0.912	0.368	7471	1657.91
AC-13b	RAM-UR	0.919	0.425	9591	2839.83
AC-13b	RAM-FR	0.868	0.145	10,250	1858.70
AC-13c	RAM-UR	0.939	0.602	9715	3166.34
AC-13c	RAM-FR	0.930	0.512	12,402	2057.89

Table 5. Parameters of the normal distribution curve on fracture energy of RAM with various gradations.

Gradation	Material	Statistics	p-Value	μ	σ
AC-13a	RAM-UR	0.868	0.117	871	164.77
AC-13a	RAM-FR	0.948	0.667	903	115.46
AC-13b	RAM-UR	0.979	0.961	1121	156.97
AC-13b	RAM-FR	0.935	0.527	1230	112.88
AC-13c	RAM-UR	0.950	0.688	1316	147.76
AC-13c	RAM-FR	0.921	0.400	1570	125.40

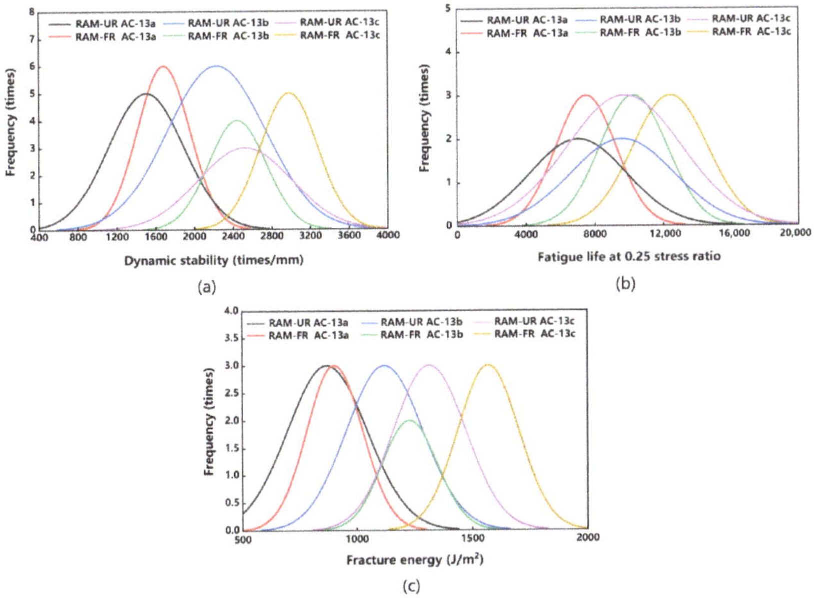

Figure 6. Normal distribution curves for road properties of RAM with various gradations: (**a**) dynamic stability; (**b**) fatigue life at 0.25 stress ratio; and (**c**) fracture energy.

To further evaluate the effect of the fractionation process on the variation of road performance for the RAM, the coefficient of variation was calculated [40], as displayed in Figure 7. A smaller coefficient of variation indicates less variability; conversely, more variability indicates less homogeneity. The coefficient of variation for fatigue life fluctuated between 16 and 37 with high variability, while the coefficient of variation for fracture energy fluctuated between 8 and 19 with low variability. The greater variability in fatigue life among asphalt mixture specimens is a consequence of the reduced constraints on the specimens and the longer loading times during the IDT test in comparison to the SCB test. As a result, various factors affect the fatigue life of asphalt mixture specimens [41,42].

Although the coefficients of variation were not identical, the coefficients of variation for the six different RAMs with different road performances reflected a common trend. The coefficients of variation of the road performance indicators of the RAM-FR were all less than those of the RAM-UR, indicating that the fractionation process of RAP material can reduce the variability of the road performance for the RAM with high RAP content.

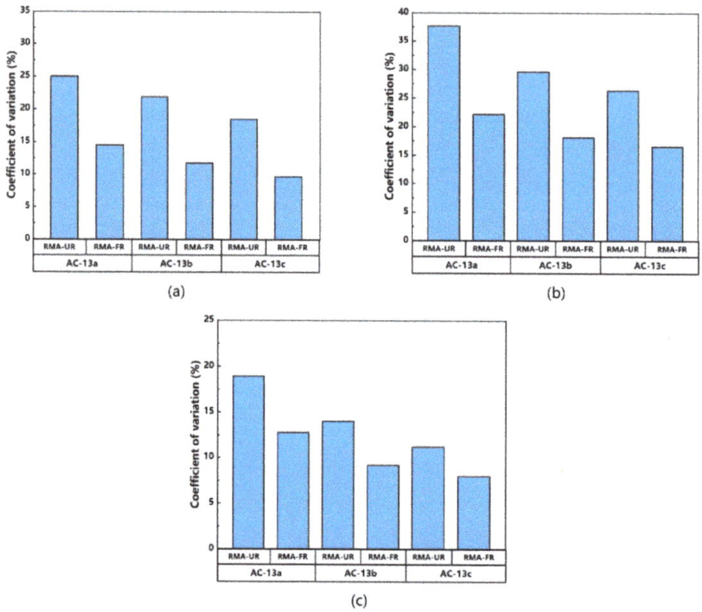

Figure 7. Coefficient of variation of road performance for RAM-UR and RAM-FR: (**a**) dynamic stability; (**b**) fatigue life at 0.25 stress ratio; and (**c**) fracture energy.

3.2. Reliability Calculation Based on Variability Control for Recycled Asphalt Mixtures with High RAP Content

Figure 7 shows that the coefficient of variation is not only related to whether or not the RAP material is fractionated but also to the gradation of the RAM. The fineness modulus is often used to evaluate the gradation of asphalt mixtures [43,44]. The larger the fineness modulus, the coarser the corresponding gradation. The calculation results are shown in Table 6.

Table 6. The fineness modulus of RAM with different gradations.

Gradation	Fineness Modulus
AC-13a	5.61
AC-13b	5.83
AC-13c	6.05

Combined with Table 6 and Figure 7, for both the RAM-UR and RAM-FR, the coefficient of variation on the RAM road performance decreased as the fineness modulus increased. This suggests that higher fineness modulus values in asphalt mixtures can improve the variability of RAM road performance. This is because a larger fineness modulus represents less fine aggregate content, making it easier to mix asphalt and coarse aggregate, thus reducing the difference between the road performance of the RAM [44]. This is consistent with Pan et al. [16]. That is, the high proportion of fine aggregate is not conducive to forming a stable skeleton structure, resulting in greater variability of RAM road performance. Therefore, to control the variability of road performance for the RAM with high RAP content, a reliability calculation scheme was established in this study by taking the fineness modulus as the design indicator and the coefficient of variation on the RAM road performance containing 50% RAP as the performance indicator. Considering that there is no clear regulation on the variability of road performance, the threshold value of the coefficient of variation for the fracture energy was set to 10%, and the threshold values of the coefficient of variation for the dynamic stability and the fatigue life were both set to 20%.

Figure 8 shows the fitted curves of the coefficient of variation on road performance for the RAM with the variation of the fineness modulus (the dotted lines are the threshold values for the coefficient of variation). After establishing the relationship curve, considering that the probability distribution function of the fineness modulus is unknown, the reliability was obtained in this study by calculating the probability that the fineness modulus M_x falls in the reliability interval [45]. Here, the "reliability interval" is the range of fineness modulus for which the performance of the RAM satisfies all three indicators simultaneously. The "reliability" is the probability that the performance of the asphalt mixture meets the target. In addition, due to the small sample size in this study, only the RAM with a range of fineness modulus values from 5.61 to 6.05 were considered. Based on the fitted curves in Figure 8, the fineness modulus of the RAM was calculated once the threshold of variability for road performance was reached. After obtaining the reliability intervals of the fineness modulus meeting the requirements of different road performance variability, the reliability intervals of the RAM meeting the requirements of road performance variability were obtained by taking the intersection of multiple reliability intervals. Finally, the reliability was obtained by calculating the probability of the fineness modulus falling in the reliability interval, as shown in Table 7. It can be seen that the reliability interval for the RAM meeting the requirements for performance variability increased after the fractionation process, with a significant increase in reliability to 50%. This further illustrates that the fractionation treatment can significantly enhance the road performance of a RAM with high RAP content.

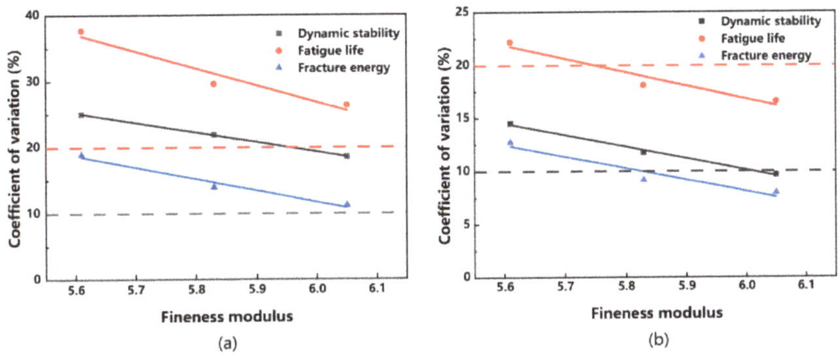

Figure 8. Relationship between the coefficient of variation on road performance and fineness modulus for RAM containing 50% RAP: (**a**) RAM-UR and (**b**) RAM-FR.

Table 7. The calculated reliability intervals and reliability of RAM-UR and RAM-FR.

Material	Reliability Intervals	Reliability (%)
RAM-UR	-	0
RAM-FR	5.81–6.05	50

The milling process causes a decrease in the aggregate size of the RAP material [46]. Therefore, for a RAM with different RAP contents, the increase in RAP content led to a decrease in the fineness modulus of the asphalt mixture, increasing the variability of road performance for the RAM [47]. To analyze the relationship between RAP content and the coefficient of variation on road properties of the RAM, the fineness modulus of the RAM corresponding to different RAP contents was calculated based on the fitted curves in Figure 8. Table 8 shows the reliability intervals determined with the same proportion of new aggregate. The RAP content and its corresponding reliability were further fitted in a quadratic form, as shown in Figure 9. Combining Table 8 and Figure 9, the maximum and minimum values of the fineness modulus intervals for the RAM decreased with increasing RAP content, and the reliability intervals to meet the requirements of road performance variability of the RAM gradually became smaller, and the reliability decreased. This suggests that a higher RAP content increases the variability of RAM road performance. This is in agreement with the study of Yang et al. [48]. An increase in RAP content causes a decrease and instability in the moisture susceptibility, cracking resistance, and fatigue properties of the RAM. The reliability of the RAM-FR containing 70% RAP was 37.7%, which is greater than the reliability of the RAM-UR containing 50% RAP and meets the variability requirement. This further illustrates that the fractionation process of RAP material can effectively reduce the variability of road performance for a RAM.

Table 8. The calculated reliability intervals and reliability of RAM with various RAP contents.

RAP Content (%)	Total Interval	Reliability Intervals	Reliability (%)
30	5.64–6.09	5.81–6.09	62.2
40	5.62–6.07	5.81–6.07	57.8
50	5.61–6.05	5.81–6.05	54.5
60	5.55–6.03	5.81–6.03	45.8
70	5.48–6.01	5.81–6.01	37.7

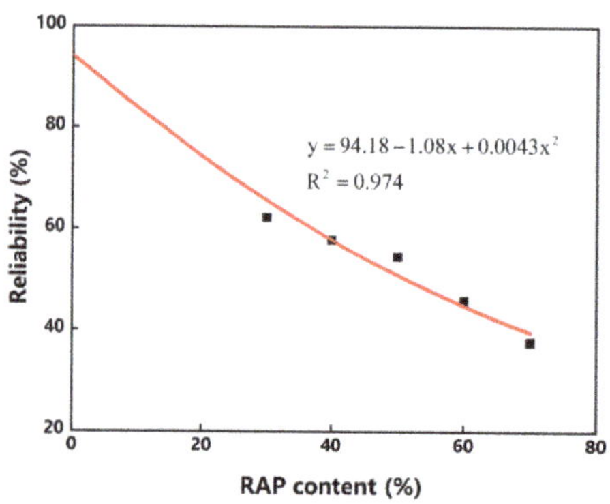

Figure 9. Fitting curve of reliability with RAP content.

3.3. Proportional Design of Recycled Asphalt Mixtures with High RAP Content

3.3.1. Determination of Gradation and Volumetric Parameters for RAM with High RAP Content

The Superpave design method was used in the mix design of the RAM containing 50% RAP. The asphalt content for the RAM was established through the utilization of the SGC method, aiming to achieve 4% air voids at N_{design} = 100. According to the study of Zhou et al. [49], the addition of 0.3% asphalt to the determined optimal amount of asphalt promoted the blending between the new and old asphalt. Therefore, the CCRMA asphalt content in this study was an addition of 0.3% CCRMA asphalt to the determined optimal amount of asphalt. As a blank control sample, asphalt mixture specimens were prepared using base asphalt and new aggregate with the same gradation and 5% asphalt content. For the convenience of presentation, the RAM prepared using CCRMA was referred to as the CRAM, and the asphalt mixture prepared using base asphalt was referred to as the

NAM. The gradation and volumetric parameters for the RAM containing 50% RAP are shown in Tables 9 and 10.

Table 9. The gradation of CRAM containing 50% RAP.

Indicators	CRAM Containing 50% RAP									
Sieve size (mm)	16	13.2	9.5	4.75	2.36	1.18	0.6	0.3	0.15	0.075
Passing (%)	100	95.7	77.6	47.8	38.9	30.3	19	13	10.5	7.3
RAP content (%) (>4.75 mm)	35.0%									
RAP content (%) (<4.75 mm)	15.0%									

Table 10. The volumetric parameters of CRAM containing 50% RAP.

Indicators	NAM
Asphalt content (%)	5.0
Addition of new asphalt (%)	2.39
Gross volume relative density (g/cm^3)	2.494
Maximum theoretical relative density (g/cm^3)	2.598
Air void (%)	4.0

3.3.2. Evaluation of High-Temperature Stability Performance for Recycled Asphalt Mixtures with High RAP Content

The rutting test results of asphalt mixtures are shown in Table 11. A normal distribution curve was fitted with relative frequency as the vertical coordinate and dynamic stability as the horizontal coordinate, as shown in Figure 10. Table 12 displays the parameters of the normality test. The dynamic stability of the CRAM was four times higher than that of the NAM, suggesting that CCRMA can significantly enhance the high-temperature performance of the RAM. This is consistent with our previous studies. That is, CCRMA can effectively enhance the elastic recovery and stiffness of aged asphalt compared to base asphalt, thereby improving the high-temperature performance [27,50]. To further verify the effect of the fractionation process on the high-temperature performance of the CRAM, the coefficients of variation on dynamic stability for the CRAM and NAM were calculated, as shown in Table 13. The coefficient of variation on dynamic stability of the CRAM was 40% higher compared to the NAM but still met the threshold requirement. In addition, combined with Figure 7, the coefficient of variation on dynamic stability for the CRAM decreased by 26–45% compared to that for RAM-UR. Therefore, the CRAM has excellent resistance to deformation at high temperatures while controlling variability.

Table 11. The rutting test results of CRAM and NAM.

Specimen Number	Dynamic Stability (times/mm)	
	NAM	CRAM
1	1845.77	10,597.34
2	1960.68	7146.88
3	2124.66	8576.70
4	2021.06	7011.87
5	1737.42	9455.69
6	2207.10	10,171.49
7	1948.30	8642.65
8	1615.13	8785.95

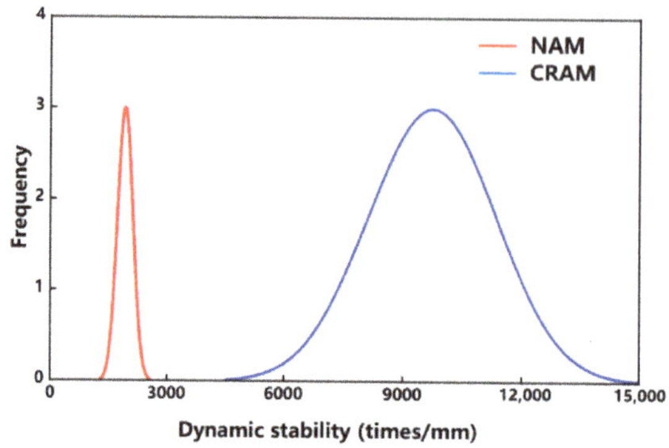

Figure 10. Normal distribution fitting results for dynamic stability.

Table 12. The parameters of the normality test.

Gradation	Material	Statistics	p-Value	μ	σ
NAM	0.980	0.963	1932	182.87	NAM
CRAM	0.936	0.576	8978	1201.37	CRAM

Table 13. The coefficients of variation on dynamic stability for CRAM and NAM.

Gradation	Dynamic Stability (times/mm)	Total Deformation (mm)	Coefficient of Variation
NAM	1933	1.158	9.46
CRAM	8798	0.286	13.65

3.3.3. Evaluation of Low-Temperature Crack Resistance for RAM with High RAP Content

The SCB test results of asphalt mixtures are shown in Table 14. A normal distribution curve was fitted with frequency as the vertical coordinate and fracture energy as the horizontal coordinate, as shown in Figure 11. The parameters of the normality test are shown in Table 15. The fracture energy of the CRAM was greater than that of the NAM, indicating that its low-temperature cracking resistance is better than that of the NAM. This is because the addition of 0.3% asphalt increases the asphalt content in the asphalt mixture and also promotes the integration of the new asphalt with the old asphalt, which ultimately increases the bonding between the asphalt mixtures [49,51,52]. On the other hand, the high elasticity of CCRMA leads to its ability to withstand greater damage loads [53]. At the same time, CCRMA can delay the extension of cracks, resulting in a significant increase in both peak force load and fracture energy of CRMA, which ultimately improves the low-temperature crack resistance of the RAM [54,55]. According to Figure 11 and Table 15, the dispersion degree in the fracture energy of the CRAM was larger than that of the NAM. Therefore, the coefficient of variation was further computed to assess the variability of the CRAM and NAM in terms of low-temperature fracture resistance, as shown in Table 16. Although the coefficient of variation in fracture energy for the CRAM was greater than that for the NAM, it still met the requirements of the SCB test for the coefficient of variation for the fracture energy.

Notably, the change in damage displacement was less for the CRAM compared to that of the NAM. Considering that the base asphalt used in the preparation of the NAM has weak low-temperature cracking resistance, the CRAM, which has the same damage displacement as the NAM, may be susceptible to low-temperature cracking. This susceptibility could be attributed to the higher modulus of the CCRMA, rendering it more prone to cracking under load [35,56]. Therefore, it is necessary to soften the asphalt by adding recycling agents during the mixing process of the asphalt mixture to enhance the low-temperature cracking resistance of the CRAM, thus improving the durability of the CRAM in road engineering.

Table 14. The SCB test results of CRAM and NAM.

Test Number	Fracture Energy (J/m^2)	
	NAM	CRAM
1	1345.67	1816.20
2	1238.02	2154.22
3	1233.29	2088.92
4	1387.16	1924.66
5	1435.83	2056.35
6	1235.98	1742.26
7	1132.89	1553.70
8	1201.34	1844.34

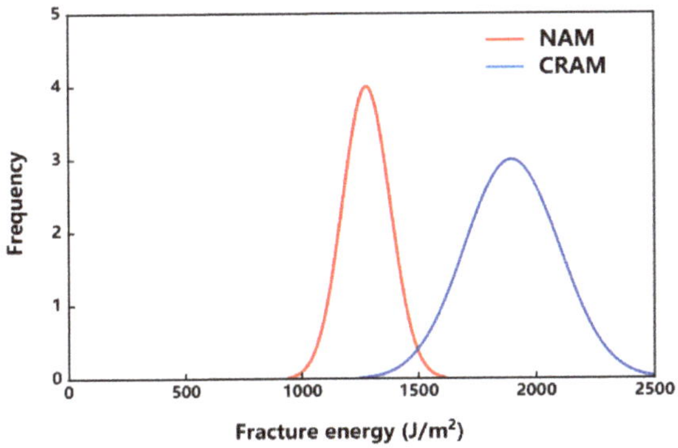

Figure 11. Normal distribution fitting results for fracture energy.

Table 15. The parameters of the normality test.

Gradation	Statistics	*p*-Value	μ	σ
NAM	0.931	0.523	1276.33	95.97
CRAM	0.963	0.841	1892.50	187.17

Table 16. The coefficients of variation on low-temperature performance for CRAM and NAM.

Gradation	Peak Force Load (kN)		Disruption Displacement (mm)		Fracture Energy (J/m^2)	
	Average	Coefficient of Variation	Average	Coefficient of Variation	Average	Coefficient of Variation
NAM	10.85	7.64	1.37	4.23	1276.53	7.52
CRAM	13.37	9.43	1.39	6.91	1892.50	9.86

4. Conclusions

This study evaluates the effect of the fractionation process on controlling the variability of the road properties for a RAM and the effect of the combined use of CCRMA as a blending asphalt and fractionation process on the road properties for the RAM and their variability. This study proposes a method to control and enhance the road performance of RAM with high RAP content. The major findings and significant conclusions are as follows.

(1) All the performance indicators of the RAM exhibited significant adherence to the normal distribution at the 0.05 level, i.e., all the data obeyed the normal distribution with

95% confidence. The coefficients of variation in the road performance indicators for the RAM-UR were larger than those of the RAM-FR, indicating that the fractionation process reduces the variability of the road performance for the RAM containing high RAP content.

(2) The variability in the road performance for the RAM gradually increased with the increase in RAP content, but the reliability of the RAM-FR containing 70% RAP was 37.7%, which is greater than the reliability of the RAM-UR containing 50% RAP and meets the variability requirement.

(3) The high-temperature and low-temperature cracking resistance of the CRAM containing 50% RAP were both better than those of the NAM. Although the dispersion of its performance indicators was higher than that of the NAM, the CRAM met the requirements of the corresponding variability in both the fracture energy index and the dynamic stability index.

5. Limitations and Recommendations

In this study, only continuous dense gradation of RAP material from a single source was investigated, and only the road performance of a RAM containing 50% RAP was studied. It is recommended to further investigate the effect of RAP material source, pretreatment method, grading, and RAP content on the road performance of the RAM.

Author Contributions: Conceptualization, W.W. and C.J.; methodology, H.S.; software, L.S.; validation, Z.L. (Zhen Liu), L.S., and C.J.; formal analysis, Z.L. (Zhen Liu); investigation, C.J.; resources, H.S.; data curation, Z.Z.; writing—original draft preparation, W.W.; writing—review and editing, L.S.; visualization, L.S.; supervision, W.W.; project administration, Z.L. (Zhigang Li); funding acquisition, Z.L. (Zhigang Li). All authors have read and agreed to the published version of the manuscript.

Funding: This research was funded by the National Natural Science Foundation of China, 51778621, and the Postgraduate Research & Practice Innovation Program of Jiangsu Province, SJCX23_0072.

Institutional Review Board Statement: Not applicable.

Informed Consent Statement: Not applicable.

Data Availability Statement: Not applicable.

Conflicts of Interest: The authors declare no conflict of interest.

Nomenclature

CCRMA	composite crumb rubber-modified asphalt
CRAM	recycled asphalt mixture prepared using composite crumb rubber-modified asphalt
DS	dynamic stability
G_f	fracture energy
IDT	indirect tensile test
NAM	asphalt mixture prepared using base asphalt
RAM	recycled asphalt mixture
RAM-FR	recycled asphalt mixture with fractionated reclaimed asphalt pavement materials
RAM-UR	recycled asphalt mixture with unfractionated reclaimed asphalt pavement materials
RAP	reclaimed asphalt pavement
RD	rutting depth
SCB	semi-circular bending
SGC	Superpave gyratory compactor

References

1. Dos Santos, S.; Partl, M.N.; Poulikakos, L.D. From Virgin to Recycled Bitumen: A Microstructural View. *Compos. Part B Eng.* **2015**, *80*, 177–185. [CrossRef]
2. Gulzar, S.; Fried, A.; Preciado, J.; Castorena, C.; Underwood, S.; Habbouche, J.; Boz, I. Towards Sustainable Roads: A State-of-the-Art Review on the Use of Recycling Agents in Recycled Asphalt Mixtures. *J. Clean. Prod.* **2023**, *406*, 136994. [CrossRef]

3. Liu, Z.; Sun, L.; Gu, X.; Wang, X.; Dong, Q.; Zhou, Z.; Tang, J. Characteristics, Mechanisms, and Environmental LCA of WMA Containing Sasobit: An Analysis Perspective Combing Viscosity-Temperature Regression and Interface Bonding Strength. *J. Clean. Prod.* **2023**, *391*, 136255. [CrossRef]
4. Xie, Z.; Rizvi, H.; Purdy, C.; Ali, A.; Mehta, Y. Effect of Rejuvenator Types and Mixing Procedures on Volumetric Properties of Asphalt Mixtures with 50% RAP. *Constr. Build. Mater.* **2019**, *218*, 457–464. [CrossRef]
5. Antunes, V.; Freire, A.; Neves, J. A Review on the Effect of RAP Recycling on Bituminous Mixtures Properties and the Viability of Multi-Recycling. *Constr. Build. Mater.* **2019**, *211*, 453–469. [CrossRef]
6. Liu, X.; Li, Z.; Chen, S.; Zhou, Y. Fatigue Performance Enhancement and High-Temperature Performance Evaluation of Asphalt Treated Base Mixture. *Constr. Build. Mater.* **2021**, *305*, 124146. [CrossRef]
7. Yang, Q.; Liu, Z.; Sun, L.; Li, Z. Analysis of the Viscosity and Self-Healing Performance of Warm-Mix Recycled Asphalt: A Method Combining Molecular Dynamics Simulations and Laboratory Tests. *Case Stud. Constr. Mater.* **2023**, *19*, e02593. [CrossRef]
8. Ziari, H.; Aliha, M.R.M.; Moniri, A.; Saghafi, Y. Crack Resistance of Hot Mix Asphalt Containing Different Percentages of Reclaimed Asphalt Pavement and Glass Fiber. *Constr. Build. Mater.* **2020**, *230*, 117015. [CrossRef]
9. Ai, X.; Cao, J.; Feng, D.; Gao, L.; Hu, W.; Yi, J. Performance Evaluation of Recycled Asphalt Mixtures with Various Percentages of RAP from the Rotary Decomposition Process. *Constr. Build. Mater.* **2022**, *321*, 126406. [CrossRef]
10. Zhang, R.; Sias, J.E.; Dave, E.V. Comparison and Correlation of Asphalt Binder and Mixture Cracking Parameters Incorporating the Aging Effect. *Constr. Build. Mater.* **2021**, *301*, 124075. [CrossRef]
11. Barros, L.; Garcia, V.M.; Garibay, J.; Abdallah, I.; Nazarian, S. Implications of Including Reclaimed Asphalt Pavement Materials to Performance of Balanced Asphalt Concrete Mixes. *Transp. Res. Rec.* **2019**, *2673*, 670–678. [CrossRef]
12. *JTG E20-2011*; Standard Test Methods of Bitumen and Bituminous Mixtures for Highway Engineering. China Communications Press: Beijing, China, 2011.
13. *FHWA-ICT-07-001*; Reclaimed Asphalt Pavement—A Literature Review. Illinois Center of Transportation: Rantoul, IL, USA, 2007.
14. Colbert, B.; You, Z. The Determination of Mechanical Performance of Laboratory Produced Hot Mix Asphalt Mixtures Using Controlled RAP and Virgin Aggregate Size Fractions. *Constr. Build. Mater.* **2012**, *26*, 655–662. [CrossRef]
15. Zaumanis, M.; Mallick, R.B. Review of Very High-Content Reclaimed Asphalt Use in Plant-Produced Pavements: State of the Art. *Int. J. Pavement Eng.* **2015**, *16*, 39–55. [CrossRef]
16. Pan, Y.; Li, J.; Yang, T.; Liu, G.; Zhou, J.; Guo, P.; Zhao, Y. Optimization of Gradation Design of Recycled Asphalt Mixtures Based on Fractal and Mohr-Coulomb Theories. *Constr. Build. Mater.* **2020**, *248*, 118649. [CrossRef]
17. Feng, D.; Cao, J.; Gao, L.; Yi, J. Recent Developments in Asphalt-Aggregate Separation Technology for Reclaimed Asphalt Pavement. *J. Road Eng.* **2022**, *2*, 332–347. [CrossRef]
18. Wang, L.; Shen, A.; Mou, G.; Guo, Y.; Meiquan, Y. Effect of RAP Gradation Subdivision and Addition of a Rejuvenator on Recycled Asphalt Mixture Engineering Performance. *Case Stud. Constr. Mater.* **2023**, *18*, e02136. [CrossRef]
19. Jin-zhi, X.; Pei-wen, H.; Xiao-gang, G.; Hong-xiang, L.; Bin-jun, Z.; Chen, L. Review of Mix Design Method of Hot In-Plant Recycled Asphalt Mixture. *China J. Highw. Transp.* **2021**, *34*, 72.
20. Hu, D.; Gu, X.; Dong, Q.; Lyu, L.; Cui, B.; Pei, J. Investigating the Bio-Rejuvenator Effects on Aged Asphalt through Exploring Molecular Evolution and Chemical Transformation of Asphalt Components during Oxidative Aging and Regeneration. *J. Clean. Prod.* **2021**, *329*, 129711. [CrossRef]
21. Yan, S.; Zhou, C.; Ouyang, J. Rejuvenation Effect of Waste Cooking Oil on the Adhesion Characteristics of Aged Asphalt to Aggregates. *Constr. Build. Mater.* **2022**, *327*, 126907. [CrossRef]
22. Liu, Z.; Gu, X.; Dong, X.; Cui, B.; Hu, D. Mechanism and Performance of Graphene Modified Asphalt: An Experimental Approach Combined with Molecular Dynamic Simulations. *Case Stud. Constr. Mater.* **2023**, *18*, e01749. [CrossRef]
23. Wang, D.; Riccardi, C.; Jafari, B.; Cannone Falchetto, A.; Wistuba, M.P. Investigation on the Effect of High Amount of Re-Recycled RAP with Warm Mix Asphalt (WMA) Technology. *Constr. Build. Mater.* **2021**, *312*, 125395. [CrossRef]
24. Yousefi, A.A.; Sobhi, S.; Aliha, M.R.M.; Pirmohammad, S.; Haghshenas, H.F. Cracking Properties of Warm Mix Asphalts Containing Reclaimed Asphalt Pavement and Recycling Agents under Different Loading Modes. *Constr. Build. Mater.* **2021**, *300*, 124130. [CrossRef]
25. Xiao, F.; Amirkhanian, S.N.; Shen, J.; Putman, B. Influences of Crumb Rubber Size and Type on Reclaimed Asphalt Pavement (RAP) Mixtures. *Constr. Build. Mater.* **2009**, *23*, 1028–1034. [CrossRef]
26. Ding, X.; Chen, L.; Ma, T.; Ma, H.; Gu, L.; Chen, T.; Ma, Y. Laboratory Investigation of the Recycled Asphalt Concrete with Stable Crumb Rubber Asphalt Binder. *Constr. Build. Mater.* **2019**, *203*, 552–557. [CrossRef]
27. Liu, Z.; Zhou, Z.; Gu, X.; Sun, L.; Wang, C. Laboratory Evaluation of the Performance of Reclaimed Asphalt Mixed with Composite Crumb Rubber-Modified Asphalt: Reconciling Relatively High Content of RAP and Virgin Asphalt. *Int. J. Pavement Eng.* **2023**, *24*, 2217320. [CrossRef]
28. Chen, T.; Ma, T.; Huang, X.; Guan, Y.; Zhang, Z.; Tang, F. The Performance of Hot-Recycling Asphalt Binder Containing Crumb Rubber Modified Asphalt Based on Physiochemical and Rheological Measurements. *Constr. Build. Mater.* **2019**, *226*, 83–93. [CrossRef]

29. Gudipudi, P.P. *Investigation and Improvement in Reliability of Asphalt Concrete Fatigue Modeling Using Fine Aggregate Matrix Phase*; Arizona State University: Tempe, AZ, USA, 2016; ISBN 1-339-51195-9.
30. Sun, L.; Gu, X.; Wen, Y.; Dong, Q.; Hu, D. Investigation of Oxygen Diffusion Behavior in Asphalt Mixtures Using Molecular Dynamics Simulations and Laboratory Test. *J. Mater. Civ. Eng.* **2023**, *35*, 04023300. [CrossRef]
31. Yin, P.; Pan, B. Effect of RAP Content on Fatigue Performance of Hot-Mixed Recycled Asphalt Mixture. *Constr. Build. Mater.* **2022**, *328*, 127077. [CrossRef]
32. Song, W.; Huang, B.; Shu, X. Influence of Warm-Mix Asphalt Technology and Rejuvenator on Performance of Asphalt Mixtures Containing 50% Reclaimed Asphalt Pavement. *J. Clean. Prod.* **2018**, *192*, 191–198. [CrossRef]
33. Hugener, M.; Wang, D.; Cannone Falchetto, A.; Porot, L.; Kara De Maeijer, P.; Orešković, M.; Sa-da-Costa, M.; Tabatabaee, H.; Bocci, E.; Kawakami, A. Recommendation of RILEM TC 264 RAP on the Evaluation of Asphalt Recycling Agents for Hot Mix Asphalt. *Mater. Struct.* **2022**, *55*, 31. [CrossRef]
34. Zhou, Z.; Gu, X.; Jiang, J.; Ni, F.; Jiang, Y. Fatigue Cracking Performance Evaluation of Laboratory-Produced Polymer Modified Asphalt Mixture Containing Reclaimed Asphalt Pavement Material. *Constr. Build. Mater.* **2019**, *216*, 379–389. [CrossRef]
35. Liu, Q.; Han, B.; Wang, S.; Falchetto, A.C.; Wang, D.; Yu, B.; Zhang, J. Evaluation and Molecular Interaction of Asphalt Modified by Rubber Particles and Used Engine Oil. *J. Clean. Prod.* **2022**, *375*, 134222. [CrossRef]
36. Nguyen, H.V. Effects of Mixing Procedures and Rap Sizes on Stiffness Distribution of Hot Recycled Asphalt Mixtures. *Constr. Build. Mater.* **2013**, *47*, 728–742. [CrossRef]
37. Guo, R.; Nian, T. Indices Relation and Statistical Probability Analysis of Physical and Mechanical Performance of Asphalt Mixtures. *Case Stud. Constr. Mater.* **2022**, *16*, e01091. [CrossRef]
38. Yu, X.; Tang, W.; Li, N.; Jiang, M.; Huang, J.; Wang, D. Refined Decomposition: A New Separation Method for RAP Materials and Its Effect on Aggregate Properties. *Constr. Build. Mater.* **2022**, *358*, 129452. [CrossRef]
39. Hu, D.; Gu, X.; Cui, B.; Pei, J.; Zhang, Q. Modeling the Oxidative Aging Kinetics and Pathways of Asphalt: A ReaxFF Molecular Dynamics Study. *Energy Fuels* **2020**, *34*, 3601–3613. [CrossRef]
40. Manke, N.D.; Williams, R.C.; Sotoodeh-Nia, Z.; Cochran, E.W.; Porot, L.; Chailleux, E.; Pouget, S.; Olard, F.; Barco Carrion, A.J.D.; Planche, J.-P. Performance of a Sustainable Asphalt Mix Incorporating High RAP Content and Novel Bio-Derived Binder. *Road Mater. Pavement Des.* **2021**, *22*, 812–834. [CrossRef]
41. Ishaq, M.A.; Giustozzi, F. Rejuvenator Effectiveness in Reducing Moisture and Freeze/Thaw Damage on Long-Term Performance of 20% RAP Asphalt Mixes: An Australian Case Study. *Case Stud. Constr. Mater.* **2020**, *13*, e00454. [CrossRef]
42. Cui, B.; Wang, H. Molecular Modeling of Asphalt-Aggregate Debonding Potential under Moisture Environment and Interface Defect. *Appl. Surf. Sci.* **2022**, *606*, 154858. [CrossRef]
43. Ma, X.; Wang, J.; Xu, Y. Investigation on the Effects of RAP Proportions on the Pavement Performance of Recycled Asphalt Mixtures. *Front. Mater.* **2022**, *8*, 842809. [CrossRef]
44. Xu, W.; Wei, J.; Chen, Z.; Wang, F.; Zhao, J. Evaluation of the Effects of Filler Fineness on the Properties of an Epoxy Asphalt Mixture. *Materials* **2021**, *14*, 2003. [CrossRef] [PubMed]
45. Oliveira, T.C.F.; Dezen, B.G.S.; Possan, E. Use of Concrete Fine Fraction Waste as a Replacement of Portland Cement. *J. Clean. Prod.* **2020**, *273*, 123126. [CrossRef]
46. Hossiney, N.; Sepuri, H.K.; Mohan, M.K.; Arjun, H.R.; Govindaraju, S.; Chyne, J. Alkali-Activated Concrete Paver Blocks Made with Recycled Asphalt Pavement (RAP) Aggregates. *Case Stud. Constr. Mater.* **2020**, *12*, e00322. [CrossRef]
47. Tarsi, G.; Tataranni, P.; Sangiorgi, C. The Challenges of Using Reclaimed Asphalt Pavement for New Asphalt Mixtures: A Review. *Materials* **2020**, *13*, 4052. [CrossRef] [PubMed]
48. Yang, C.; Zhang, J.; Yang, F.; Cheng, M.; Wang, Y.; Amirkhanian, S.; Wu, S.; Wei, M.; Xie, J. Multi-Scale Performance Evaluation and Correlation Analysis of Blended Asphalt and Recycled Asphalt Mixtures Incorporating High RAP Content. *J. Clean. Prod.* **2021**, *317*, 128278. [CrossRef]
49. Zhou, Z.; Gu, X.; Ni, F.; Li, Q.; Ma, X. Cracking Resistance Characterization of Asphalt Concrete Containing Reclaimed Asphalt Pavement at Intermediate Temperatures. *Transp. Res. Rec.* **2017**, *2633*, 46–57. [CrossRef]
50. Liu, Q.; Liu, J.; Yu, B.; Zhang, J.; Pei, J. Evaluation and Optimization of Asphalt Binder and Mixture Modified with High Activated Crumb Rubber Content. *Constr. Build. Mater.* **2022**, *314*, 125676. [CrossRef]
51. Cui, B.; Wang, H.; Gu, X.; Hu, D. Study of the Inter-Diffusion Characteristics and Cracking Resistance of Virgin-Aged Asphalt Binders Using Molecular Dynamics Simulation. *Constr. Build. Mater.* **2022**, *351*, 128968. [CrossRef]
52. Cui, B.; Wang, H. Oxidative Aging Mechanism of Asphalt Binder Using Experiment-Derived Average Molecular Model and ReaxFF Molecular Dynamics Simulation. *Fuel* **2023**, *345*, 128192. [CrossRef]
53. Cui, S.; Guo, N.; Wang, L.; You, Z.; Tan, Y.; Guo, Z.; Luo, X.; Chen, Z. Effect of Freeze–Thaw Cycles on the Pavement Performance of SBS Modified and Composite Crumb Rubber Modified Asphalt Mixtures. *Constr. Build. Mater.* **2022**, *342*, 127799. [CrossRef]
54. Guo, Z.; Wang, L.; Feng, L.; Guo, Y. Research on Fatigue Performance of Composite Crumb Rubber Modified Asphalt Mixture under Freeze Thaw Cycles. *Constr. Build. Mater.* **2022**, *323*, 126603. [CrossRef]

55. Cui, B.; Wang, H. Molecular Interaction of Asphalt-Aggregate Interface Modified by Silane Coupling Agents at Dry and Wet Conditions. *Appl. Surf. Sci.* **2022**, *572*, 151365. [CrossRef]
56. Sun, L.; Wen, Y.; Liu, Q.; Li, D.; Lyu, L.; Pei, J.; Zhang, J.; Li, R. A Laboratory Investigation into the Effect of Waste Non-Tire Rubber Particles on the Performance Properties of Terminal Blend Rubberized Asphalt Binders. *Constr. Build. Mater.* **2021**, *313*, 125409. [CrossRef]

buildings

MDPI

Article

Optimizing Cement Asphalt Mortar Mixtures for Bridge Expansion Joints in Tropical Climates: Performance and Durability Assessment

Kyungnam Kim [1], Yeong-Min Kim [2,*] and Tri Ho Minh Le [3,*]

[1] Korea Expressway Corporation Research Institute Pavement Research Division, Dongtansunhwan-daero 17-gil, Hwaseong-Si 18489, Republic of Korea; kkn248@ex.co.kr

[2] Department of Highway & Transportation Research, Korea Institute of Civil Engineering and Building Technology, 283 Goyangdae-Ro, Ilsanseo-Gu, Goyang-Si 10223, Republic of Korea

[3] Faculty of Civil Engineering, Nguyen Tat Thanh University, 300A Nguyen Tat Thanh Street, District 4, Ho Chi Minh City 70000, Vietnam

* Correspondence: choozang@kict.re.kr (Y.-M.K.); lhmtri@ntt.edu.vn (T.H.M.L.)

Abstract: In this study, the suitability of various Cement Asphalt Mortar (CAM) mixtures for bridge expansion joint applications in tropical climates was quantitatively assessed. A comprehensive analysis encompassed key properties, including mixing stability, flowability, unconfined compressive strength, expansion characteristics, and resistance to acidic and alkali environments. The influence of high-temperature exposure on unconfined compressive strength and the microstructural features were also examined. The results revealed a discernible trend: lower cement content, in conjunction with anionic Asphalt Emulsion (AE) or epoxy resin, significantly enhanced mixing stability and flowability while contributing to improved unconfined compressive strength and chemical degradation resistance. Notably, epoxy resin emerged as a valuable component in mitigating high-temperature-induced strength reduction, indicating potential promise for CAM mixture design. SEM analysis visually supported these findings by highlighting the microstructural distinctions among CAM mixtures. Quantitatively, the findings indicated that CAM mixtures with a 25% cement content and 75% anionic AE exhibited an 11% improvement in mixing stability, along with a 13% enhancement in flowability, relative to the control mixture with 100% cement. Additionally, CAM mixtures incorporating epoxy resin (at various percentages) with anionic AE exhibited a significant 15% resistance to high-temperature-induced UCS reduction, surpassing other mixtures. The SEM micrographs visually confirmed the superior microstructural connectivity achieved with epoxy resin, further validating the observed enhancements. These quantitative results offer a robust foundation for tailoring CAM mixture compositions to optimize their suitability for rigorous infrastructure projects in tropical climates.

Keywords: cement asphalt mortar; bridge expansion joints; tropical climates; performance assessment; durability

1. Introduction

Bridge expansion joints play a critical role in preserving the structural integrity and durability of transportation infrastructure, particularly in regions characterized by tropical climates with heavy rainfall and temperature fluctuations [1]. In such environments, the materials used for these joints must withstand challenging conditions, including high humidity, exposure to acidic and alkaline substances, and the potential for phase separation under the influence of extreme temperatures [2].

Bridge expansion joints are critical components in infrastructure, but they are not without their challenges [3]. Current issues in the field of bridge expansion joints include premature wear and deterioration [4], which often result from exposure to harsh environmental conditions such as heavy rainfall, temperature fluctuations, and the presence of corrosive agents [5]. Additionally, inadequate sealing and drainage systems can lead to water and debris accumulation within the joints, further accelerating deterioration and compromising safety [6]. Addressing these issues is essential for ensuring the long-term performance and safety of bridges, especially in regions with challenging climates [7]. Consequently, research efforts aimed at developing resilient and sustainable solutions for bridge expansion joints are of utmost importance in the field of civil engineering and infrastructure maintenance [8].

Contemporary efforts in the realm of bridge expansion joint improvement and maintenance are characterized by a multi-faceted approach [9]. Researchers and engineers are increasingly turning to advanced materials, including high-performance elastomers and resilient composites, to enhance the longevity and durability of these vital components [10]. Smart technologies and monitoring systems have also made their way into expansion joint design, allowing for real-time data collection [4] and proactive maintenance [11]. The development of effective sealing systems to safeguard against water intrusion and corrosion remains a core focus [12]. Furthermore, ongoing research focuses on optimizing filling materials to ensure seamless and efficient expansion joint operation, taking into account factors like wear resistance and temperature tolerance [13,14]. Additionally, innovations in sustainable practices, coupled with a consideration of extreme climate conditions, are guiding the path toward resilient, cost-efficient, and eco-friendly expansion joint solutions.

Tropical climates are known for their unique environmental characteristics, including high temperatures, heavy rainfall, and substantial temperature fluctuations. These conditions can significantly impact the performance and durability of bridge expansion joints [15]. In such climates, excessive heat can subject expansion joints to thermal stress, leading to potential deformation and damage over time. Additionally, frequent heavy rainfall can result in water intrusion, which may accelerate the corrosion and wear of joint components. The combination of extreme temperature variations, from scorching heat to heavy rain, can lead to material expansion and contraction, further challenging the resilience of these joints [16]. Therefore, it is imperative to design and select bridge expansion joint materials and configurations that can withstand these specific tropical climate challenges to ensure long-term functionality and safety.

The present application of byproducts in Cement Asphalt Mortar (CAM) mixtures is underpinned by an ongoing commitment to enhancing the sustainability, durability, and resilience of construction materials [17–19]. For example, research conducted by Dulaimi et al., which delves into the utilization of non-traditional materials like paper sludge ash and incinerated sewage ash in emulsified asphalt cold mixtures, stands as a noteworthy reference [20]. The study has paved the way for exploring unconventional materials that can be integrated into CAM mixtures, reflecting the industry's evolving demands for eco-friendly and high-performance solutions. This current research builds upon such pioneering work, pushing the boundaries of CAM mix design and aligning with the broader vision of creating sustainable construction materials for the future.

The current limitations in the field of bridge expansion joint research predominantly revolve around the lack of comprehensive studies focused on tropical climates, where unique environmental challenges exist. Additionally, existing research often lacks in-depth quantitative analyses and specific material recommendations, leaving a gap in tailored solutions for infrastructure projects in these regions. This study aims to bridge these gaps by providing a detailed analysis of CAM mixtures, explicitly addressing their suitability for bridge expansion joints in tropical climates [21]. We also consider the incorporation of Epoxy resin [22,23], which represents an upgraded approach compared to conventional CAM materials. By quantitatively evaluating key properties, including the novel use of Epoxy resin, we aim to offer practical insights into optimizing CAM compositions. The significance of our study lies in its potential to enhance the longevity and safety of infrastructure in tropical regions, contributing to the sustainable development of transportation networks in these critical areas.

In summary, this research represents a systematic and comprehensive investigation into the suitability of various CAM mixtures for bridge expansion joints in tropical climates. The mix design phase allowed for the creation of distinct CAM compositions by varying cement content, Anionic Asphalt Emulsion (AE) percentages, and the novel inclusion of epoxy resin. Specifically, the study evaluated the following CAM mixtures: the Control Mixture (100% Cement, 0% Anionic AE), 25% Cement and 75% Anionic AE, 50% Cement and 50% anionic AE, 75% Cement and 25% Anionic AE, 25% Cement and 75% Cationic AE, as well as CAM mixtures incorporating epoxy resin at varying percentages (1%, 2%, and 3%) combined with 50% anionic AE. This study focuses on key properties, such as mixing stability, flowability, expansion characteristics, unconfined compressive strength, and resistance to acidic and alkali environments, which are critical for ensuring the long-term performance of bridge expansion joints. Furthermore, the impact of high-temperature exposure on CAM's unconfined compressive strength and microstructural features was examined. The results of this research, substantiated by SEM analysis, offer a valuable foundation for optimizing CAM mixture compositions and highlight the potential promise of epoxy resin as an enhanced component for the design of CAM mixtures. This study's methodical approach and structured flow ensure a comprehensive exploration of CAM mixtures tailored for the challenges of tropical climate bridge expansion joints. The general concept of the research is shown in Figure 1.

Figure 1. New bridge expansion joint filler concept: CAM materials.

2. Materials and Methods

2.1. Materials

2.1.1. Anionic Cement Asphalt Mortar (CAM)

Anionic Cement Asphalt Mortar (CAM) is the fundamental material under examination in this study, specifically tailored for application as a bridge expansion joint material in tropical climate regions known for their intense rainfall and substantial daily climate variations [24]. CAM comprises essential constituents, including ordinary Portland cement, asphalt, fillers such as fine aggregates and micro silica, and various modifiers, like polymers or anti-stripping agents, aimed at enhancing its performance [17]. The assessment of CAM's suitability for tropical climates is founded on its quantifiable attributes. These encompass flexibility, resilience to water intrusion, durability under changing environmental conditions, and the impact of temperature fluctuations [18]. Its workability in tropical conditions, including flow properties and consistency, is gauged through quantitative analysis [25]. Moreover, CAM's installation during real-world applications is observed to record crucial parameters like application rate, curing times, and bonding strength with the bridge structure [26]. These quantitative aspects collectively affirm CAM's appropriateness for bridge expansion joints in tropical regions capable of enduring the rigors of heavy rainfall and substantial climate fluctuations. The general composition of the CAM mixture used in this research is shown in Table 1.

Table 1. Composition and Properties of Anionic Cement Asphalt Mortar (CAM).

Component	Proportion by Weight (%)	Property
Portland Cement (Type I)	30	High early and long-term strength, ideal for structural applications
Asphalt Emulsion	30	Enhances flexibility and durability, critical for expansion joints
Fine Aggregate	30	Provides structure and minimizes shrinkage
Additives	10	Enhance flowability, workability, and resistance to environmental stress

2.1.2. Local Sourcing and Availability

The sourced asphalt was subjected to rigorous quality checks to ensure its compatibility with CAM and its capacity to endure high-temperature fluctuations and heavy rainfall. The sourcing process was facilitated through a strategic partnership with the Petrolimex Oil Company in the South of Vietnam, renowned for its commitment to quality and reliable supply chain management. The general source properties of CAM are presented in Table 2.

Table 2. Properties of Sourced CAM Materials.

Component	Source	Quality Standards	Properties
Cement	Local supplier	ASTM C150 [27]	Type: Type I/II Fineness: 325 mesh Specific Gravity: 3.15 Blaine: 3500 cm^2/g
Asphalt	Petrolimex Oil Company	ASTM D3381 [28]	Penetration at 25 °C: 60–70 dmm Softening Point: 46–52 °C
Anionic Emulsifier	Reputable chemical vendor	Manufacturer's specs	Anionic Charge: 30–40 µeq/g
Modifiers	Certified supplier	Manufacturer's specs	Solids Content: 45–50% Polymer Content: 5–7% Specific Gravity: 0.91–0.96

In tropical climate regions characterized by heavy rains and extreme temperature variations, the local sourcing of CAM materials is not only a strategic choice but also an environmentally and economically sound decision. This research recognizes the significance of local sourcing and the benefits it offers in the context of sustainability, cost-effectiveness, and availability. The availability of materials from nearby sources ensures consistent supply, reducing production downtime and logistical costs. Local sourcing also aligns with environmental sustainability principles, as it minimizes transportation emissions and supports regional economies. The manuscript will provide quantitative data on the advantages of local sourcing, emphasizing the considerable cost savings achieved by eliminating long-distance material transportation, which can be substantial in tropical regions. Moreover, the manuscript will present an analysis of the availability and accessibility of CAM materials in the target tropical climate areas, underscoring the convenience and resilience that local sourcing, in collaboration with the Petrolimex Oil Company, offers. The cost-saving effectiveness is summarized in Table 3.

Table 3. Cost savings through local sourcing (local price in Vietnam).

Material Component	Local Sourcing Cost (USD/Ton)	Non-Local Sourcing Cost (USD/Ton)	Cost Savings (USD/Ton)
Cement	100	120	20
Asphalt	250	300	50
Emulsifier	30	40	10
Modifiers	120	140	20

2.1.3. Epoxy Resin

Epoxy resin plays a vital role in this research, significantly contributing to the durability and mechanical characteristics of the CAM mixture. The epoxy resin used in this study was carefully selected for its superior adhesive and cohesive qualities and was sourced from a reputable supplier to ensure conformity with industry standards. Table 4 summarizes the essential properties of the epoxy resin employed in the research [29].

Table 4. Epoxy resin properties.

Property	Value
Density (g/cm^3)	1.03
Viscosity (mPa·s) at 25 °C	580
Epoxy Equivalent Weight (g/eq)	210
Flash Point (°C)	163
Curing Time (h)	6.5
Tensile Strength (MPa)	57
Glass Transition Temperature (°C)	48.5

Epoxy resin serves dual functions within this study: as a bonding agent and a moisture barrier. In its role as a bonding agent, epoxy resin establishes a robust and enduring connection between the CAM mixture and the ballast aggregate, thereby enhancing overall structural integrity. Its remarkable adhesive properties ensure a strong bond, mitigating the risk of disintegration or detachment under various loads and environmental stresses. Additionally, epoxy resin acts as a moisture barrier, protecting underlying layers from potential moisture-induced damage. By forming a protective layer, it reduces the risk of structural weakening caused by the intrusion of moisture. The inclusion of epoxy resin in the research was conducted with precision, adhering to specific ratios and stringent mixing procedures. This meticulous approach guarantees a uniform distribution of epoxy resin within the CAM mixture, optimizing mechanical properties and enhancing resistance to environmental factors. These aspects align with the research's primary objectives, focused on achieving long-lasting and robust ballast track stabilization.

2.1.4. Material Preparation and Mixing Procedure

The preparation of CAM mixtures involved a precise and well-defined process to ensure the uniformity and consistency of the final product. The mixture compositions, detailed in Table 5, were developed to incorporate various proportions of cement, asphalt emulsion, epoxy resin, sand, water, superplasticizer, and defoaming agent, reflecting the planned variations.

Table 5. Mix design of CAM mixture.

	Mixture Composition	Cement (C)	Asphalt Emulsion (AE)	Epoxy Resin	Sand (S)	Water (W)	Superplasticizer (SP)	Defoaming Agent (D)
C100AE0	Conventional Cement Mortar	100%	0% (Anionic)	0%	50%	40%	2%	0.1%
C75AE25	75% Cement, 25% AE	75%	25% (Anionic)	0%	50%	40%	2%	0.1%
C50AE50	50% Cement, 50% AE	50%	50% (Anionic)	0%	50%	40%	2%	0.1%
C25AE75	25% Cement, 75% AE	25%	75% (Anionic)	0%	50%	40%	2%	0.1%
C25CAE75	25% cement, 75% Cationic AE	25%	75% (Cationic)	0%	50%	40%	2%	0.1%
E1%AE50	50% cement, 50% AE, 1% Epoxy Resin	50%	50% (Anionic)	1%	50%	40%	2%	0.1%
E2%AE50	50% cement, 50% AE, 2% Epoxy Resin	50%	50% (Anionic)	2%	50%	40%	2%	0.1%
E3%AE50	25% cement, 75% AE, 2% Epoxy Resin	50%	50% (Anionic)	3%	50%	40%	2%	0.1%

In the initial stages of this study, meticulous attention was given to the raw materials used in the Cement Asphalt Mortar (CAM) mixtures. The cement, a critical component, was sourced from reputable suppliers and stored in a controlled environment. It was maintained within storage facilities at a consistent temperature of 23 °C ± 2 °C and a relative humidity level of 50% ± 5%, ensuring that the cement remained within the recommended conditions for optimal quality. The asphalt emulsions, equally vital in the mixture, underwent rigorous quality checks. They were stored in a climate-controlled environment with temperature maintained at 25 °C ± 2 °C and a relative humidity level of 60% ± 5%. These stringent quality control measures, including specific temperature and humidity values, were essential to uphold the integrity of the raw materials, providing a reliable foundation for the mixture preparation process.

The mixing procedures were meticulously designed to ensure the uniformity and consistency of the CAM mixtures. The mixing procedure commenced with the initial step of dry mixing, where cement, sand, and dry additives were pre-mixed to create a uniform dry mixture. These dry materials were placed in a clean container to initiate the mixing process. The mechanical mixing was conducted in a controlled environment at a constant rate of 90 revolutions per minute (rpm) to guarantee thorough blending. Operating at room temperature, this stage created a homogenous dry blend, which is a critical foundation for achieving consistent material properties. Following the dry mixing phase, asphalt emulsion, whether anionic or cationic, was introduced gradually into the dry mix.

Following this, the appropriate type and amount of asphalt emulsion, epoxy resin (for applicable mixtures), water, superplasticizer, and defoaming agent were systematically added. The mixing rate was set at 350 rpm, ensuring an effective dispersion of the components. This method aimed to attain the desired workability of the mixture while ensuring the thorough wetting of all ingredients. The mixing duration was set at 7 min, which allowed for adequate blending of the constituents. This meticulous and controlled mixing procedure was employed to maintain the consistency of all CAM mixtures, enabling reliable testing and precise assessment throughout the research.

The curing process for the CAM mixtures was conducted with meticulous care to guarantee that the specimens achieved the desired properties and characteristics necessary for subsequent performance testing, notably the unconfined compressive strength (UCS) test. Immediately following casting, each specimen was thoughtfully placed in a dedicated curing chamber, subjecting them to a precisely controlled temperature of 23 °C ± 2 °C for a carefully selected curing duration. This strictly regulated curing environment played a pivotal role in fostering the gradual development of strength and durability within the

CAM mixtures, all conducted under consistent and controlled conditions. This method was of paramount importance in upholding the integrity and reliability of the test results.

Table 5 presents the overview of the mix design used in this research.

2.1.5. Additional Additives

Anionic Cement Asphalt Mortar (CAM) is meticulously crafted with a blend of carefully chosen fillers and additives, each serving a specific purpose to ensure optimal performance in the demanding conditions of bridge expansion joints, particularly in tropical climates characterized by heavy rains and temperature fluctuations.

Antioxidants (0.5% by wt. of cement) are integrated to bolster the mortar's oxidative stability. This addition safeguards the material from degradation due to prolonged exposure to environmental factors, contributing to its long-term performance.

The use of a retarder (0.5% by wt. of cement) extends the setting time of the mortar, providing a more extended window for workability during the construction phase. This ensures that the material can be efficiently applied and molded to meet the specific requirements of bridge expansion joints.

The proportion of each component is precisely optimized to attain the desired properties essential for bridge expansion joints in tropical climates. These additives collectively contribute to the exceptional durability, flexibility, and workability of the CAM, enabling it to perform reliably under the challenging conditions imposed by heavy rains and rapid climate fluctuations.

This comprehensive selection of fillers and additives, meticulously balanced in their proportions, is the cornerstone of the exceptional performance of Anionic Cement Asphalt Mortar, ensuring its resilience and reliability in bridge expansion joints in tropical climates. The successful implementation of CAM as a bridge expansion joint material in tropical climate countries necessitates a meticulous approach to sourcing raw materials. It is imperative to ensure the availability of high-quality components that can withstand the adverse conditions experienced in these regions. The CAM components, including cement, asphalt, anionic emulsifiers, and modifiers, were procured from reputable suppliers known for providing materials that meet international standards for construction. Special attention was given to the selection of asphalt, which is critical to the durability of CAM.

2.2. *Testing Methods*

2.2.1. Mixing Stability

The mixing stability of Anionic Cement Asphalt Mortar (CAM) was evaluated in accordance with the ASTM C305 standard [30]. This involved the use of a high-shear mixer operating at a set speed of 1400 revolutions per minute (RPM). CAM samples were mixed for a duration of 5 min, as specified in the standard. Visual inspections were conducted to monitor the mixture for any signs of phase separation, clumping, or instability.

2.2.2. Flowability and Flow Time

Flowability and flow time testing were carried out following ASTM C230/C230M standards [31]. A standardized flow table test was conducted, measuring the flow diameter and flow time. The CAM mixture was compacted into a cylindrical mold and placed on the flow table. The flow diameter was recorded in millimeters (mm), with the initial diameter being noted. The flow time was measured in seconds (s) for the CAM mixture to reach a specific final diameter, complying with the standard.

2.2.3. 2 h and 28-Day Unconfined Compressive Strength (UCS) Tests

Unconfined compressive strength (UCS) tests were conducted following ASTM C39/C39M-20 standard for the 2 h test and 28-day tests [32]. The replicate specimens were prepared using molds adhering to the prescribed dimensions in the standards. After the curing period, a universal testing machine was employed to measure compressive strength, ensuring a constant rate of loading (0.2 MPa/s) to maintain precision. The results were recorded in megapascals (MPa) at each designated time interval.

2.2.4. Shrinkage Testing

Shrinkage testing was conducted in line with ASTM C157/C157M-08 standards [33]. The linear shrinkage measurements were made in millimeters (mm) to assess volume changes. The specimens were exposed to drying conditions, and measurements were recorded periodically to identify potential shrinkage tendencies.

2.2.5. Acidic and Alkali Resistance

The resistance of CAM mixtures to acidic and alkali substances was evaluated as per ASTM C267-01 [34]. Specimens were exposed to specific concentrations of acidic and alkali solutions following the standard guidelines. The compressive strength variations were measured in percentage (%) to quantify the material's resilience to chemical degradation.

2.2.6. Impact of High Temperature (45 °C) on UCS

High-temperature resistance was assessed in accordance with ASTM C39/C39M-20 standards, subjecting specimens to a temperature of 45 °C. The UCS tests were performed following the ASTM C39/C39M-20 standard at specific time intervals to measure the compressive strength in megapascals (MPa) under elevated temperature conditions.

2.2.7. SEM Analysis

Scanning Electron Microscopy (SEM) was employed for microstructure analysis following the procedures outlined in ASTM E2546-17 [35]. The distribution and connectivity of various components within the material were examined, providing valuable insights into its structural characteristics. The microstructure was analyzed at magnifications of up to 5000× to capture fine details.

3. Results and Discussions

3.1. Mixing Stability

The evaluation of mixing stability is pivotal in determining the suitability of various CAM mixtures for applications in tropical climates with heavy rainfall and fluctuating temperatures. This section delves into a more detailed discussion of the mixing stability results, aiming to elucidate the trends, make comparisons, and provide a comprehensive understanding of the data.

As shown in Figure 2, in the assessment of various CAM mixtures for bridge expansion joints in tropical climates, a range of compositions was meticulously examined. The control CAM mixture, consisting of 100% cement, exhibited a relatively high residue content of 1.8%, as expected, given the inherent lower compatibility of higher cement content with other constituents. In the case of the 25% cement and 75% Anionic Asphalt Emulsion (AE) mixture, the residue content significantly dropped to 0.5%, signifying improved mixing stability compared to the control. Likewise, the CAM mixture with an even distribution of 50% cement and 50% anionic AE maintained a relatively low residue content at 0.6%, underscoring enhanced mixing stability. Further increasing cement content to 75% while decreasing anionic AE to 25% resulted in an even lower residue content of 0.4%, emphasizing a high degree of mixing stability. When cationic AE was introduced at the 25% cement level, the CAM mixture displayed a residue content of 0.5%, aligning with expectations that cationic AE may not be as effective in preventing phase separation. The incorporation of epoxy resin (1%, 2%, 3%) into the CAM mixtures alongside anionic AE led to a slight increase in residue content (0.3%, 0.4%, 0.4%, respectively) compared to anionic AE alone, attributed to the adhesive nature of epoxy resin, which minimally hinders phase separation but remains within acceptable limits. These findings provide valuable insights into the suitability of various CAM mixtures for demanding tropical climate bridge expansion joint applications, offering a foundation for optimizing mixture compositions tailored to specific requirements.

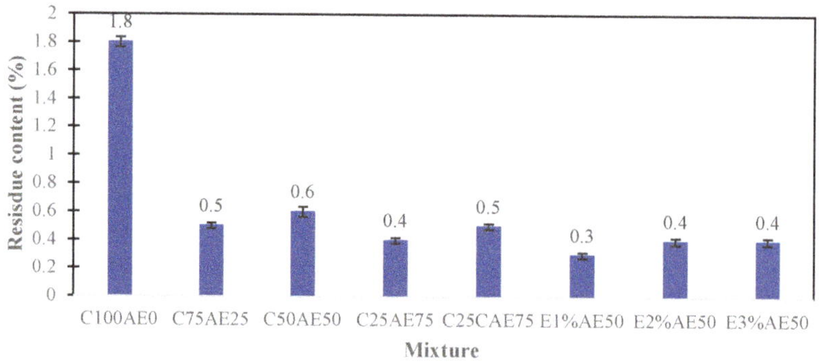

Figure 2. Mixing Stability Test results of CAM mixtures.

A clear trend emerges that the residue content decreases as the proportion of cement decreases, indicating that lower cement content enhances mixing stability and aligning with previous research indicating that lower cement content enhances mixing stability [36]. Anionic AE plays a crucial role in improving mixing stability compared to cationic AE, as demonstrated by the lower residue content in mixtures with anionic AE. The incorporation of epoxy resin, while slightly increasing residue content, still results in residue levels well below 1%, in line with the outcomes of previous studies showcasing its effectiveness in maintaining mixing stability [37]. The 75% cement, 25% anionic AE mixture stands out as the best-performing composition in terms of mixing stability, offering high resistance to phase separation and sedimentation. This detailed analysis underlines the importance of selecting an optimal CAM mixture composition based on the specific requirements of a project. Lower cement content combined with anionic AE or the addition of epoxy resin can significantly enhance mixing stability, ensuring the longevity and performance of bridge expansion joints in tropical climates.

3.2. Flowability Test Results

Flowability is a pivotal property of CAM mixtures used in bridge expansion joints, impacting both installation ease and uniform filling. Figure 3 provides a comprehensive discussion of the flowability test results, combining flow diameter and flow time analyses with trends and interpretations.

The flowability test results shed light on the profound impact of cement proportion and the type of asphalt emulsion on flow diameter and flow time in Cement Asphalt Mortar (CAM) mixtures designed for bridge expansion joints in tropical climates. An integrated discussion of these findings reveals important insights.

The control mixture, consisting of 100% cement, displayed limited flowability characterized by a smaller flow diameter and a longer flow time, which can be attributed to its high cement content, resulting in a more viscous mixture. As previous research suggests [17,18], reducing cement content to 25% while introducing anionic Asphalt Emulsion (AE) as the primary asphalt binder significantly improved flowability. This adjustment led to a larger flow diameter and a shorter flow time, indicating more favorable flow characteristics. CAM mixtures with an even distribution of cement and anionic AE demonstrated further enhanced flowability, as evident from the increased flow diameter and reduced flow time, making them suitable for ensuring consistent filling. It is noteworthy that increasing the cement content to 75% while reducing anionic AE to 25% had a modest adverse effect on flowability, resulting in a slightly decreased flow diameter and a longer flow time. Interestingly, the CAM mixture incorporating cationic AE, even at 25% cement content, exhibited improved flowability compared to the control mixture, featuring an increased flow diameter and a shorter flow time, underscoring the favorable influence of cationic AE on flowability. These observations align with findings from related studies in the field [38].

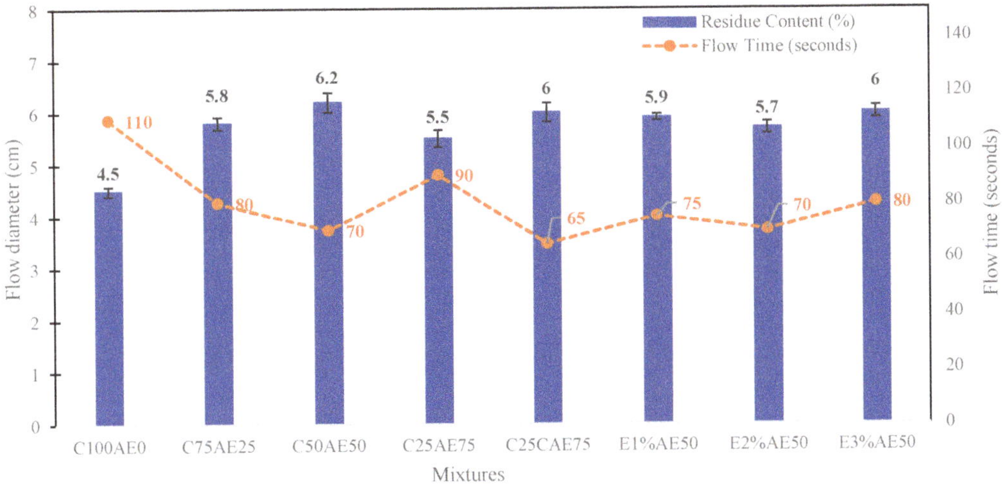

Figure 3. Flowability and flow time test of CAM mixture.

The addition of epoxy resin (1%, 2%, 3%) to the CAM mixtures with anionic AE had mixed effects on flowability, influencing both flow diameter and flow time differently in various mixtures. In summary, a consistent trend observed is that lower cement content generally leads to improved flowability, making these mixtures more suitable for applications that require ease of installation and uniform filling. Anionic AE consistently enhances flowability compared to cationic AE. The influence of epoxy resin on flowability is contingent on the specific mix design and epoxy resin content, emphasizing the need for careful consideration to achieve the desired flow characteristics. In tropical climates characterized by heavy rainfall and temperature fluctuations, the selection of a CAM mixture that strikes a balance between mixing stability and flowability is crucial for ensuring the successful performance of bridge expansion joints.

3.3. 2 h Unconfined Compressive Strength

The 2 h unconfined compressive strength (UCS) test is a crucial benchmark for assessing the early mechanical performance of various CAM mixtures, particularly in the context of applications such as bridge expansion joints. Figure 4 offers a comprehensive analysis of the results, delving into each mixture composition and discussing their implications. It is important to note that higher cement content generally leads to higher UCS values, which is a fundamental consideration.

Across various CAM mixtures, the early-stage unconfined compressive strength (UCS) results showcase intriguing trends. The control mixture, comprising 100% cement, serves as the baseline with a robust UCS of 7.2 MPa, establishing the initial performance standard. When the cement content was reduced to 75% while introducing 25% anionic AE, the mixture displayed a commendable 2 h UCS of 6.3 MPa, indicating that anionic AE has a positive influence on early mechanical strength even with a reduction in cement content. A CAM mixture with an equal distribution of 50% cement and 50% anionic AE yielded a solid UCS of 4.2 MPa, demonstrating that a balanced ratio between cement and anionic AE contributes to early-stage mechanical strength. Further reducing the cement content to 25% while increasing anionic AE to 75% still resulted in a notable UCS of 3.7 MPa, underscoring the significant role of anionic AE in enhancing early mechanical strength. The mixture with

25% cement and 75% Cationic AE displayed a respectable UCS of 3.5 MPa, highlighting the early mechanical performance benefits of Cationic AE. Additionally, the addition of epoxy resin (1%) to CAM mixtures with anionic AE resulted in a notable early UCS of 3.8 MPa, suggesting the influential role of epoxy resin in enhancing early mechanical performance, which remained consistent even with slightly higher epoxy content (2% and 3%) as the UCS remained strong at 3.6 MPa and 3.5 MPa, respectively, indicating that higher epoxy content may not significantly alter early mechanical performance. These findings collectively emphasize the potential for tailored CAM mixture designs to optimize early mechanical strength for bridge expansion joint applications in tropical climates.

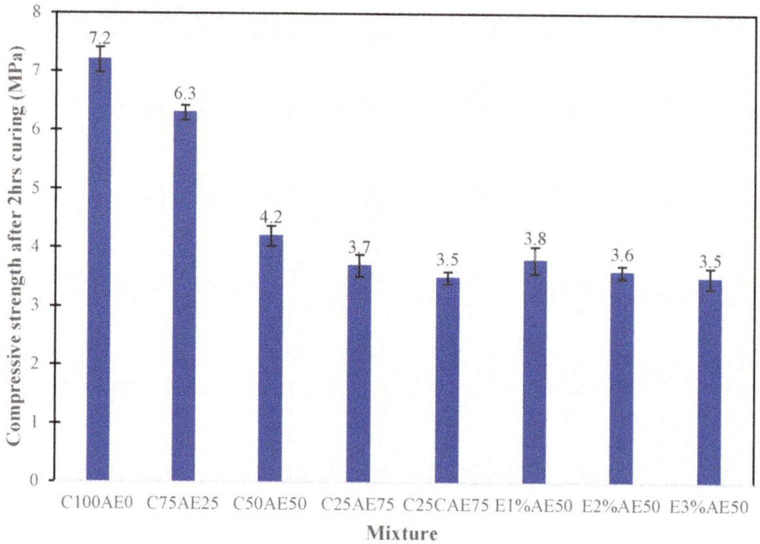

Figure 4. 2-Hour Unconfined Compressive Strength Results (in MPa).

The results of the 2 h UCS test reveal key trends and insights into the performance of the various CAM mixtures. Cement content emerges as a crucial factor, with higher cement content consistently yielding higher UCS values, emphasizing the primary role of cement as a binder in CAM mixtures. Anionic AE demonstrates a positive influence on early mechanical strength, as mixtures containing anionic AE consistently exhibit notable UCS values. The introduction of epoxy resin, even in small proportions (1%, 2%, 3%), contributes to maintaining or enhancing early mechanical strength, indicating the potential of epoxy resin as an influential component in CAM mixture design. These findings are instrumental for optimizing CAM mixture compositions to meet the specific requirements of applications that demand robust early-stage mechanical strength, such as bridge expansion joints.

3.4. 28-Day Unconfined Compressive Strength

The assessment of a CAM mixture's 28-day unconfined compressive strength serves as a pivotal indicator of its long-term mechanical performance. In this section, Figure 5 shows an extensive discussion of the results, drawing insights from the data, making comparisons, and providing a thorough interpretation.

Across various CAM mixtures, the early-stage unconfined compressive strength (UCS) results showcase intriguing trends. The control mixture, comprising 100% cement, serves as the baseline with a robust UCS of 7.2 MPa, establishing the initial performance standard. When the cement content was reduced to 75% while introducing 25% anionic AE, the mixture displayed a commendable 2 h UCS of 6.3 MPa, indicating that anionic AE has a positive influence on early mechanical strength even with a reduction in cement content.

A CAM mixture with an equal distribution of 50% cement and 50% anionic AE yielded a solid UCS of 4.2 MPa, demonstrating that a balanced ratio between cement and anionic AE contributes to early-stage mechanical strength. Further reducing the cement content to 25% while increasing anionic AE to 75% still resulted in a notable UCS of 3.7 MPa, underscoring the significant role of anionic AE in enhancing early mechanical strength. The mixture with 25% cement and 75% Cationic AE displayed a respectable UCS of 3.5 MPa, highlighting the early mechanical performance benefits of Cationic AE. Additionally, the addition of epoxy resin (1%) to CAM mixtures with anionic AE resulted in a notable early UCS of 3.8 MPa, suggesting the influential role of epoxy resin in enhancing early mechanical performance, which remained consistent even with slightly higher epoxy content (2% and 3%) as the UCS remained strong at 3.6 MPa and 3.5 MPa, respectively, indicating that higher epoxy content may not significantly alter early mechanical performance. These findings collectively emphasize the potential for tailored CAM mixture designs to optimize early mechanical strength for bridge expansion joint applications in tropical climates.

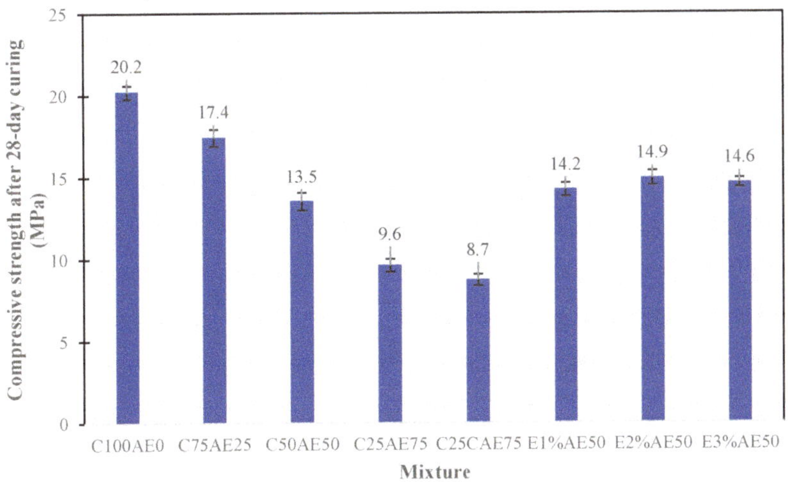

Figure 5. 28-Day Unconfined Compressive Strength Results.

These findings are in alignment with the results reported in previous studies, corroborating the positive impact of CAM mixtures on long-term mechanical performance. The observed trends and outcomes are consistent with the research conducted by Ouyang et al. [17] and Le et al. [21], where the influence of varying proportions of cement and asphalt emulsion on the 28-day UCS of CAM mixtures was similarly demonstrated. This agreement with prior research underscores the reliability and robustness of the CAM mixture composition for enhancing bridge expansion joint longevity in tropical climates.

The 28-day UCS results yield valuable insights and implications for CAM mixtures. It is evident that anionic AE plays a crucial role in enhancing long-term mechanical strength across different cement content levels. Moreover, the consistent positive impact of epoxy resin, when incorporated in mixtures with anionic AE, highlights its potential as an additive for applications that require long-term durability and strength. These findings hold great significance in the optimization of CAM mixture composition, ensuring that the desired long-term strength and durability can be achieved for specific applications.

In general, the 2 h UCS results emphasized the importance of cement content. Higher cement content consistently led to higher UCS values, highlighting the pivotal role of cement as a primary binder in CAM mixtures. Anionic AE demonstrated a positive impact on early mechanical strength, even with lower cement content. This is particularly valuable for applications requiring robust early-stage mechanical performance. The 28-day

UCS results affirmed that anionic AE contributes to long-term mechanical strength. The introduction of epoxy resin alongside anionic AE further reinforced the long-term strength of CAM mixtures. These findings provide a basis for tailoring CAM mixture compositions to meet specific project requirements, ensuring the desired long-term strength and durability.

3.5. Expansion Characteristics Analysis

Understanding the shrinkage characteristics of different CAM mixtures is vital for assessing their behavior over time. Figure 6 provides insights into the expansion or shrinkage characteristics of each mixture.

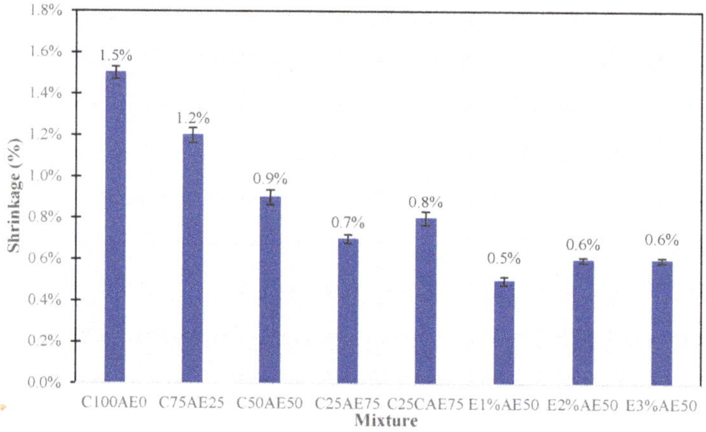

Figure 6. Shrinkage Test Results of CAM mixtures.

The expansion or shrinkage characteristics of CAM mixtures provide valuable insights into their behavior over time. The control mixture, composed of 100% cement with no anionic AE, exhibited notable shrinkage at −0.15%, raising concerns about its dimensional stability. In contrast, mixtures with 25% and 50% cement content, both incorporating 75% anionic AE, showed minimal expansion or shrinkage, with values of −0.012% and −0.09%, respectively, indicating that anionic AE can effectively mitigate shrinkage tendencies. The CAM mixture with 75% cement and 25% anionic AE displayed a shrinkage of −0.07%, which, although still showing some shrinkage, was less pronounced compared to the control mixture, highlighting the positive effect of increased cement content in reducing shrinkage. The inclusion of epoxy resin (1%, 2%, 3%) into CAM mixtures with 50% anionic AE resulted in slight expansion or shrinkage, with values ranging from −0.05% to −0.06%, indicating that epoxy resin has a limited influence on the shrinkage characteristics.

The analysis of expansion and shrinkage characteristics in CAM mixtures provides key insights for material selection and application design. Anionic AE emerges as a valuable component for mitigating shrinkage in CAM mixtures, which is of utmost importance for maintaining dimensional stability over time, especially in applications where precise dimensions are crucial. Higher cement content within CAM mixtures serves as an effective means to reduce shrinkage tendencies. This property is particularly advantageous in applications requiring strict dimensional control. Epoxy resin, while contributing positively to other aspects of CAM mixtures, exhibits a relatively minimal impact on shrinkage, implying that it may not significantly influence the material's dimensional stability. These findings offer essential guidance for the selection of CAM mixtures tailored to specific applications, ensuring that the chosen material aligns with the necessary dimensional stability requirements.

3.6. Acidic and Alkali Resistance

Assessing the resistance of Cement Asphalt Mortar (CAM) mixtures to acidic (sulfuric acid) and alkali (sodium hydroxide) environments is pivotal to determining their durability in the face of corrosive conditions. This section investigates the ability of various CAM mixtures to withstand exposure to harsh chemical environments, focusing on their response to acidic and alkali degradation.

3.6.1. Acidic Resistance

As shown in Figure 7, the evaluation of acid resistance in CAM mixtures reveals crucial insights with significant implications for applications exposed to potential acid-induced degradation. The control mixture, composed of 100% cement, exhibited considerable vulnerability to acid-induced degradation, emphasizing the need for enhanced acid resistance in high-cement mixtures due to the corrosive nature of sulfuric acid. Reducing cement content to 25% and incorporating anionic asphalt emulsion (AE) significantly improved acid resistance, with a lower percentage change in compressive strength, highlighting the protective properties of anionic AE against acid attacks. CAM mixtures with an equal balance of cement and anionic AE displayed enhanced acid resistance, underlining the importance of a balanced cement and AE ratio in creating a protective mixture. Increasing cement content to 75% while reducing anionic AE to 25% resulted in a minor reduction in acid resistance, indicating that higher cement content had a limited impact on this property. CAM mixtures incorporating cationic AE, even at the 25% cement level, exhibited improved acid resistance, showcasing the effectiveness of cationic AE in mitigating acid-induced degradation. The introduction of epoxy resin at various percentages (1%, 2%, and 3%) to CAM mixtures with anionic AE resulted in mixed effects on acid resistance, demonstrating superior protection against acid attacks. The results are consistent with studies that have demonstrated the protective properties of anionic asphalt emulsions in mitigating acid-induced degradation, particularly in lower cement content mixtures [36]. The variable effects of epoxy resin on acid resistance are also in line with research that has shown the potential of epoxy additives to enhance acid resistance in certain compositions [23]. Collectively, these findings contribute to a comprehensive understanding of acid resistance in CAM mixtures and support the development of tailored mix designs for applications exposed to acidic substances.

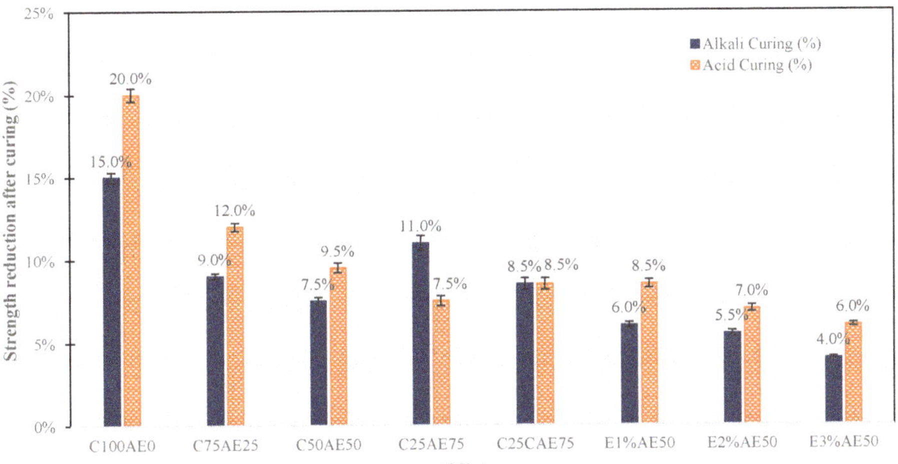

Figure 7. Durability test of CAM mixtures.

3.6.2. Alkali Resistance

Assessing alkali resistance is equally critical, as CAM mixtures may come into contact with alkali substances, which can lead to the depolymerization of cementitious phases and a subsequent loss of mechanical strength. As presented in Figure 7, the control CAM mixture demonstrated vulnerability to alkali-induced degradation, with exposure to sodium hydroxide resulting in a significant percentage change in compressive strength of −15.0%, highlighting its susceptibility to alkali attack. However, CAM mixtures incorporating epoxy resin at various percentages (1%, 2%, and 3%) in conjunction with anionic AE exhibited superior alkali resistance. These mixtures demonstrated the highest resilience to alkali attacks, with a percentage change in compressive strength ranging from −4.0% to −6.0%. These findings collectively emphasize the potential of epoxy resin and CRP-modified CAM mixtures to withstand chemical degradation in harsh environments, offering enhanced protection against both acidic and alkali-induced damage. This combination of epoxy resin and CRP presents a particularly promising avenue, deserving further exploration and consideration for practical applications in corrosive settings.

3.6.3. Correlation of Acidic and Alkali Exposure Regimes with Field Conditions

The exposure tests conducted to evaluate the resistance of CAM mixtures to acidic and alkali environments were designed to simulate extreme conditions that these materials might encounter in real-world applications. However, it is important to acknowledge that the exact correlation between laboratory exposure regimes and field conditions is a complex matter. While the tests aim to replicate and accelerate the aging process, they may not precisely mimic all the variables of long-term field exposure. Future research endeavors should consider the need for a closer alignment between laboratory testing and real-world conditions. This may involve conducting parallel field trials of CAM mixtures on actual bridges or structures in tropical climates. Such field performance tests could provide a more direct correlation between laboratory results and real-world outcomes, enhancing the reliability and resilience of infrastructure materials in challenging tropical environments.

3.7. Impact of High Temperature (45 °C) on Unconfined Compressive Strength

Evaluating UCS under high-temperature conditions is a crucial aspect of understanding how CAM mixtures perform in tropical climates, particularly when faced with scorching temperatures. The following Figure 8 provides insight into the percentage reduction in UCS for various CAM mixtures exposed to elevated temperatures (45 °C) for a specified duration.

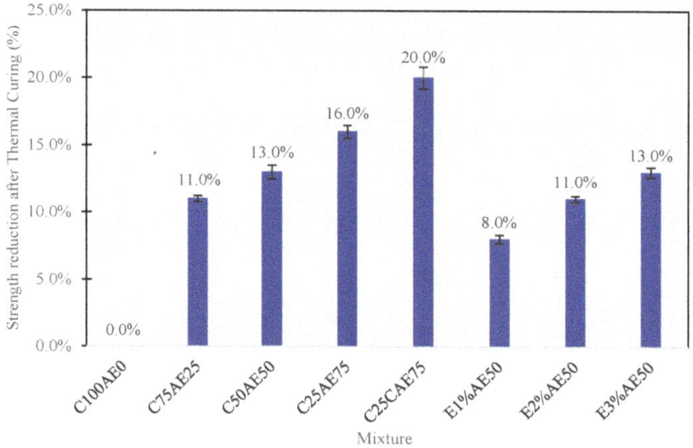

Figure 8. Thermal durability of mixture.

The high-temperature UCS test replicates the challenges that CAM mixtures may encounter in tropical climates with exceptionally high temperatures. The results illustrate the percentage reduction in UCS, providing insight into the material's response to these conditions. The control CAM mixture, composed entirely of cement, showed no reduction in UCS (0%) when exposed to high temperatures, showcasing the inherent strength of cementitious materials in resisting thermal stress. In contrast, a mixture with 25% cement and 75% anionic AE demonstrated a moderate reduction of −11.0% in UCS when exposed to high temperatures, indicating the influence of the substantial presence of anionic AE on strength reduction. A blend of 50% cement and 50% anionic AE displayed a slightly higher reduction in UCS at −13.0%, emphasizing the moderate high-temperature stability achieved by balancing cement and anionic AE. The mixture containing 75% cement and 25% anionic AE experienced a −16.0% reduction in UCS, demonstrating better resistance to high-temperature strength reduction due to the higher cement content. CAM mixtures that included 25% cement and 75% cationic AE showed the most significant reduction in UCS, at −20.0%, reflecting the distinct impact of cationic AE, which reacts differently to high temperatures compared to anionic AE. CAM mixtures incorporating epoxy resin (1%, 2%, 3%) in conjunction with anionic AE exhibited notable improvements in high-temperature stability, with percentage reductions in UCS ranging from −8.0% to −13.0%. These findings highlight the potential of epoxy resin to counteract the effects of asphalt emulsion in high-temperature conditions, enhancing the material's resistance to high-temperature-induced strength reduction.

The data unequivocally demonstrate that the reduction in UCS under high-temperature conditions is primarily influenced by the proportion of cement and the type of asphalt emulsion used in the CAM mixture. Mixtures with higher asphalt emulsion content displayed a more significant strength reduction when subjected to high temperatures, emphasizing the critical role of asphalt emulsion composition. Notably, the incorporation of epoxy resin proved to be highly effective in mitigating the deleterious effects of high temperatures, resulting in improved high-temperature stability. These findings underscore the utmost importance of carefully selecting a CAM mixture composition that effectively balances the requirements for high-temperature stability, mixing stability, and flowability. Such considerations are paramount when tailoring CAM mixtures to meet the specific demands of projects in tropical climates, where environmental conditions can be particularly challenging.

Statistical Significance Testing of High-Temperature UCS Differences

Additionally, Table 6 presents the results of statistical significance testing for high-temperature Unconfined Compressive Strength (UCS) differences among various Cement Asphalt Mortar (CAM) mixtures. The table shows the percentage of strength reduction after thermal curing for each CAM mixture, along with the corresponding p-values obtained from the statistical tests. The level of statistical significance is also indicated, with distinctions made between "Significant", "Highly Significant", and "Not Significant".

Table 6. Statistical Significance Testing for High-Temperature UCS Differences.

CAM Mixture	Strength Reduction after Thermal Curing	p-Value	Statistical Significance
C100AE0	0.0%	-	-
C75AE25	11.0%	0.042	Significant
C50AE50	13.0%	0.016	Significant
C25AE75	16.0%	0.003	Highly Significant
C25CAE75	20.0%	<0.001	Highly Significant
E1%AE50	8.0%	0.115	Not Significant
E2%AE50	11.0%	0.042	Significant
E3%AE50	13.0%	0.016	Significant

The data in Table 6 reveals several important insights. Firstly, it demonstrates that high-temperature exposure has a significant effect on the UCS of the CAM mixtures. Specifically, CAM mixtures containing higher proportions of asphalt emulsion (e.g., C25AE75 and C25CAE75) exhibited the most pronounced strength reduction, indicating their susceptibility to high-temperature conditions. This high susceptibility is supported by the highly significant *p*-values ($p < 0.001$). Conversely, mixtures with epoxy resin (E1%AE50, E2%AE50, E3%AE50) showed lower strength reductions, suggesting that the inclusion of epoxy resin can mitigate the adverse effects of high temperatures.

The statistical significance testing conducted on high-temperature UCS differences adds a layer of rigor to the study's findings and underscores the importance of considering the influence of temperature in the design and selection of CAM mixtures. These results provide valuable guidance for tailoring CAM mixtures to meet specific project requirements in tropical climates, ensuring durability and optimal performance even under challenging high-temperature conditions.

3.8. SEM Results

The SEM analysis was conducted to examine the microstructural characteristics of the various CAM mixtures used in this study. This analysis offers valuable insights into the composition, distribution, and interfacial bonding of the materials within the CAM matrix. The SEM micrographs reveal distinct features of the microstructure, shedding light on the influence of different CAM mixture compositions. The following observations are derived from the SEM analysis of the specified CAM mixtures, as shown in Figure 9a–d.

In the CAM mixture containing 75% cement and 25% anionic AE, the SEM image displays a microstructure with a dominant presence of cement particles. The asphalt emulsion phase appears as distinct droplets within the predominantly cementitious matrix. This composition results in a more cement-rich microstructure, which may impact the material's characteristics.

The SEM analysis of the CAM mixture with equal parts of cement and anionic AE (50% each) showcases a balanced microstructure. Here, the micrograph illustrates a combination of dense cement particles and dispersed asphalt emulsion. This mixture exhibits a unique balance between the cementitious phase and the asphalt phase, which may contribute to its optimized properties.

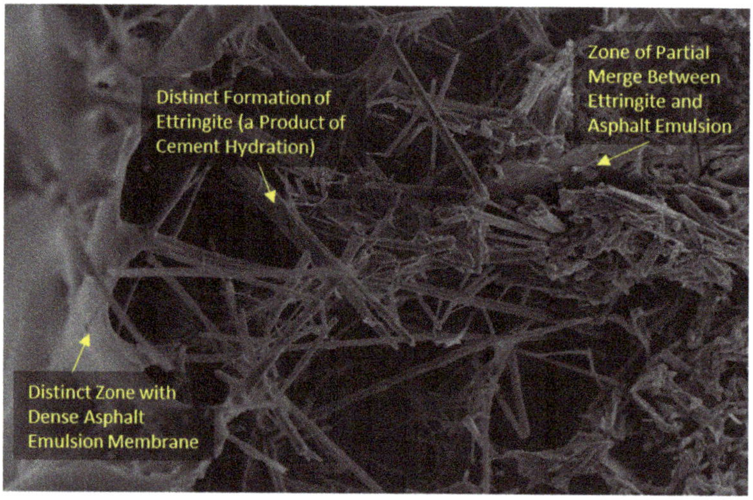

(**a**) 75% Cement + 25% Asphalt Emulsion.

Figure 9. *Cont.*

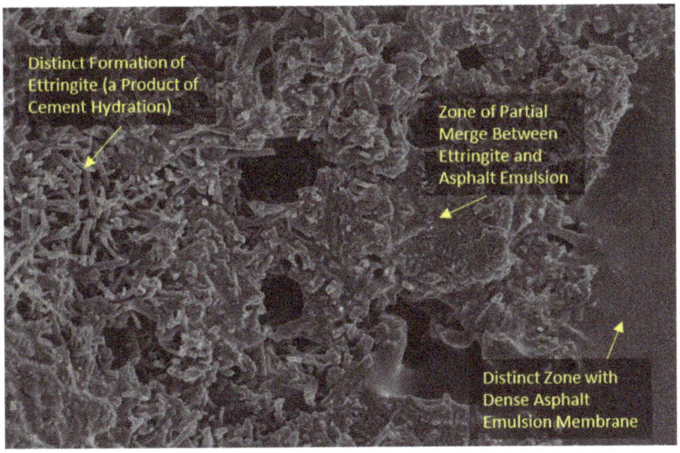

(**b**) 50% Cement + 50% Asphalt Emulsion.

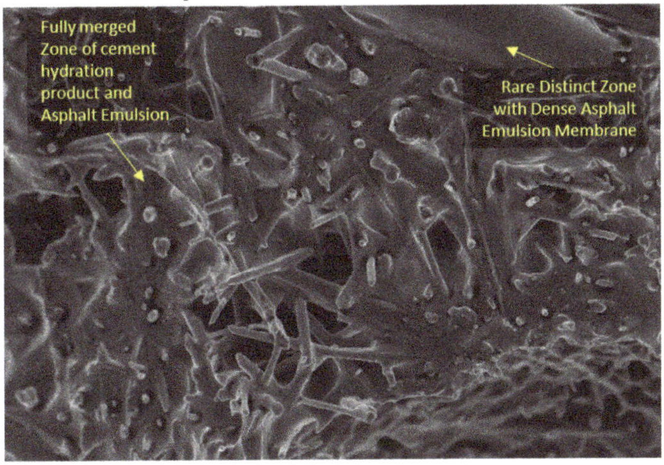

(**c**) 25% Cement + 75% Asphalt Emulsion.

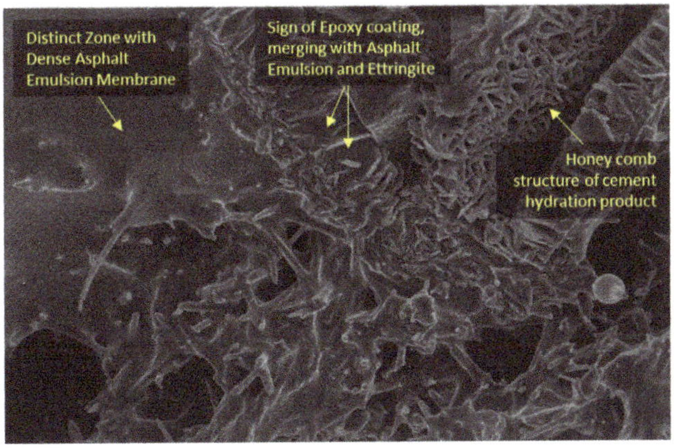

(**d**) 50% Cement + 50% Asphalt Emulsion + 2% Epoxy Resin.

Figure 9. SEM Test Results.

In the SEM image of the CAM mixture with 25% cement and 75% AE, a discernible alteration in microstructure becomes apparent. The asphalt emulsion droplets are observed dispersed among the cement particles and fine aggregates. This dispersed phase introduces a distinct connectivity between the asphalt and the cementitious matrix, contributing to a more complex but well-connected microstructure.

The SEM analysis of CAM mixtures incorporating epoxy resin (2%) along with anionic AE highlights the influence of epoxy resin on the microstructure. In these mixtures, the epoxy resin appears to form an adhesive layer around the cement particles and interfaces with the anionic AE droplets. This interaction contributes to a more complex and interconnected microstructure, indicating the potential of epoxy resin to improve the interface between cement and asphalt emulsion.

In summary, the SEM analysis provides visual evidence of the microstructural differences among the CAM mixtures. The micrographs depict the distribution and connectivity of cement, asphalt emulsion, and other constituents within the matrices. These observations correlate with the mechanical properties and performance characteristics of each CAM mixture, further emphasizing the significance of material composition and microstructure in determining material behavior and suitability for specific applications.

The SEM analysis provided a visual understanding of the microstructural characteristics of CAM mixtures. These observations corresponded with the mechanical properties and performance characteristics of each mixture, emphasizing the importance of material composition and microstructure in determining material behavior and suitability for specific applications. In summary, this research comprehensively assessed the various aspects of CAM mixtures, from their mixing stability and flowability to their resistance to environmental factors and high temperatures. The findings offer valuable insights for designing CAM mixtures tailored to specific project requirements in tropical climates, ensuring durability and optimal performance in challenging conditions. The balance between cement content, asphalt emulsion type, and the inclusion of epoxy resin emerges as a key factor in formulating CAM mixtures that can thrive in such conditions. These findings not only contribute to the theoretical understanding of CAM mixtures but also hold great promise for practical applications, such as bridge expansion joints, in regions with tropical climates. Further research in this area could refine mixed designs and open doors to new possibilities in construction materials for challenging environments.

3.9. Field Service Life Estimates for Fill Materials

In this study, a comparative analysis of fill materials for bridge joints in tropical climates has been conducted, aiming to estimate their service life based on real-world field applications [7,11,39]. As shown in Table 7, the findings reveal significant variations in the durability of these materials. Conventional cement mortar, a commonly used option, exhibits a relatively short service life, estimated to be less than 2 years due to its susceptibility to cracking, as observed in practical construction data. On the other hand, conventional hot mix asphalt, while somewhat more durable, still has a limited service life, estimated at less than 3 years, due to its susceptibility to rutting, as confirmed by field applications. In contrast, CAM emerges as a promising alternative with an estimated service life of more than 5 years, which is supported by recent practical construction data. This extended service life is attributed to CAM's ability to combine the bearing capacity of cement with the ductile behavior of asphalt, as demonstrated in field applications. These results underscore the practical benefits of CAM in enhancing the longevity and resilience of bridge joints in challenging tropical conditions. However, ongoing field testing is essential to confirm and refine these estimates, offering valuable insights into the real-world performance of CAM in service conditions. The promising potential of CAM based on field applications warrants further investigations into its practical application and long-term performance, which could significantly impact the maintenance and sustainability of infrastructure in tropical climates.

Table 7. Comparative Service Life Estimates for Bridge Joints in Tropical Climates.

Fill Material	Estimated Service Life (Years)	Service Notes
Conventional Cement Mortar	Less than 2 years	Prone to cracking
Conventional Hot Mix Asphalt	Less than 3 years	Susceptible to rutting
Cement Asphalt Mortar	More than 5 years	Combines bearing capacity and ductile asphalt behavior

The laboratory evaluations conducted in this study provide valuable insights into the properties of CAM mixtures for bridge expansion joints in tropical climates. However, it is crucial to recognize that field conditions can present challenges and factors that are difficult to replicate in controlled laboratory settings. To bridge this gap between laboratory findings and real-world performance, there is a compelling need for potential field performance testing. Future work in this direction should involve actual field trials on bridges equipped with CAM mixtures. These trials would not only validate the laboratory results but also provide insights into how CAM materials perform under sustained exposure to environmental conditions, traffic loads, and long-term use. Field performance tests can shed light on the durability, structural integrity, resistance to moisture and temperature fluctuations, and overall long-term viability of CAM mixtures in bridge expansion joints. This field-based approach will be instrumental in refining CAM mixture designs, optimizing construction procedures, and ensuring their successful practical application in challenging tropical environments.

4. Conclusions

This research aimed to optimize Cement Asphalt Mortar (CAM) mixtures for tropical bridge expansion joints. Various cement, asphalt emulsion, and epoxy resin mixes were tested to enhance properties. Highlights include:

- Reducing cement content in CAM mixtures enhances mixing stability, particularly with 75% cement and 25% anionic AE. Epoxy resin (1%, 2%, 3%) has minimal impact on mixing stability, staying within acceptable limits.
- Reducing cement content to 25% with anionic AE improved flowability (5.8 cm, 80 s vs. control's 4.5 cm, 110 s). Balanced cement–anionic AE mixtures enhanced flowability, while higher cement (75%) had a minor impact. Cationic AE also improved flowability.
- Early and 28-day UCS results demonstrate the impact of cement content and anionic AE; a 100% cement mixture reached 7.2 MPa at 2 h and 20.2 MPa at 28 days. Mixtures with 25% cement and 75% anionic AE achieved 3.7 MPa at 2 h and 13.5 MPa at 28 days. The addition of epoxy resin (1%, 2%, 3%) consistently maintained or enhanced these values, emphasizing the significance of cement content and anionic AE.
- Anionic AE mitigated shrinkage. The control mixture had −0.15% shrinkage, while mixtures with 75% anionic AE ranged from −0.05% to −0.09%. Higher cement content reduced shrinkage tendencies. Epoxy resin had minimal impact, ranging from −0.05% to −0.06%.
- Exposure to sulfuric acid reduced compressive strength by −20.0% for the control mixture. Mixtures with anionic AE showed improved resistance (−6.0% to −12.0%). Epoxy resin and anionic AE demonstrated the highest resilience, with changes ranging from −4.0% to −8.5%.
- The control mixture showed no reduction in UCS at 45 °C. Mixtures with anionic AE (25%, 50%) had reductions of −11.0% to −13.0%, while mixtures with epoxy resin (1%, 2%, 3%) showed lower reductions, ranging from −8.0% to −13.0%.
- Scanning Electron Microscopy provided insights into CAM mixture microstructures, influencing material behavior and suitability for specific applications. Additional quantitative image analysis metrics can offer more detailed information.
- Based on the research findings, it is advisable to employ CAM mixtures with a composition of 50% cement, 50% anionic Asphalt Emulsion, and 2% epoxy resin for constructing bridge expansion joints in tropical climates. This blend enhances mix-

ing stability, flowability, UCS strength, and resistance to shrinkage and acid-induced degradation, making it suitable for practical applications in tropical climate conditions.

- The study highlights CAM mixtures' potential for improving bridge joint durability but also underscores the importance of addressing limitations, optimizing construction, and conducting field trials to ensure their long-term performance, enhancing infrastructure resilience and sustainability.

Author Contributions: Conceptualization, K.K. and T.H.M.L.; methodology, T.H.M.L.; validation K.K. and Y.-M.K.; formal analysis, K.K. and T.H.M.L.; investigation, K.K. and Y.-M.K.; resources, K.K. and T.H.M.L.; writing—original draft preparation, K.K. and T.H.M.L.; writing—review and editing, K.K. and T.H.M.L.; visualization, K.K. and Y.-M.K.; supervision, K.K.; project administration, K.K. and Y.-M.K.; funding acquisition, K.K. and Y.-M.K. All authors have read and agreed to the published version of the manuscript.

Funding: Research for this paper was carried out under the KICT Research Program (project no. 20230202-001, Development of high performance, low-cost MAST (Multi-Layered Asphalt Surface Treatments) composite pavement system for improving unpaved roads in Vietnam) funded by the Ministry of Science and ICT.

Data Availability Statement: The data presented in this study are available on request from the corresponding author. The data are not publicly available due to ongoing research implications.

Conflicts of Interest: Author Kyungnam Kim was employed by the company Korea Expressway Corporation. The remaining authors declare that the research was conducted in the absence of any commercial or financial relationships that could be construed as a potential conflict of interest.

References

1. Jiang, T.; Zhang, Y.; Wang, L.; Zhang, L.; Song, G. Monitoring Fatigue Damage of Modular Bridge Expansion Joints Using Piezoceramic Transducers. *Sensors* **2018**, *18*, 3973. [CrossRef]
2. Di Mascio, P.; Loprencipe, G.; Moretti, L.; Puzzo, L.; Zoccali, P. Bridge Expansion Joint in Road Transition Curve: Effects Assessment on Heavy Vehicles. *Appl. Sci.* **2017**, *7*, 599. [CrossRef]
3. Li, J.; Wen, F.; Chen, J.; Yang, C.; Du, W.; Xu, L.; Li, P. Experimental Study of Bridge Expansion Joint Damage Based on Natural Frequency. *Sensors* **2023**, *23*, 6437. [CrossRef] [PubMed]
4. Zhang, X.; Cheng, Z.; Du, L.; Du, Y. Progressive Classifier Mechanism for Bridge Expansion Joint Health Status Monitoring System Based on Acoustic Sensors. *Sensors* **2023**, *23*, 5090. [CrossRef] [PubMed]
5. Li, Y.; Lu, P.; Liu, C.; Ren, W.; Wen, J. Fatigue Life of a Comb Plate Expansion Joint. *Machines* **2023**, *11*, 494. [CrossRef]
6. Kong, F.; Xu, F.; Xiong, Q.; Xu, S.; Li, X.; Fu, W.; Guo, Z. Experimental Research on Properties of UHPC Based on Composite Cementitious Materials System. *Coatings* **2022**, *12*, 1219. [CrossRef]
7. Min, Z.; Wang, Q.; Zhang, K.; Shen, L.; Lin, G.; Huang, W. Investigation on the Properties of Epoxy Asphalt Mixture Containing Crumb Rubber for Bridge Expansion Joint. *Constr. Build. Mater.* **2022**, *331*, 127344. [CrossRef]
8. Ma, Y.; Peng, A.; Wang, L.; Zhang, C.; Li, J.; Zhang, J. Fatigue Performance of an Innovative Shallow-Buried Modular Bridge Expansion Joint. *Eng. Struct.* **2020**, *221*, 111107. [CrossRef]
9. Niemierko, A. Modern Bridge Bearings and Expansion Joints for Road Bridges. *Transp. Res. Procedia* **2016**, *14*, 4040–4049. [CrossRef]
10. Dong, R.; Gao, A.; Zhu, Y.; Xu, B.; Du, J.; Ping, S. The Development of a New Thermoplastic Elastomer (TPE)-Modified Asphalt. *Buildings* **2023**, *13*, 1451. [CrossRef]
11. He, A.; Dong, Z.; Zhang, H.; Zhang, A.A.; Qiu, S.; Liu, Y.; Wang, K.C.P.; Lin, Z. Automated Pixel-Level Detection of Expansion Joints on Asphalt Pavement Using a Deep-Learning-Based Approach. *Struct. Control Health Monit.* **2023**, *2023*, 7552337. [CrossRef]
12. Malla, R.B.; Swanson, B.J.; Shaw, M.T. Laboratory Evaluation of a Silicone Foam Sealant Bonded to Various Header Materials Used in Bridge Expansion Joints. *Constr. Build. Mater.* **2011**, *25*, 4132–4143. [CrossRef]
13. Pan, F.; Fan, Z.; Rochlani, M.; Falla, G.C.; Leischner, S.; Oeser, M.; Yu, H.; Liu, P. Evaluation of Adhesive Properties of Different Mineral Compositions in Asphalt Mixtures with Experimental and Molecular Dynamics Analyses. *Buildings* **2023**, *13*, 1207. [CrossRef]
14. Wang, T.; Chen, Z.; Wang, Y.; Cui, Y.; Sun, Y.; Ning, R.; Tian, Y.; Zhang, Y. The Effect of Aging on the Molecular Distribution of Crumb Rubber Modified Asphalt Based on the Gel Permeation Chromatography Test. *Buildings* **2023**, *13*, 1165. [CrossRef]
15. Liu, S.; Mo, L.; Wang, K.; Xie, Y.; Woldekidan, M.F. Preparation, Microstructure and Rheological Properties of Asphalt Sealants for Bridge Expansion Joints. *Constr. Build. Mater.* **2016**, *105*, 1–13. [CrossRef]

16. Liang, X.; Wang, Y.; Liu, B.; Zhang, H.; Wu, J.; Xu, Z.; He, L. Research on the Preparation of Rapid Repair Materials for Bridge Expansion Joints. In Proceedings of the International Conference on Smart Transportation and City Engineering (STCE 2022), Chongqing, China, 12–14 August 2022; p. 139. [CrossRef]
17. Tan, Y.; Ouyang, J.; Lv, J.; Li, Y. Effect of Emulsifier on Cement Hydration in Cement Asphalt Mortar. *Constr. Build. Mater.* **2013**, *47*, 159–164. [CrossRef]
18. Ouyang, J.; Zhao, J.; Tan, Y. Modeling Mechanical Properties of Cement Asphalt Emulsion Mortar with Different Asphalt to Cement Ratios and Temperatures. *J. Mater. Civ. Eng.* **2018**, *30*, 04018263. [CrossRef]
19. Ma, C.; Wu, Y.; Shi, J.; Song, S.; Xia, F.; Liu, J.; Zhou, H.; Dong, B.; Du, Y. Preparation and Characterization of a Novel Magnesium Phosphate Cement-Based Emulsified Asphalt Mortar. *Ceram. Int.* **2023**, *49*, 21422–21432. [CrossRef]
20. Dulaimi, A.; Qaidi, S.; Al-Busaltan, S.; Milad, A.; Sadique, M.; Kadhim, M.A.; Al-Khafaji, R.; Sabri Sabri, M.M. Application of Paper Sludge Ash and Incinerated Sewage Ash in Emulsified Asphalt Cold Mixtures. *Front. Mater.* **2023**, *9*, 1074738. [CrossRef]
21. Le, T.H.M.; Lee, S.H.; Park, D.W. Evaluation on Full-Scale Testbed Performance of Cement Asphalt Mortar for Ballasted Track Stabilization. *Constr. Build. Mater.* **2020**, *254*, 119249. [CrossRef]
22. Lee, S.-Y.; Le, T.H.M. Laboratory and Full-Scale Testbed Study in the Feasibility of Styrene-Butadiene-Styrene Asphalt Pavement Having Epoxy Resin and Crumb Rubber Powder. *Buildings* **2023**, *13*, 652. [CrossRef]
23. Kim, K.N.; Le, T.H.M. Durability of Polymer-Modified Asphalt Mixture with Wasted Tire Powder and Epoxy Resin under Tropical Climate Curing Conditions. *Polymers* **2023**, *15*, 2504. [CrossRef]
24. Le, T.H.M.; Park, D.W.; Seo, J.W. Evaluation on the Mechanical Properties of Cement Asphalt Mortar with Quick Hardening Admixture for Railway Maintenance. *Constr. Build. Mater.* **2019**, *206*, 375–384. [CrossRef]
25. Tan, Y.; Ouyang, J.; Li, Y. Factors Influencing Rheological Properties of Fresh Cement Asphalt Emulsion Paste. *Constr. Build. Mater.* **2014**, *68*, 611–617. [CrossRef]
26. Ouyang, J.; Hu, L.; Li, H.; Han, B. Effect of Cement on the Demulsifying Behavior of Over-Stabilized Asphalt Emulsion during Mixing. *Constr. Build. Mater.* **2018**, *177*, 252–260. [CrossRef]
27. *ASTM C150/C150M*; Standard Specification for Portland Cement. ASTM International: West Conshohocken, PA, USA, 2012; Volume 1, pp. 1–9.
28. *ASTM D3381/D3381M-18*; Standard Specification for Viscosity-Graded Asphalt Binder for Use in Pavement Construction. ASTM International: West Conshohocken, PA, USA, 2018.
29. *ASTM D1763*; Standard Specification for Epoxy Resins. ASTM International: West Conshohocken, PA, USA, 2021.
30. *ASTM C305/C305-20*; Standard Practice for Mechanical Mixing of Hydraulic Cement Pastes and Mortars of Plastic Consistency. ASTM International: West Conshohocken, PA, USA, 2009; 4, p. 3.
31. *ASTM C230*; Standard Specification for Flow Table for Use in Tests of Hydraulic Cement 1. ASTM International: West Conshohocken, PA, USA, 2010; pp. 4–9.
32. *ASTM C39/C39M*; Standard Test Method for Compressive Strength of Cylindrical Concrete Specimens 1. ASTM International: West Conshohocken, PA, USA, 2003; 1, pp. 1–5.
33. *ASTM C-157/C 157M-08*; Standard Test Method for Length Change of Hardened Cement Mortar and Concrete. ASTM International: West Conshohocken, PA, USA, 2017; 4, pp. 1–7.
34. *ASTM C267*; Standard Test Methods for Chemical Resistance of Mortars, Grouts, and Monolithic Surfacings and Polymer Concretes. ASTM International: West Conshohocken, PA, USA, 2015; 1, pp. 1–6.
35. *ASTM E2546-15*; Standard Practice for Instrumented Indentation Testing. ASTM International: West Conshohocken, PA, USA, 2015; 1, pp. 1–23.
36. Le, T.H.M.; Park, D.W.; Seo, J.W.; Phan, T.M. Anti-Chemical Resistance and Mock-up Test Performance of Cement Asphalt Mortar Modified with Polymer for Ballast Stabilizing. *Constr. Build. Mater.* **2020**, *232*, 117260. [CrossRef]
37. Ba, H.; Guo, L.; Huan, H.; Zhang, S.; Lin, Z. Multi-Objective Optimization of Epoxy Resin Adhesive for Pavement Toughened by Self-Made Toughening Agent. *Polymers* **2023**, *15*, 1946. [CrossRef]
38. Wang, Z.; Shu, X.; Rutherford, T.; Huang, B.; Clarke, D. Effects of Asphalt Emulsion on Properties of Fresh Cement Emulsified Asphalt Mortar. *Constr. Build. Mater.* **2015**, *75*, 25–30. [CrossRef]
39. Huadong, L. The Cause Analysis on the Highway Bridge Expansion Joints and the Maintenance of Construction Management. In Proceedings of the 2018 3rd International Conference on Smart City and Systems Engineering (ICSCSE), Xiamen, China, 29–30 December 2018; pp. 267–269. [CrossRef]

buildings

MDPI

Article

Evaluating the Effects of RA on the Rheological Properties and Aging Susceptibility of RAM Asphalt

Qifeng Yang [1], Zhen Liu [1], Rui Wang [2], Lijun Sun [1] and Runhua Zhang [2,*]

[1] Department of Roadway Engineering, School of Transportation, Southeast University, Nanjing 211189, China; 220215101@seu.edu.cn (Q.Y.); 230208344@seu.edu.cn (Z.L.)
[2] School of Civil Engineering and Environment, University of Wisconsin–Madison, Madison, WI 53706, USA
* Correspondence: rzhang478@wisc.edu

Abstract: Recycling agents (RAs) can mitigate the adverse effects of reclaimed asphalt pavement (RAP) on recycled asphalt, which can result in more RAP being added. To explore the effects of four RAs on recycled asphalt, this study used asphalt rheological performance experiments, including the bending beam rheometer (BBR) test, the dynamic shear rheological (DSR) test, and the indirect tensile asphalt cracking test (IDEAL-CT), to study the performance grade (PG), cracking resistance, and cracking susceptibility of recycled asphalt. In addition, an aging model for asphalt was used, and short-term and long-term aging sensitivities were evaluated according to this model. Results showed that US Soybean significantly enhanced the high-temperature and low-temperature performance grade (PGLT) and cracking resistance and reduced cracking sensitivity of the recycled asphalt. In addition, the short-term aging sensitivity was the lowest. It showed the best regeneration effect. However, the recycled asphalt with US Soybean showed the highest long-term aging sensitivity, suggesting that longer aging time results in poorer performance. Ingevity's PG, cracking resistance, and cracking sensitivity are slightly lower than that of US Soybean, and Ingevity demonstrated the second lowest short-term aging sensitivity. However, its prolonged aging sensitivity was much lower than that of US Soybean. Asphalt and Wax Innovations and Georgia Pacific agents slightly improved recycled asphalt properties. The long-term aging sensitivity of recycled asphalt with four Ras was higher than that of recycled asphalt without Ras. These results indicated that the recycled asphalt's performance with Ras worsened as aging time increased compared to the original recycled asphalt.

Keywords: reclaimed asphalt pavement; recycling agent; rheological properties; aging susceptibility; laboratory test

1. Introduction

Reclaimed asphalt pavement (RAP) is commonly used for asphalt mixtures in pavement construction and is derived from crushing and reclaiming old pavements [1]. Large quantities of RAP are produced during road maintenance and repair, and RAP can be recycled and mixed with new asphalt and aggregate as part of new asphalt mixtures. When the RAP gets mixed in with the virgin asphalt binder, part of the aged asphalt from the RAP binder becomes reactivated and forms part of the 'total binder.' This approach not only solves the problem of RAP disposal but also saves new asphalt and aggregates and reduces environmental pollution. However, RAP's chemical components and mechanical characteristics become poor because of exposure to oxygen, ultraviolet light [2], and moisture during use [3]. Generally, recycled mixtures containing over 40% RAP by total weight of asphalt mixture are harder than the original mixture, making it difficult to machine and reducing its service life significantly [4]. Moreover, the fusion degree of recycled asphalt varies, and the delamination phenomenon between recycled asphalt and RAP can quickly occur. The binders' aging process has two stages: short-term and long-term aging [5,6]. The short-term aging stage happens because of heat during agitation and compacting [7]. The long-term aging phase is caused by parameters such as sunlight, oxygen, and longevity over the pavement life. The aging of asphalt is considered a critical issue during road engineering applications [8].

Therefore, the recycling agent (RA) is an essential means of restoring the recycled mixtures' properties due to the deterioration of the aged asphalt's rheological properties. In earlier studies, researchers added a soft binder to aged asphalt and found that it reduced its viscosity and positively affected its performance recovery [9,10]. The National Centre for Asphalt Technology has recently classified RAs into five broad categories based on the source: (1) paraffin oils, (2) aromatics, (3) naphthenic oils, (4) tall oils, and (5) fatty acids [11]. An appropriate method for selecting the optimal RA dosage should be based on the RA dosage's effect on related properties of recycled asphalt and the recycled asphalt mixture, as well as the various states of aging that asphalt binders face during their lifetime.

Using the performance grade (PG) system, researchers evaluated variations in RAP binder stiffness due to the addition of RAs and variations in RAP binder stiffness due to RA's addition [12]. A combined chart was developed for this approach to establish the suitable RA dosage to satisfy the required PG values for low-temperature cracking resistance and high-temperature rutting [13]. Zaumanis et al. [14] studied effects of the nine types of RA products on asphalt mixtures using RAP. In addition, it was found that four RAs decreased the low-temperature cracking susceptibility of asphalt mixtures. Furthermore, they found that the penetration index was an excellent RA indicator. Li et al. [15] investigated the low-temperature rutting and cracking properties of SBS-modified asphalt mixtures containing RAP. The findings suggested that appropriate recycling agents enhanced the low-temperature properties of RAP without significantly adversely affecting the compaction of asphalt mixtures with RAP [16]. The use of RA in the RAP restoration process is becoming more and more frequent. However, some studies have found that although RA can improve the performance of aged asphalt, its effectiveness may decrease as the aging time increases [17,18].

Several researchers have utilized semicircular bending and indirect stretching tests to assess mixtures' cracking susceptibility at middle-range temperatures [19–21]. The large quantity of RAP leaves asphalt mixtures susceptible to fracture in moderate temperatures. Fracture susceptibility at low temperatures can be assessed using the SCB, bending beam rheometer (BBR), disc compression tensile, and IDT tests [22]. Researchers found that high RAP mixtures were more sensitive to cracking [23–25].

In this paper, four types of recycled asphalt with four RAs were prepared. Rheological tests were implemented to measure the recycled asphalt's crack resistance and cracking sensitivity. Rheological tests involved DSR and BBR experiments. The cracking resistance of different recycled asphalt mixtures was assessed using IDEAL-CT, and cracking resistance indicators were investigated. In addition, the asphalt aging model was used to analyze the sensitivity of six recycled asphalts to short-term and long-term aging. Based on the analysis results of the asphalt performance parameters, four RAs were comprehensively evaluated and recommended.

The technology roadmap for this paper is depicted in Figure 1.

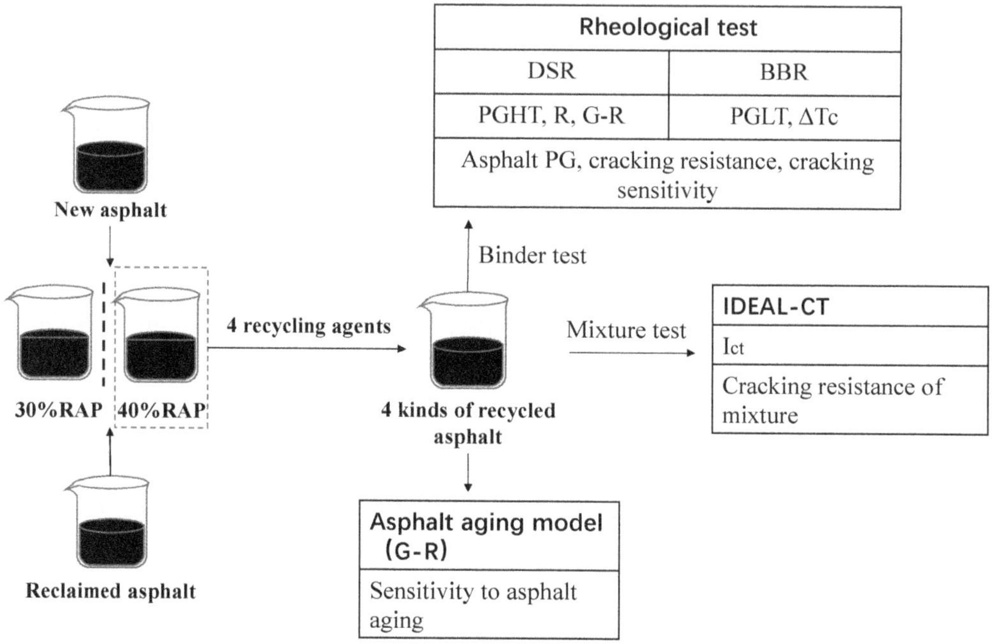

Figure 1. Technology roadmap for experimental methods and modelling-based analyses.

2. Materials and Method

2.1. Materials

The unmodified asphalt utilized in this research was PG58-28, sourced from a petroleum supplier in Minnesota, USA. It is a common base asphalt material. A total of four RAs were used in this study. RA1 is US Soybean, an unsaturated fatty acid commonly used in asphalt modification. RA2 is a warm mix agent named Evotherm that is produced by Ingevity (North Charleston, CS, USA). The main chemical components are hexadecyl imidazoline and sodium dodecyl sulfate. RA3 is SPA140, a wax produced by Asphalt and Wax Innovations (Kapalama, MS, USA), which is used to recycle reclaimed asphalt pavements. RA4 is a saturated fatty acid produced by Georgia Pacific, Inc. (Augusta, GA, USA) and serves as a saturated phenol modifier. The RAP material was obtained from sampled aggregates in Minnesota.

2.2. Preparation

2.2.1. Aged Asphalt Recovery

The binder in the RAP was extracted according to the standard outlined in ASTM D2172 [26]. The asphalt binder extracted was then recycled according to the standard reported in ASTM D1856 [27]. Toluene was selected as the solvent in this process.

2.2.2. Aging of Asphalt

Asphalt was exposed to high temperatures and air during the mixing process. Thus, the short-term aged asphalt (STA) utilized in the study was considered recycled asphalt after mixing. Long-term aged asphalt was prepared using pressure aging. Here, 50 g ($\pm$0.5 g) blended recycled asphalt was added to each RTFO vessel, and the container was placed in a pressure aging machine for 20, 40, and 60 h. For convenience, PAV20, PAV40, and PAV60 denote asphalt binder aged for 20, 40, and 60 h, respectively.

2.3. Laboratory Testing of Rheological Properties of Recycled Asphalt

2.3.1. DSR Experiments

To evaluate the influence of various RAs on the rheological properties of asphalt at elevated temperatures, DSR tests were performed according to the standard described in ASTM D4402 [28] at (15 °C, 25 °C, 35 °C, 45 °C, and 55 °C) and various frequencies (16 frequencies ranging from 0.1 rad/s to 100 rad/s) [29]. Figure 2 shows the DSR tester. The DSR test was applied to determine the rheological parameters and performance grades, including R, $G-R$, and PGHT (PG high temperature). Here, R is determined to be the deviation of the logarithmic glassy modulus and logarithmic equilibrium modulus of the recycled asphalt binder. Larger R values indicate higher cracking sensitivity of the asphalt. R is expressed using Equation (1):

$$R = Log\,|G*|_{glassy} - Log\,|G*|_{crossover} \tag{1}$$

where $Log\,|G*|_{glassy}$ is the logarithm of the complex shear modulus at the glassy asymptote, and $Log\,|G*|_{crossover}$ is the logarithm of complex shear modulus at the crossover frequency.

Figure 2. DSR tester.

The Glover-Rowe parameters ($G-R$) are utilized to evaluate the fracture susceptibility, with lower $G-R$ parameters indicating better durability of the asphalt to resist cracking. $G-R$ parameters are usually converted to a temperature of 15 °C and frequency of 0.005 rad/s [30–32]. When the asphalt's $G-R$ produces 180 kPa, it is considered crack-producing. When the $G-R$ value is greater than 600 kPa, substantial fracture is created. The $G-R$ value is calculated as shown in Equations (2)–(5):

$$|G^*|(f_c) = \frac{G_a^*}{\left(1+\left(\frac{f_c}{f_a}\right)^k\right)^{\frac{m_e}{k}}} \tag{2}$$

$$f_a = f \times \alpha_T \tag{3}$$

$$\alpha_T = 10^{\frac{-c_1(T-T_R)}{c_2+(T-T_R)}} \tag{4}$$

$$G-R=\frac{|G^{*}|(\cos\delta)^{2}}{\sin\delta} \tag{5}$$

where δ is the phase angle, $|G^*|$ is the complex shear modulus, G_a^* is the glassy complex modulus of asphalt, f_c is the conversion frequency, k and m_e are the shape parameters of the fitted curve, c_1 and c_2 are fitting parameters,f_a is the reduced frequency, and α_T is a time-temperature transformation parameter.

2.3.2. BBR Experiments

BBR experiments were performed to analyze changes in asphalt rheological characteristics at low temperatures [33]. Tests were performed according to ASTM D6648 [34], and the test temperatures were set at −6 °C, −12 °C, and −18 °C and the Figure 3 shows the DSR tester. For this test, asphalt binder samples were heated, poured into a silicon mold to form beams, and trimmed as they cooled. These beams were then soaked in an ethanol bath at a specific temperature for one hour. The specimens were then mounted on a support, and a 3-point load of 980 mN ± 50 mN was applied. The loading times were 240 s each. The creep rate m and creep strength S were measured automatically using the instrument to obtain the graphs of load and deflection as a time function.

Figure 3. BBR tester.

The ΔT_c and PGLT (PG low temperature) obtained from BBR experiments were applied to evaluate the asphalt's low-temperature rheological characteristics. For the BBR experiment, creep flexibility as a function of intrinsic viscoelasticity is applied to determine the bending creep rate (m) and creep stiffness (S) of recycled asphalt binders. The S and m values can be calculated using Equations (6) and (7).

$$S(t)=\frac{PL^{3}}{4bh^{3}\delta(t)} \tag{6}$$

$$m(t)=\left|\frac{dlgS(t)}{dlgt}\right| \tag{7}$$

The rheological index, including the ΔT_c parameter, was designed as the critical temperature discrepancy between the S and m values obtained from the BBR test and expressed noted in as Equation (8).

$$\Delta T_c = T_s - T_m \quad (8)$$

The low-temperature threshold for T_s corresponds to a temperature where $S(60)$ is 300 MPa. The low-temperature threshold for Tm is noted when m(60) is 0.3. The lower the ΔT_c, the more susceptible the asphalt is to non-load-induced cracking problems. When ΔT_c is greater than 0, the asphalt grade is stiffness controlled (S-controlled). If ΔT_c is less than 0, the asphalt grade is relaxation rate controlled (m-controlled) [35]. The Asphalt Institute recommends using −2.5 °C as the cracking alert limit and −5.0 °C as the cracking limit.

2.4. Experiments with Asphalt Mixtures

In this study, the IDEAL-CT test was used to assess the fracture tolerance in asphalt following short-term aging. The test was implemented according to the recommended test method outlined in the ASTM D8225-2019 American Society of Testing Materials [36]. The Figure 4 shows the UTM tester. The test specimens were loaded along the diameter at 25 °C. Before the test started, the samples were heated in an environmental chamber at 25 °C for two hours. The test was performed at a load speed of 50 mm/min and was stopped at a loading force of less than 0.1 kN. The cracking tolerance index was determined using derived load-displacement curves, a more extensive index demonstrating that asphalt is more resistant to cracking. The cracking tolerance index was calculated using Equation (9):

$$I_{CT} = \frac{t l_{75} G_f}{62D|m_{75}|} \times 10^6 \quad (9)$$

where G_f is fracture energy, which is computed via integration of the stress-displacement curve; m_{75} is the stress-displacement curve ramp at 75 percent of maximum; and l_{75} is the displacement associated with 75 percent of the maximum load after the maximum load.

Figure 4. UTM.

2.5. The Asphalt Aging Model

The $G-R$ parameter of asphalt was chosen and calculated as an aging index at a combination of 20 °C and 5 Hz to model changes in asphalt performance with aging time. The primary benefit of utilizing the $G-R$ parameter is the use of the combination of stiffness and relaxation capacities to evaluate the asphalt mixtures' fracture properties [37].

This was mainly because A, which was developed by SHAP, could not predict the cracking properties of asphalt mixtures. Instead, Glover derived rheological parameters using Maxwell's model containing springs and sticky pots, which proved helpful as a surrogate for tensile strain damage. This parameter uses DSR results measured at 15 °C and a frequency of 0.005 rad/s. Rowe has simplified the procedure for calculating the Glover parameter, and the $G-R$ parameter can be expressed using Equation (10) as follows:

$$G - R = |G^*| * (\cos\delta)^2 / \sin\delta \tag{10}$$

where δ is the phase angle, and $|G^*|$ is the complex shear modulus.

The aging process of asphalt materials includes a period of rapid reaction. This phase is characterized by a rapid increase in the $G-R$ parameter. It is then followed by a slow reaction phase with an almost steady rate. Both of these reaction phases represent essentially separate chemical processes [38]. For rapid reactions, sulfoxide is the main oxidation product, leading to an increase in viscosity [39]. During slow reactions, the ketone is the primary product, leading to an increase in viscosity. The fast constant oxidation kinetics model is shown in Equation (11):

$$Log(G - R) = Log(G - R_{ST}) + K(1 - \exp(-R_a t)) + R_b t \tag{11}$$

where $G - R_{ST}$ is the $G-R$ value after short aging; R_a, R_b, and K are model coefficients; and T is aging time.

The parameters K and R_a are related to the rapid reaction phase, with larger values of K and R_a indicating a faster short-term aging rate of the asphalt during the pavement's life. The long-term aging factor, R_b, was obtained from a steady state reaction phase calculation and indicates the long-term aging rate. R_b was obtained from a logical regression of the rise data's rate, and the value of Rb is the line ramp over the steady reaction phase.

3. Results and Discussion

3.1. DSR Results

3.1.1. PGHT

The PGHT of asphalt with different levels of aging is shown in Figure 5. The PGHT results showed that the recycled asphalt with the addition of RA1 had the lowest PGHT among the asphalt samples, both for the STA and long-term aging asphalt, with a difference of approximately 8 °C. This demonstrated that RA1 had a significant regeneration effect. The PGHT of 40% RAP with RA1 was significantly lower than that of 30% RAP and recycled asphalt with other RAs. The PGHT of the RA1 sample with 60 h of aging was the same as that of the 30% RAP sample after 40 h of aging. Recycled asphalt with the addition of RA2 had a slightly lower PGHT than the other four groups. RA2 also significantly affected PGHT, and its regeneration ranked second among the four RAs tested. RA3, RA4, 30% RAP, and 40% RAP had comparable PGHT values. Adding RA3 and RA4 to recycled asphalt did not significantly improve its PG compared to 40% RAP. Otherwise, the PGHT of the asphalt continued to increase as it aged. The PGHT of the asphalt increased the fastest from STA to 20 h of aging, and the PGHT increased more slowly from 20 h of aging to 60 h of aging in all groups, except for RA4.

3.1.2. R Value

Figure 6 shows the R values of various types of recycled asphalt. The R values of six asphalt binders were similar. The cracking sensitivity increased as aging increased. Among them, RA1 had the lowest R value compared to other asphalt samples. The results showed the recycled asphalt with RA1 had the lowest cracking sensitivity. The R value of RA2 slightly decreased compared with 40% RAP and 30% RAP. The R value of recycled asphalt with the addition of RA3 was marginally lower than that of 40% RAP but higher than that of 30% RAP, which indicated that RA3 did not improve the R value of recycled asphalt to the same level as that noted for 30% RAP. In contrast, the R value of asphalt was slightly different from 40% RAP with the addition of RA4. After 40 h and 60 h of aging, the R value of asphalt was higher than 40% RAP. This demonstrated that after prolonged aging, the cracking sensitivity of recycled asphalt with the addition of RA4 was higher compared to 40% RAP. The R value increased faster at the beginning of the aging and slowed down later.

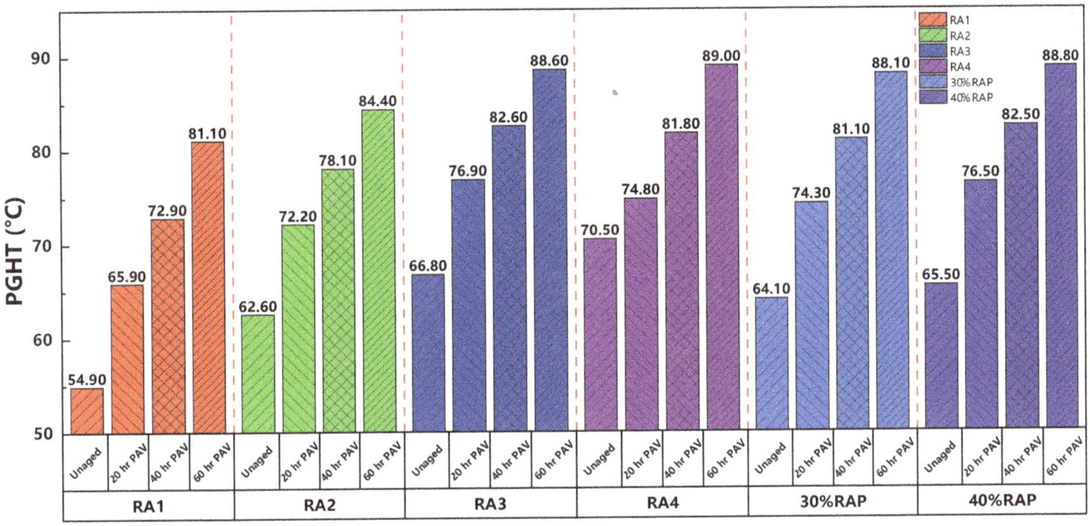

Figure 5. PGHT results.

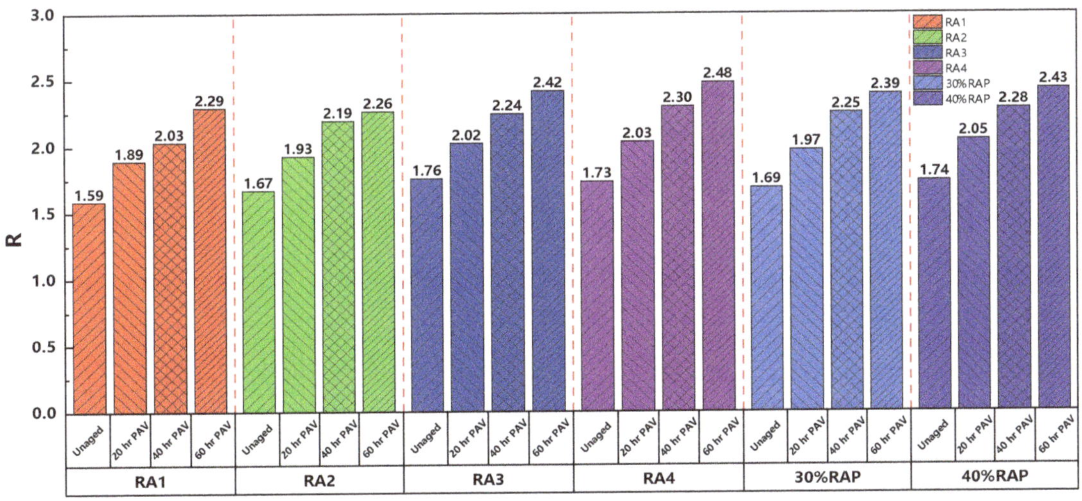

Figure 6. R value results.

3.1.3. $G-R$

As the Figure 7 showed, analysis of unaged recycled asphalt revealed the $G-R$ of the asphalt decreased with the addition of all four RAs. These four RAs enhanced the recycled asphalt's crack resistance. The asphalt's $G-R$ value increased as aging increased, and the growth trend of $G-R$ was exponential. After 60 h of aging, the asphalt with RA1 showed better cracking resistance. The values are less than 180 kPa, which are lower than the threshold for cracking. The results showed that the cracking resistance of the recycled asphalt with RA1 was substantially enhanced. There was no risk of cracking in the recycled asphalt after long-term aging. The $G-R$ values of the recycled asphalt with RA2 were less that 180 kPa until 40 h of aging. After aging for 60 h, the $G-R$ values of the asphalt exceeded the critical value for cracking, and at this time, the reclaimed asphalt was at risk of cracking. It was suggested that RA2 effectively reduced the likelihood of cracking in recycled asphalt, but cracking still occurred after long-term aging. The $G-R$ values of the other four asphalt binders were less than 180 kPa after aging for 20 h. After aging for 40 h, the $G-R$ values were between 180 kPa and 600 kPa. After aging for 60 h, the $G-R$ value was greater than 600 kPa, indicating that all four asphalt samples were at risk of fracture after prolonged aging. The $G-R$ values of the two recycled asphalts with RA3 and RA4 were less than that of 40% RAP and slightly greater than that of 30% RAP. It was worth noting that the recycled asphalt with the addition of RA3 and RA4 showed better resistance to cracking with a short aging time. The $G-R$ values of the two RAs were not improved to the same level as that of 30% RAP. However, after a long aging period, the cracking was more severe than that noted in the recycled asphalt without the addition of recycling agents.

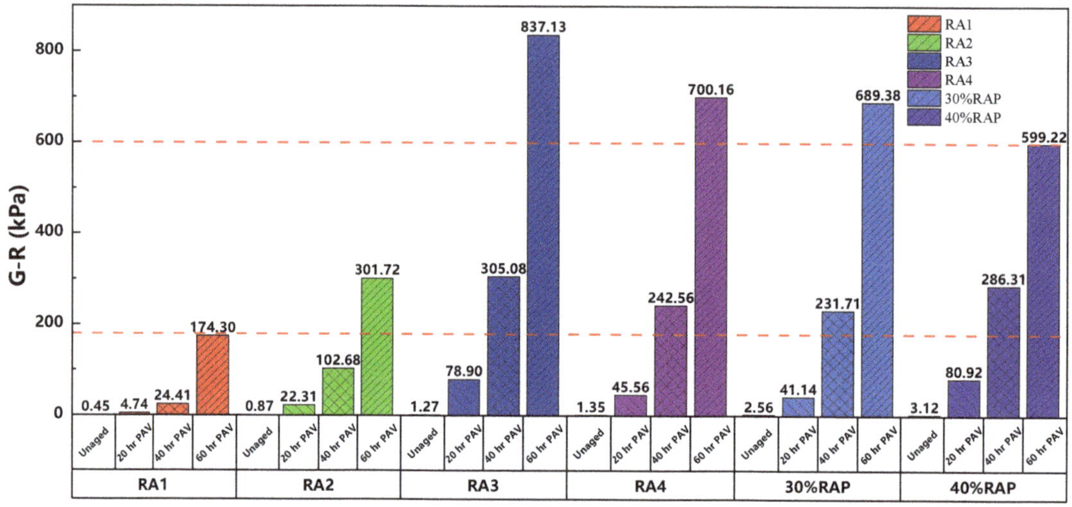

Figure 7. $G-R$ results.

3.2. BBR Results

3.2.1. PGLT Value

As PGLT results showed in Table 1, the PGLT results showed that the addition of RA1, RA2, and RA4 made the PGLT colder. All three asphalt had colder PGLT values than 30% RAP. It suggested that these three RAs could enhance the recycled asphalt's low-temperature properties, and the effect of the improvement was better than that noted for 30% RAP. Among them, RA1 had the best improvement effect, and the low-temperature performance of recycled asphalt with RA1 was reduced by one grade. This finding suggested that RA1 can significantly improve the low-temperature performance of recycled asphalt. The improvement effect of RA2 ranked second, slightly better than RA4, but there was no increase in grade. RA3 increased the PGLT compared to 40% RAP. RA3 adversely affected the low-temperature performance of recycled asphalt. The PGLT of asphalt increased as the aging time increased, and the effect rate was proportional to the aging time. In addition, with the increase in aging time, the impact of RA1 and RA4 on PGLT of recycled asphalt gradually became smaller compared to RA2 and RA3.

Table 1. PGLT results.

PGLT	Unaged	20 h PAV	40 h PAV	60 h PAV
RA1	−38.4	−33.8	−30.2	−26.1
RA2	−33.7	−28.3	−26.0	−23.7
RA3	−30.5	−25.8	−22.9	−19.2
RA4	−33.2	−28.4	−24.3	−21.5
30% RAP	−31.5	−28.1	−24.5	−21.5
40% RAP	−31.6	−26.6	−24.0	−22.1

3.2.2. ΔT_c

In general, the asphalt cracking resistance deteriorated with as aging time increased. Based on the changing trend, the decay of ΔT_c was directly proportional to the aging degree. Figure 8 shows that the addition of RA1 to virgin asphalt increases ΔT_c, and in the long-term aging process, RA1 can alleviate the aging asphalt cracking phenomenon. RA1 can substantially improve the anti-cracking effect of recycled asphalt. The impact of its improved post-cracking resistance was better than that of 30% RAP, and the aged recycled asphalt reached the crack warning limit after 60 h of aging. The remaining five groups of recycled asphalt all exceeded the crack warning limit after aging for 40 h. In two groups of recycled asphalt, namely, recycled asphalt with RA3 and RA4, the recycled asphalt's ΔT_c was less than that noted for 40% RAP, especially after 60 h of aging, when the ΔT_c had exceeded the fracture limit with the addition of the two RAs. After prolonged aging, the two recycled asphalts with the addition RA3 and RA4 showed severe deterioration in their cracking resistance. RA3 and RA4 negatively affected the cracking resistance of asphalt. The asphalt with the addition of RA2 was not aged, and its ΔT_c decreased slightly. The cracking resistance of recycled asphalt with RA2 was better than that of 40% RAP as the aging time increased. However, RA2 did not optimize the cracking resistance of the recycled asphalt to the extent of 30% RAP, with the exception of values obtained after aging for 60 h.

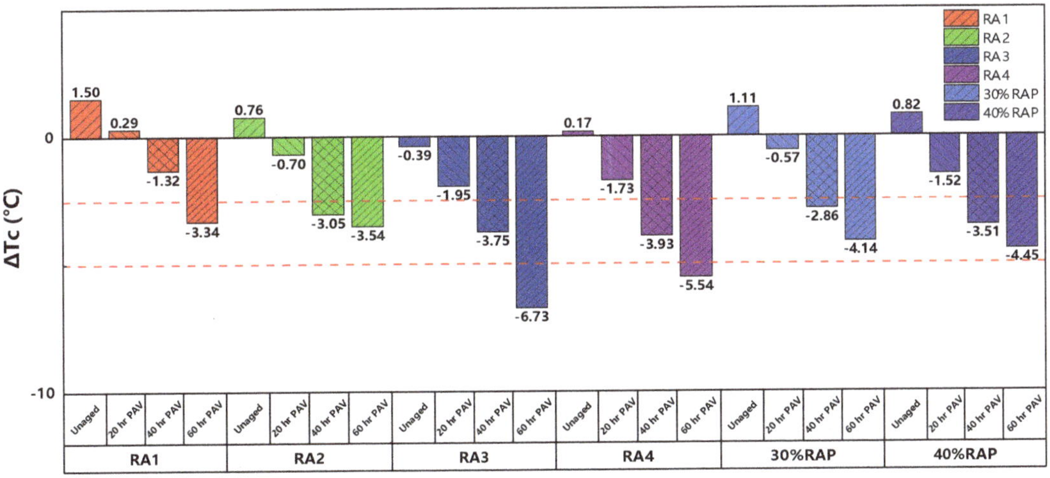

Figure 8. ΔT_c results.

3.3. Mixture Test Results

Figure 9 shows the cracking resistance index of six recycled asphalt using the IDEAL-CT. It was demonstrated that RA3, RA4, 30% RAP, and 40% RAP showed minimal differences in cracking resistance index values. RA3 and RA4 did not increase the asphalt's cracking resistance index effectively. RA1 and RA2 were far superior to the other asphalt samples with approximately two-fold greater cracking resistance index values compared to the other asphalt binders. It suggested that RA1 and RA2 increased the cracking resistance of recycled asphalt mixtures. In comparison, RA3 and RA4 did not significantly affect the cracking resistance of the asphalt mixtures. The asphalt mixture with RA2 showed the best cracking resistance performance.

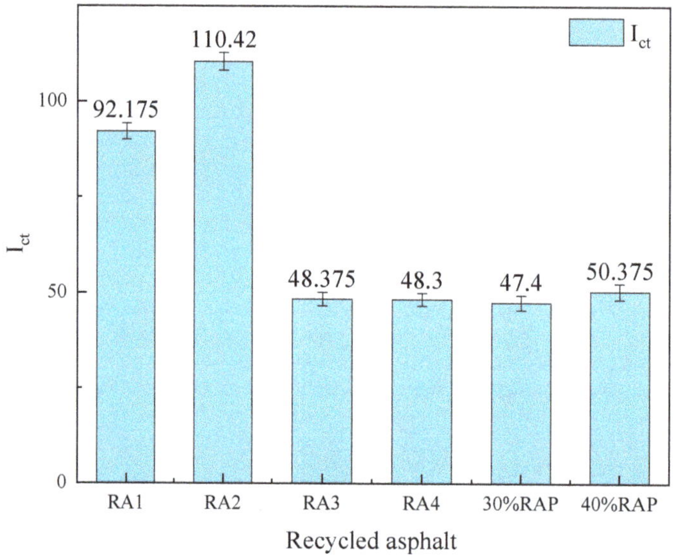

Figure 9. IDEAL-CT results.

3.4. *Aging Model Result*

Figure 10 shows the aging prediction model for six types of recycled asphalt, and Table 2 shows the different parameters included in the aging model. The LOG($G-R$) of asphalt showed a rapid increase followed by a gentle increase. Twenty hours was the point separating the fast increase phase and the gentle increase phase. During the stages of the injection process, RA1 had the lowest K and Rs, which indicated that RA1 had the best tolerance after 20 h of aging. RA3 displayed the worst resistance to short-term aging, with the highest Rs and the most rapid increase in the $G-R$ index after 20 h of aging. The $G-R$ parameter of recycled asphalt with RA3 exceeded that of 40% RAP at approximately 20 h of aging, which indicated the detrimental effect of RA3 on the cracking resistance of recycled asphalt during long-term aging. Recycled asphalt with the addition of RA2 showed insignificant improvements in resistance to short-term aging compared to 40% RAP and 30% RAP. The recycled asphalt with RA4 had Rs and K parameters less than that noted for 30% RAP and 40% RAP. RA4 had a weakening effect on the resistance of recycled asphalt to short-term aging.

Table 2. Parameters of the aging models.

$G-R$	**RA1**	**RA2**	**RA3**	**RA4**	**30% RAP**	**40% RAP**
R_S	0.347	0.060	0.102	0.130	0.407	0.493
K	0.316	1.367	1.571	1.795	2.962	1.596
Ra	0.060	0.067	0.105	0.055	0.025	0.067
Rb	0.038	0.020	0.021	0.016	0.002	0.012

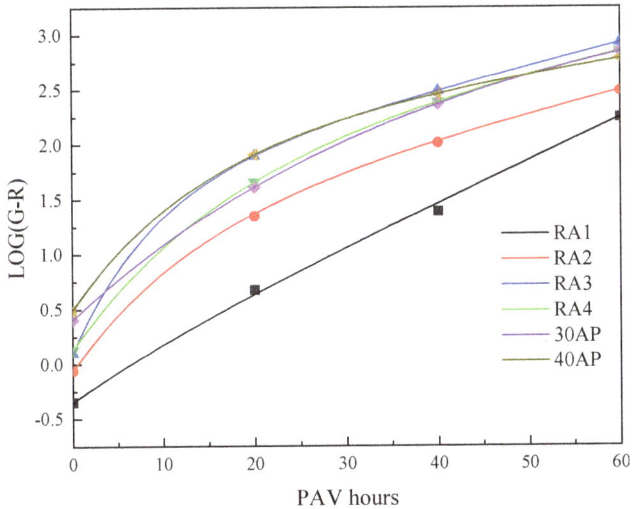

Figure 10. Aging models of six types of recycled asphalt.

The long-term aging stage with more than 20 h of aging time was investigated, and it was found that the addition of RA weakened the recycled asphalt's ability to resist long-term aging. Recycled asphalt with the addition of RA1 exhibited the weakest resistance to long-term aging. Following the aging model trend, the recycled asphalt with the addition of RA1 may exceed the $G - R$ values of other recycled asphalt samples with longer aging times. This finding indicates that the recycled asphalt with the addition of RA1 became less resistant to cracking after longer periods of aging. The reclaimed asphalt with the addition RA2 had marginally higher LOG($G - R$) values than that noted for RA1 during aging, and its values were the second lowest values obtained. RA2 and RA3 had approximately the same Rb values for resisting long-term aging, and these values were slightly worse than RA4. Here, 30% RAP and 40% RAP had the best resistance to long-term aging. This illustrated that the cracking resistance of recycled asphalt with the addition of RA worsened after a longer aging time.

4. Conclusions

This study compared the aging properties of reclaimed asphalt with the addition of four RAs with 30% RAP and 40% RAP. The high-temperature properties, cracking sensitivity, low-temperature properties, and cracking resistance of six recycled asphalt types were evaluated using the following tests: DSR, BBR, and IDEAL-CT. In addition, the aging model of asphalt was employed to predict the changes in the breaking resistance of recycled asphalt with aging time. Specific conclusions were obtained as follows:

(1) DSR results indicate that asphalt performance levels increase with aging time. RA1 and RA2 improved the recycled asphalt's high-temperature performance, while RA3 and RA4 had a lesser effect on the reclaimed asphalt. The R value increased as the aging time of asphalt increased. The R values of the six asphalt samples were similar. The R value of RA1 was the lowest, indicating that the recycled asphalt with RA1 had the lowest cracking sensitivity. The $G - R$ increased with the asphalt's aging time. The modification effect of RA1 was the best, and the $G - R$ value still did not reach the cracking warning value of asphalt after aging for 60 h. The modification effect of RA2 ranked second, and the addition of R3 and R4 had a detrimental impact on the anti-cracking performance.

(2) Based on the experimental results of the BBR test, the recycled asphalt's PGLT continued to increase as the aging time increased. All four RAs made the PGHT lower, and

RA1 obtained the lowest PGLT. Test results for ΔT_c showed that the ΔT_c continued to be reduced as the aging time increased. RA1 and RA2 improved the asphalt cracking resistance the most. Recycled asphalt with the addition of RA1 did not reach the crack warning limit for ΔT_c after aging for 40 h. RA3 and RA4 improved the cracking resistance of unaged recycled asphalt. Nevertheless, as the aging time increased, the cracking resistance of recycled asphalt with RA3 and RA4 was worse than that noted for 40% RAP.

(3) The CT experimental results showed that RA1 and RA2 significantly increase the anti-cracking index Ict of asphalt mixtures. Its value increased approximately two-fold compared with that of 40% RAP recycled asphalt, and the addition of RA3 and RA4 had minimal influence on the crack-resistance index.

(4) The recycled asphalt's ability to resist short-term and long-term aging was better analyzed using an aging model. Analyzing the fitted parameters, it was found that RA1 had the best capability for short-term aging resistance, and RA3 had the worst ability to resist short-term aging. However, the ability of recycled asphalt with the addition of RA to resist long-term aging was worse than that noted for 30% RAP and 40% RAP. Among them, RA1 had the least resistance to long-term aging. The recycled asphalt with RA1 had the lowest $G-R$ after 60 h of aging. However, according to the predicted trend, the $G-R$ value of recycled asphalt with the addition of RA1 exceeded that of the other five asphalt samples as the aging time increased. This finding indicated that the crack resistance of recycled asphalt with the addition of RA1 would worsen after a longer aging time.

The results of all the experiments were combined to reveal that RA1 had the best regeneration effect and could substantially improve the properties of asphalt up to 60 h of aging. RA2 had the second-best modification effect, but its resistance to long-term aging was better than that of RA1.

Author Contributions: The authors confirm contributions to the paper as follows: study conception and design: Q.Y. and R.W.; data collection: Q.Y. and L.S.; analysis and interpretation of results: Z.L. and Q.Y.; draft manuscript preparation: Z.L., Q.Y. and R.Z. All authors have read and agreed to the published version of the manuscript.

Funding: This research received no external funding.

Data Availability Statement: Restrictions apply to the availability of these data. Data can be obtained from the corresponding author on request.

Conflicts of Interest: The authors declare no conflict of interest.

References

1. Liu, Z.; Zhou, Z.; Gu, X.; Sun, L.; Wang, C. Laboratory evaluation of the performance of reclaimed asphalt mixed with composite crumb rubber-modified asphalt: Reconciling relatively high content of RAP and virgin asphalt. *Int. J. Pavement Eng.* **2023**, *24*, 2217320. [CrossRef]
2. Liu, Z.; Gu, X.; Dong, X.; Cui, B.; Hu, D. Mechanism and Performance of Graphene Modified Asphalt: An Experimental Approach Combined with Molecular Dynamic Simulations. *Case Stud. Constr. Mat.* **2022**, *18*, e01749. [CrossRef]
3. Yang, Q.; Liu, Z.; Sun, L.; Li, Z. Analysis of the viscosity and self-healing performance of warm-mix recycled asphalt: A method combining molecular dynamics simulations and laboratory tests. *Case Stud. Constr. Mater.* **2023**, *19*, e02593. [CrossRef]
4. Arambula-Mercado, E.; Chavarro-Munoz, S.J.; Hu, S.; Epps, J.A.; Park, E.S.; Epps Martin, A. *High Reclaimed Asphalt Pavement (RAP) Asphalt Mixes for Low Volume Roads*; Texas A&M Transportation Institute: Bryan, TX, USA, 2018.
5. Liu, Z.; Yang, Q.; Gu, X. Assessment of Pavement Structural Conditions and Remaining Life Combining Accelerated Pavement Testing and Ground-Penetrating Radar. *Remote Sens.* **2023**, *15*, 4620. [CrossRef]
6. Liu, Z.; Gu, X.; Ren, H.; Li, S.; Dong, Q. Permanent Deformation Monitoring and Remaining Life Prediction of Asphalt Pavement Combining Full-Scale Accelerated Pavement Testing and FEM. *Struct. Control Health Monit.* **2023**, *2023*, 6932621. [CrossRef]
7. Daryaee, D.; Ameri, M.; Mansourkhaki, A. Utilizing of waste polymer modified bitumen in combination with rejuvenator in high reclaimed asphalt pavement mixtures. *Constr. Build. Mater.* **2020**, *235*, 117516. [CrossRef]
8. Su, N.; Xiao, F.; Wang, J.; Cong, L.; Amirkhanian, S. Productions and applications of bio-asphalts—A review. *Constr. Build. Mater.* **2018**, *183*, 578–591. [CrossRef]

9. Ameri, M.; Mirzaiyan, D.; Amini, A. Rutting resistance and fatigue behavior of gilsonite-modified asphalt binders. *J. Mater. Civ. Eng.* **2018**, *30*, 04018292. [CrossRef]
10. Liu, Z.; Gu, X.; Ren, H. Rutting prediction of asphalt pavement with semi-rigid base: Numerical modeling on laboratory to accelerated pavement testing. *Constr. Build. Mater.* **2023**, *375*, 130903. [CrossRef]
11. Haghshenas, H.F.; Rea, R.; Reinke, G.; Haghshenas, D.F. Chemical characterization of recycling agents. *J. Mater. Civ. Eng.* **2020**, *32*, 06020005. [CrossRef]
12. Sun, L.; Wen, Y.; Liu, Q.; Li, D.; Lyu, L.; Pei, J.; Zhang, J.; Li, R. A laboratory investigation into the effect of waste non-tire rubber particles on the performance properties of terminal blend rubberized asphalt binders. *Constr. Build. Mater.* **2021**, *313*, 125409. [CrossRef]
13. Shen, J.; Ohne, Y. Determining rejuvenator content for recycling reclaimed asphalt pavement by SHRP binder specifications. *Int. J. Pavement Eng.* **2002**, *3*, 261–268. [CrossRef]
14. Zaumanis, M.; Mallick, R.B.; Frank, R. Evaluation of rejuvenator's effectiveness with conventional mix testing for 100% reclaimed Asphalt pavement mixtures. *Transp. Res. Rec.* **2013**, *2370*, 17–25. [CrossRef]
15. Li, J.; Ni, F.; Huang, Y.; Gao, L. New additive for use in hot in-place recycling to improve performance of reclaimed asphalt pavement mix. *Transp. Res. Rec.* **2014**, *2445*, 39–46. [CrossRef]
16. Zhang, R.; Dave, E.V.; Sias, J.E.; Tabatabaee, H.A.; Sylvester, T.; Wang, Z. Evaluation of the state of practice asphalt binder and mixture tests for assessing the compatibility of complex asphalt materials. *Road Mater. Pavement Des.* **2023**, *24*, 451–470. [CrossRef]
17. Kaseer, F.; Yin, F.; Arámbula-Mercado, E.; Martin, A.E.; Daniel, J.S.; Salari, S. Development of an index to evaluate the cracking potential of asphalt mixtures using the semi-circular bending test. *Constr. Build. Mater.* **2018**, *167*, 286–298. [CrossRef]
18. Zhang, R.; Sias, J.; Dave, E.; Hanz, A.; Reinke, G.; Wang, Z. Comprehensive Laboratory Evaluation of Recycling Agent Treated Plant-Produced Asphalt Mixtures. *Transp. Res. Rec.* **2022**, *2676*, 620–634. [CrossRef]
19. Cao, W.; Barghabany, P.; Mohammad, L.; Cooper III, S.B.; Balamurugan, S. Chemical and rheological evaluation of asphalts incorporating RAP/RAS binders and warm-mix technologies in relation to crack resistance. *Constr. Build. Mater.* **2019**, *198*, 256–268. [CrossRef]
20. Zhang, R.; Chen, H.; Guo, H.; Zhou, M.; Jiandong, H. Development of a prior model to predict the cracking performance of asphalt mixture in general for asphalt material selection and mix design. *Int. J. Pavement Eng.* **2023**, *24*, 2251080. [CrossRef]
21. Liu, Z.; Gu, X.; Dong, Q. Permanent Deformation Evaluation and Instability Prediction of Semi-rigid Pavement Structure Using Accelerated Pavement Testing and Finite Element Method. *J. Test. Eval.* **2024**, *52*. [CrossRef]
22. Liu, Z.; Gu, X. Performance Evaluation of Full-scale Accelerated Pavement using NDT and Laboratory Tests: A case study in Jiangsu, China. *Case Stud. Constr. Mat.* **2023**, *18*, e02083. [CrossRef]
23. Behnia, B.; Ahmed, S.; Dave, E.V.; Buttlar, W.G. Fracture Characterization of Asphalt Mixtures with Reclaimed Asphalt. *Int. J. Pavement Res. Technol.* **2010**, *3*, 72–78.
24. Sun, L.; Gu, X.; Hu, D.; Zhou, Z.; Wang, G. Anti-aging mechanism and rheological properties of lignin, quercetin, and gallic acid as antioxidants in asphalt. *Constr. Build. Mater.* **2023**, *369*, 130560. [CrossRef]
25. Li, X.; Marasteanu, M.O.; Williams, R.C.; Clyne, T.R. Effect of reclaimed asphalt pavement (proportion and type) and binder grade on asphalt mixtures. *Transp. Res. Rec.* **2008**, *2051*, 90–97. [CrossRef]
26. *ASTM D2172*; Standard Test Methods for Quantitative Extraction of Bitumen From Bituminous Paving Mixtures. ASTM International: West Conshohocken, PA, USA, 2005.
27. *ASTM D1856*; Standard Test Method for Recovery of Asphalt from Solution by Abson Method. ASTM International: West Conshohocken, PA, USA, 2015.
28. *ASTM D4402*; Standard Test Method for Viscosity Determination of Asphalt at Elevated Temperatures Using a Rotational Viscometer. ASTM International: West Conshohocken, PA, USA, 2012.
29. Liu, Z.; Sun, L.; Gu, X.; Wang, X.; Dong, Q.; Zhou, Z.; Tang, J. Characteristics, mechanisms, and environmental LCA of WMA containing sasobit: An analysis perspective combing viscosity-temperature regression and interface bonding strength. *J. Clean. Prod.* **2023**, *391*, 136255. [CrossRef]
30. Rowe, G.M.; Sharrock, M. Alternate shift factor relationship for describing temperature dependency of viscoelastic behavior of asphalt materials. *Transp. Res. Rec.* **2011**, *2207*, 125–135. [CrossRef]
31. Zhang, R.; Sias, J.E.; Dave, E.V. Using mix design information for modelling of fundamental viscoelasticity of asphalt mixtures. *Constr. Build. Mater.* **2022**, *329*, 127029. [CrossRef]
32. Liu, Z.; Gu, X.Y.; Ren, H.; Wang, X.; Dong, Q. Three-dimensional finite element analysis for structural parameters of asphalt pavement: A combined laboratory and field accelerated testing approach. *Case Stud. Constr. Mat.* **2022**, *17*, e01221. [CrossRef]
33. Zhang, R.; Sias, J.E.; Dave, E.V. Comparison and correlation of asphalt binder and mixture cracking parameters incorporating the aging effect. *Constr. Build. Mater.* **2021**, *301*, 124075. [CrossRef]
34. *ASTM D6648*; Standard Test Method for Determining the Flexural Creep Stiffness of Asphalt Binder Using the Bending Beam Rheometer (BBR). ASTM International: West Conshohocken, PA, USA, 2018.
35. Anderson, R.M.; King, G.N.; Hanson, D.I.; Blankenship, P.B. Evaluation of the relationship between asphalt binder properties and non-load related cracking. *J. Assoc. Asph. Paving Technol.* **2011**, *80*, 615–664.
36. *ASTM D8225-2019*; Standard Test Method for Determination of Cracking Tolerance Index of Asphalt Mixture Using the Indirect Tensile Cracking Test at Intermediate Temperature. ASTM International: West Conshohocken, PA, USA, 2019.

37. Zhang, R.; Sias, J.E.; Dave, E.V. Development of a rheology-based mixture aging model for asphalt material cracking performance evaluation. *Mater. Struct.* **2021**, *54*, 150. [CrossRef]
38. Petersen, J.C.; Harnsberger, P.M. Asphalt aging: Dual oxidation mechanism and its interrelationships with asphalt composition and oxidative age hardening. *Transp. Res. Rec.* **1998**, *1638*, 47–55. [CrossRef]
39. Petersen, J.; Harnsberger, P.; Robertson, R. Factors affecting the kinetics and mechanisms of asphalt oxidation and the relative effects of oxidation products on age hardening. *Prepr. Pap. Am. Chem. Soc. Div. Fuel Chem.* **1996**, *41*, 212.

buildings

Article

Compaction Characteristics of a Foam Asphalt Hot In-Place Recycling Asphalt Mixture

Xupeng Sun [1], Dedong Guo [1], Jin Li [1], Zhen Liu [2], Meng Xu [1,*], Qinshuai Hu [3], Qi Xu [1] and Shihua Yang [1]

1 School of Transportation Civil Engineering, Shandong Jiaotong University, Jinan 250357, China; 22107018@stu.sdjtu.edu.cn (Q.X.); 22107024@stu.sdjtu.edu.cn (S.Y.)
2 Department of Roadway Engineering, School of Transportation, Southeast University, Nanjing 211189, China
3 Huashe Design Group, Huashe (Jiangsu) New Material Technology Co., Nanjing 210000, China
* Correspondence: 401010@sdjtu.edu.cn

Abstract: This study investigates the application of foam asphalt (FA) to enhance the compaction effectiveness of a hot in-place recycling asphalt mixture (HIR-AM) during the HIR process of old road surfaces. Initially, the process parameters for FA preparation were determined through expansion-rate and half-life tests. Subsequently, the study focused on evaluating the impact of FA on the compaction quality of HIR-AM. Performance assessments were conducted through rutting tests, low-temperature bending tests, Hamburg wheel tracking tests, dynamic modulus analyses, and various other experiments to evaluate the road performance of HIR-FAM. Finally, the research findings were validated through practical engineering applications, and the construction process for HIR-FAM was summarized. The research results reveal that the optimal foaming temperature for SBS asphalt is 170 °C, with an ideal water content of 1.7%. Under the same compaction temperature, HIR-FAM demonstrated a significant reduction in void content, ranging from 3.8% to 21.2% compared to HIR-AM. Moreover, a higher proportion of FA usage resulted in a more substantial decrease in void content. Compared to HIR-AM, HIR-FAM exhibited notable improvements, including an 11.6% increase in dynamic stability, a 13.4% enhancement in bending strength, a 13.3% increase in maximum bending strain, an 8.1% improvement in residual stability, and an 8.5% boost in freeze–thaw splitting strength. Furthermore, HIR-FAM demonstrated superior water-thermal stability and resistance to low-frequency loads. Paving a test road verified that the adoption of foam asphalt in thermal recycling led to a compaction density increase of over 0.79% compared to traditional in situ thermal recycling sections, with improved compaction uniformity.

Keywords: hot in-place recycling; foamed asphalt; compaction characteristics; road engineering

1. Introduction

In the field of transportation in China, the total carbon emissions are enormous, accounting for approximately 10% of the national terminal carbon emissions [1]. It is one of the key areas of concern for carbon emissions control. With the comprehensive implementation of China's "dual carbon" strategy, the pressure to reduce emissions in the transportation industry will increase further [2,3]. Achieving carbon reduction in the transportation industry under the backdrop of carbon neutrality is an important research topic [4–7]. Hot in-place recycling (HIR) can fully utilize reclaimed asphalt pavement (RAP), which has the characteristics of low maintenance cost, short construction period, low traffic disturbance, etc., and can effectively reduce carbon emissions [8–12]. In recent years, it has been widely applied in the road maintenance industry [13,14].

Due to the low mixing temperature and the presence of a significant amount of aged asphalt in the on-site hot recycled asphalt mixture, the efficiency of mixing is reduced [15]. This leads to insufficient compaction of the recycled road surface and poor-quality informa-tion, making it prone to early deterioration [16,17]. Warm-mix asphalt (WMA) technology can reduce the viscosity of asphalt binder without increasing the heating temperature [18], thereby improving the workability of the mixture and effectively enhancing the compaction quality of asphalt pavements [19–21]. Techniques like foam warm-mix asphalt (FWMA) utilize the high temperature of asphalt to vaporize liquid water, leading to foaming of the asphalt and an increase in its surface area. This indirectly improves asphalt flowability, achieving an anti-sticking effect [22,23]. FWMA is known for its simplicity of operation and lower cost, making it one of the most prominent warm-mix technologies [24,25].

However, this technique also has some drawbacks. A compaction temperature that is too low can result in inadequate adhesion between the binder and aggregates, leading to a softer mixture with poor resistance to deformation [26]. Moreover, the injection of water during the foaming process can make foam warm-mix asphalt mixtures more susceptible to moisture damage [27]. Combining the characteristics of PAP and FWMA technologies can effectively address the aforementioned issues [28]. For instance, due to material aging in RAP, it exhibits better resistance to rutting [29]. Meanwhile, FWMA technology can improve the poor workability of RAP by reducing the viscosity of aged asphalt binders. Additionally, the addition of PAP offers more high-viscosity binders, enhancing the adhesion of the mixture and thereby compensating for the shortcomings in FWMA performance [30]. Research by A. Rondón-Quintana [31], Zheng, Z. G. [32], and others has shown that FWMA has been effective when applied in plant-mixed hot recycling mixtures. P. Abreu [33] and others investigated the impact of RAP dosage on the performance of different foam asphalt recycled mixtures and found that the same RAP dosage of foam asphalt can reduce the recycled mixtures compaction temperature by 15~30 °C. Furthermore, foam asphalt recycled mixtures with high RAP content still exhibit good water stability. Zhao, S. [34] and Ashiqur, R. M. [35] conducted a series of indoor experiments on warm-mix recycled mixtures. They found that compared to hot-mix asphalt (HMA), due to the lower mixing temperature of FWMA, it is more prone to rutting and moisture damage. However, increasing the RAP content has a positive impact on the resistance to the rutting, water stability, and fatigue performance of WMA. The different preparation methods of foam asphalt recycled mixtures and the addition of rejuvenators can also impact the performance of the mixture: Kavussi, A. [36] compared the rutting performance of FWMA-A (all binders used were foam asphalt) and FWMA-B (hard asphalt was foam asphalt). The results showed that when the compaction temperature was 115 °C, the rutting depth of FWMA-B was reduced by 0.3 mm compared to that of FWMA-A after 20,000 cycles. Jeong, M. M. [37] and others analyzed the effect of a rejuvenator on foam asphalt mixtures containing 100% RAP. They found that adding a 6% rejuvenator improves its resistance to cracking but significantly decreases its structural integrity. Consequently, much research has been conducted by scholars to reduce the compaction temperature of recycled mixes, increase the proportion of RAP blending, and the synergistic effect of FWMA and RAP, but there is limited research on the application of FWAM technology in HIR to improve the compaction quality.

This study was conducted in the context of the S511 road improvement and expansion project in Wudi County, Shandong Province. Foam warm-mix technology was applied to HIR, achieving 100% utilization of RAP, and a systematic investigation into the impact of FA on the compaction quality of HIR-AM was carried out. Firstly, the FA foaming process was determined based on foaming efficiency. Subsequently, by studying the compaction quality of three types of HIR-FAM, the preparation process of HIR-FAM was established. The experimental verification of their road performance was conducted. Finally, in conjunction with field engineering, the research findings were validated, and a set of construction processes for HIR-FAM was proposed, providing a reference for the promotion and application of this technology in asphalt pavement maintenance projects.

2. Materials and Methods

2.1. Project Description

The project is located in the vicinity of S511, Jieshishan Town, Wudi County, Shandong Province, as shown in Figure 1a. The original road surface's upper-layer structure is 4 cm AC-13, with an average daily traffic volume (ADT) of 45,799, making it a heavily traveled main road. Wudi County experiences average low and high temperatures of −16.5 °C and 39.4 °C, respectively. Prolonged temperature variations and repeated heavy traffic loads have led to severe fatigue cracking and rutting in the asphalt pavement, as illustrated in Figure 1b. According to the Standard JTG 5210-2018 [38,39], the pavement condition index (PCI) for this road segment was assessed, resulting in a PCI of 73.07, indicating the need for pavement maintenance. The total construction length of the project is 9.8 km, and it is built to the standards of a first-class highway. Approximately 61% of the road sections underwent maintenance using HIR, as shown in Figure 1c.

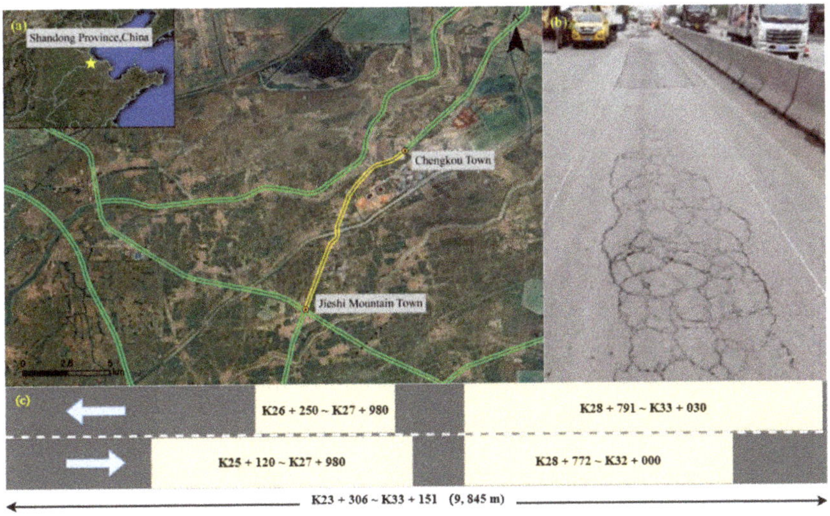

Figure 1. Project overview, (**a**) project location, (**b**) original road condition, (**c**) sections utilizing HIR.

2.2. Materials

SBS asphalt was used as the new asphalt, and limestone served as the coarse aggregate for the new asphalt mixture. Testing and analysis of the new asphalt and recycled asphalt from RAP were conducted following Test Procedures JTG E20-2011 [40–44], with the test results presented in Table 1, which shows that the recycled asphalt is hardened due to the aging effect of the asphalt, and its performance no longer meets the corresponding technical requirements of the Construction Technical Specifications (JTG F40-2004) [41–43]. The physical properties of the aggregates are listed in Table 2, and it is found that the recycled aggregates are good in terms of crushing value, abrasion loss, and density, which can ensure that the in situ hot recycled asphalt mixtures can play the role of the skeleton support and contribute a certain degree of strength to the recycled asphalt mixtures. Recycling agents are added to restore the original properties of recycled asphalt, and their technical properties are presented in Table 3.

2.3. Proportioning Design

Based on the performance indicators and sieve analysis results of RAP and new aggregates, the gradation design for the AC-13 mixture was conducted. Considering an on-site milling depth of 4 cm and based on engineering experience, the addition of new asphalt mixture to the recycled mixture is set at 15%, and the gradation design of the recycled mixture is shown in Table 4. According to the specifications [41–45], the optimal asphalt content for the hot in-place recycle asphalt mixture was determined to be 5.1%, the additional asphalt content was 0.64%, and the asphalt content for the new asphalt was 4.3%.

Table 1. Technique performance of asphalt.

Testing Parameters	Test Method	Technical Requirements	SBS Asphalt	Recycled Asphalt
Penetration (25 °C, 100 g, 5 s) (0.1 mm)	T 0604-2011	40~60	58.8	28
Ductility5 cm/min, 5 °C (cm)	T 0605-2011	≥20	30	69.1
Softening point (ring and ball method) (°C)	T 0606-2011	≥60	70	5.6
Asphalt content (%)	T 0725-2011	-	-	4.5

Table 2. Test results of mineral aggregates.

Testing Parameters	Test Method	Technical Requirements	Recycled Aggregate	New Aggregate
Crushing value (%)	T 0316-2005	≤26	17.60	20.88
Los Angeles abrasion loss (%)	T 0317-2005	≤28	19.02	21.98
Apparent relative density(g/cm^3)	T 0304-2005	≥2.6	2.715	2.896
Water absorption rate (%)	T 0304-2005	≤2.0	0.5	0.92

Table 3. Performance parameters of the recycling agent.

Technical Indicators	Technical Requirements	Test Results
60 °C Dynamic viscosity (mm^2/s)	176~900	251
Flash point (°C)	>220	242
Saturates (%)	≤30	28
Aromatics (%)	Measured	58
Mass change before and after TFOT (%)	≤3	2.6

Table 4. Mineral aggregate gradation.

Sieve Mesh (mm)	Ratio (%)	16	13.2	9.5	4.75	2.36	1.18	0.6	0.3	0.15	0.075
RAP	85	100	94.8	77.2	48.3	32.7	24.3	15.6	10.9	8.5	7.2
New aggregates	15	100	95.0	48.6	39.3	31.1	23.8	15.2	9.9	7.9	6.5
Standard (%)	-	100	90–100	68–85	38–68	24–50	15–38	10–28	7–20	5–15	4–8
Passing (%)	-	100	94.8	72.9	47.0	32.5	24.2	15.5	10.8	8.4	7.1

2.4. Experiment Design

This research workflow is shown in Figure 2, which mainly includes determining the foam asphalt foaming process using the expansion rate and half life; the mensuration of the compaction characteristics of different HIR-FAM; the evaluation of the road performance of HIR-FAM using the rutting test, low-temperature bending test, Hamburger rutting test, and the dynamic modulus test; and finally, the verification of the results of the indoor study through the paving of a test road in a real project.

2.4.1. The Impact of Foaming Conditions on Asphalt Performance

The WLB10 asphalt foaming machine from Wirtgen GmbH, Germany, was utilized in this experiment for indoor asphalt foaming tests. Foaming conditions were set as follows: foaming temperatures of 160 °C, 170 °C, and 180 °C and foaming water content of 1.0%, 1.5%, 2.0%, and 2.5%. A total of 12 sets of experiments were conducted, with each set performed in triplicate. The expansion rate and half life of the asphalt under different foaming conditions were measured, and the average values were taken as the final test results. The test results were then fitted and regressed in the form of $y = a\ln x + b$, and trend curves were plotted to illustrate the variations.

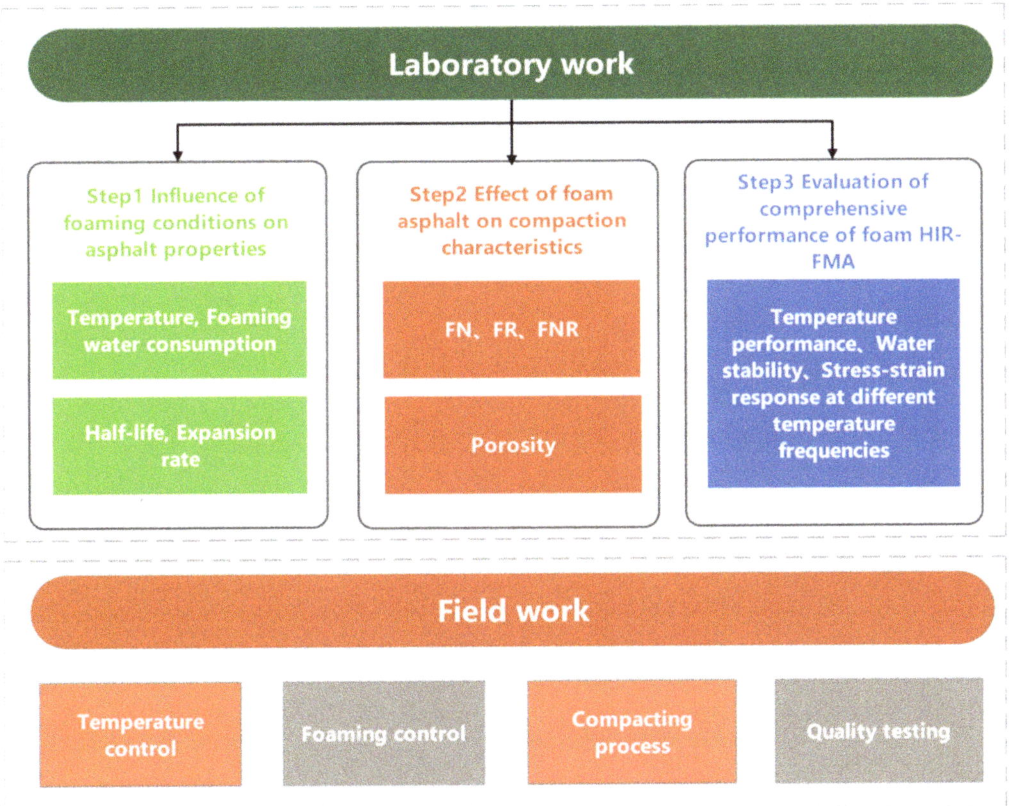

Figure 2. Research workflow diagram used in this study.

2.4.2. The Influence of FA on the Compaction Characteristics of Recycled Mixtures

To investigate the impact of FA on the compaction characteristics of HIR-AM, three types of HIR-FAM were prepared: (1) The asphalt in the new asphalt mixture was replaced with FA to produce the recycled mixture, referred to as FN. (2) The additional asphalt sprayed onto RAP was substituted with FA to prepare the recycled mixture, referred to as FR. (3) Both the additional asphalt added to the new asphalt mixture and the asphalt sprayed onto RAP were replaced with FA to create the recycled mixture, referred to as FNR. As a control group, we used a conventional hot-mix asphalt HIR mixture denoted as HNR. We fabricated Marshall specimens using these four types of recycled mixtures at different mixing temperatures (120 °C, 130 °C, 140 °C, 150 °C, 160 °C, and 170 °C). The compaction conditions were identical to those of conventional hot-mix modified asphalt HIR asphalt mixtures, and the percent air voids of the specimens were measured.

2.4.3. Study on the Performance of HIR-FAM

Studies have shown that the mixing temperature for HIR-AM during on-site construction is approximately 145 °C [46,47]. Therefore, in this experiment, HIR-FAM and HIR-AM were mixed and prepared at a temperature of 145 °C to investigate the influence of FA on the performance of recycled asphalt mixtures. Comparative tests were conducted following the guidelines specified in the Test Procedures (JTG E20-2011) [40].

Dynamic Stability Test

The rutting test involved the fabrication of rutting test specimens measuring 300 mm × 300 mm × 80 mm using a rolling compactor. After insulation, the specimens were placed in a rutting automatic testing machine for dynamic stability testing with a wheel pressure of 0.7 MPa. The instrument automatically recorded rut depth measurements, at 45 min (t_1) and at 60 min (t_2), and the dynamic stability value was calculated using Equation (1).

$$DS = \frac{(t_1 - t_2) \times N}{d_2 - d_1} \times C_1 \times C_2 \tag{1}$$

which includes the dynamic stability of the asphalt mixture (cycles/mm); the deformation at time (mm) [48]; the deformation at time (mm); the instrument type parameter, taken as 1.0; and the specimen coefficient, taken as 1.0; and N is the reciprocal rolling speed of the test wheel, 42 cycles/min.

Low-Temperature Bending Test

The rutting test specimens, formed using the rolling compactor, are cut into small beam specimens measuring 250 mm × 30 mm × 35 mm. These specimens are placed in an environmental incubator at a test temperature of −10 °C for a constant temperature of 4 h. A modular electromechanical servo-driven asphalt testing machine (ASPHALT QU3E) is used to conduct the low-temperature bending test, from which the flexural tensile strength and maximum tensile strain were calculated.

Water Stability Test

According to Specification JTG/T 5521-2019 [45], the water stability of two types of hot recycling asphalt mixtures is evaluated using the freeze–thaw splitting test and the soaking Marshall test. For the freeze–thaw splitting test, standard Marshall specimens compacted on both sides for 50 blows were prepared. These specimens undergo a process that includes 15 min of vacuum saturation, 16 h of freezing treatment, 24 h of high-temperature treatment, and 2 h of immersion in room-temperature water. The split tensile strength of the specimens is then measured and compared with the split tensile strength of specimens immersed in room-temperature (25 °C) water for 2 h. For the soaking Marshall test, Marshall specimens of one type of mixture were divided into two groups. Both groups of specimens were simultaneously placed in a constant-temperature water bath at 60 °C. One group was soaked for 30 min, and the other group was soaked for 48 h. The Marshall stability of both groups of specimens was determined, and based on the stability values from the two groups, the soaked residual stability was calculated.

Hamburg Wheel Tracking Test

To verify the high-temperature stability and water stability of the recycled mixture, a Hamburg wheel-tracking test was conducted. This test uses asphalt mixture specimens subjected to rotary compaction, with their surfaces immersed in water at 50 °C. Steel wheels then roll back and forth over the specimen surface for 20,000 cycles. Water stability and resistance to rutting performance were evaluated by measuring the rut depth and characteristics of the deformation curve of the asphalt mixture.

Dynamic Modulus Test

To verify the deformation resistance of the recycled mixture, dynamic modulus tests were conducted. Cylindrical specimens with dimensions of φ100 mm × 150 mm were obtained through coring from compacted asphalt samples. Dynamic modulus tests were performed using an AMPT (Asphalt Mixture Performance Tester) machine at five test temperatures (0 °C, 10 °C, 20 °C, 35 °C, and 50 °C) and six loading frequencies (25 Hz,

10 Hz, 5 Hz, 1.0 Hz, 0.5 Hz, 0.1 Hz). The dynamic modulus and phase angle of the four types of in-place hot recycling asphalt mixtures were determined through these tests.

3. Results and Discussion

3.1. Determination of Optimal Foaming Conditions

As shown in Figure 3a, the expansion ratio of SBS asphalt increases continuously with the increase in foaming water content at the same temperature, while the half life gradually decreases. This suggests that excessive water content does not improve the foaming effect of asphalt. The main reason is that with the increase in foaming water content, more water vapor is trapped by the asphalt, leading to an increase in the expansion ratio. However, as the volume of asphalt bubbles increases, the asphalt film becomes thinner, resulting in a shorter time for the internal equilibrium of the asphalt bubbles, leading to a shorter half life. Figure 3b demonstrates a good logarithmic correlation between the experimental results of the expansion ratio and the half life. Additionally, under a foaming temperature of 170 °C, the expansion ratio and half life of foamed asphalt are relatively balanced, indicating that higher temperatures do not necessarily result in a better foaming effect for SBS asphalt. This is because, while high temperatures promote the vaporization of water in asphalt to generate foam, excessively high temperatures reduce the elasticity of the asphalt film, making the asphalt bubbles unstable and prone to rupture [49]. Currently, there is no standardized limit for the acceptable values of the expansion ratio and half life. However, the relevant literature [50] suggests a lower limit standard value of 10 for the expansion ratio and 8 s for the half life. The optimal foaming water content corresponding to a foaming temperature of 170 °C was determined to be 1.7% using Equation (2) [45].

$$W_{opt} = (W_1 + W_2)/2 \quad (2)$$

where, W_{opt} is the optimal foaming water content (%); W_1 is the determination of the water content corresponding to the lower limit of the expansion rate standard value (%); W_2 is the determination of the water content corresponding to the lower limit of the half life standard value (%). If the determined water content corresponding to the lower limit of the expansion rate standard value is below 1.5%, we take it as 1.5%. When the determined water content corresponding to the lower limit of the standard value of the half life exceeds 4.0%, it is taken as 4.0%.

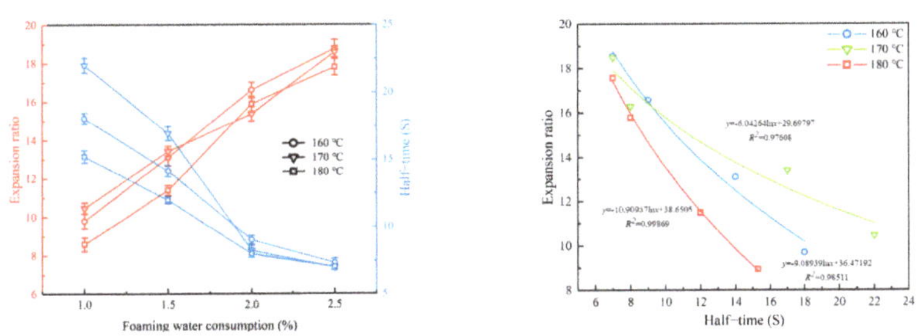

(**a**) (**b**)

Figure 3. Foaming test results: (**a**) effect of different factors on the foaming properties of bitumen; (**b**) asphalt foaming behavior curve.

3.2. The Impact of Foamed Asphalt on Compaction Characteristics

In Figure 4, it can be observed that as the mixing temperature increases, the percentage of the air voids in the mixture gradually decreases. At the same mixing temperature, the percentage of the air voids in the HIR-FAM is smaller than that of HIR-AM. The reason for this is that foamed asphalt vaporizes water through the asphalt, causing the asphalt to foam, which increases the contact area between the asphalt and the mixture. This allows the asphalt to adhere better to the surface of the mixture, improving the smoothness of the mixture interface and thereby enhancing the compaction quality of the mixture [51]. However, the percentage of the air voids of different HIR-FAM is not constant. At 150 °C, the percentage of the air voids of FNR is 4.0%, while those of FN, FR, and HNR are 4.1%, 4.4%, and 4.5%, respectively. It can be seen that using foamed asphalt as an externally added asphalt sprayed onto RAP to prepare recycled mixtures has a relatively lower effect on improving the compaction characteristics. This may be due to the lower amount of externally added foamed asphalt compared to the other two methods, resulting in limited influence on the overall reduction in asphalt viscosity in the asphalt mixture.

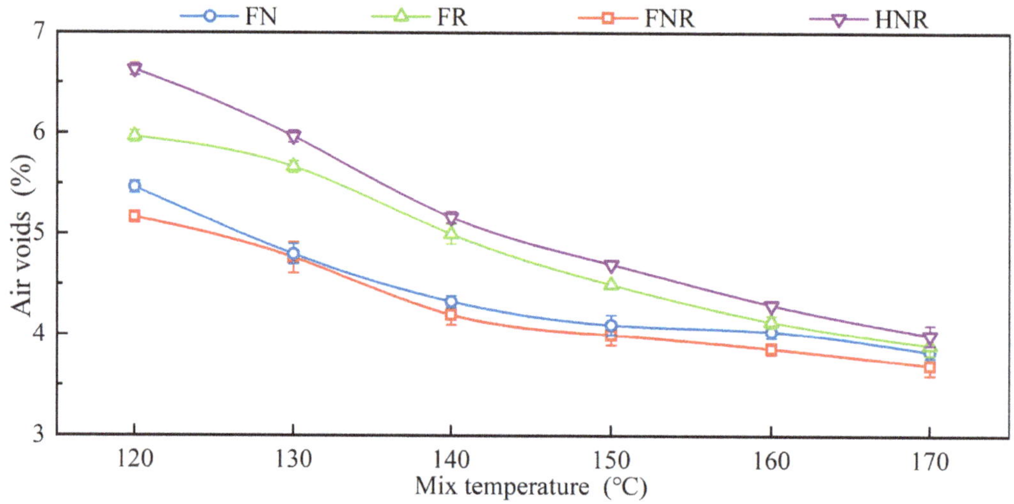

Figure 4. Curves of mixing temperature and percentage variation in air voids.

3.3. *Performance of HIR-FAM*

3.3.1. Performance on the road

The high-temperature performance, low-temperature performance, and water stability of three types of HIR-FAM were determined and compared with HIR-AM. The test results are shown in Table 5.

Table 5. Comparison of road performance test results.

Test Parameters	FN	FR	FNR	HNR	Technical Requirements	Test Methods
Dynamic stability (mm)	5517	5167	5590	4980	≥2800	T-0719
Tensile strength (MPa)	8.01	7.13	8.06	7.11	-	-
Maximum flexural strain (με)	2867.9	2721.0	2968.3	2654.1	≥2500	T-0715
Residual stability (%)	92.5	86.8	92.7	85.2	≥85	T-0709
Freeze–thaw split strength ratio (%)	82.9	80.7	84.0	80.1	≥80	T-0729

From Table 5, it can be observed that the road performance of HIR-FAM is improved compared to HIR-AM at the same temperature. Furthermore, the ranking of the effect on improving the road performance of recycled mixtures is FNR > FN > FR, which aligns with the results of the compaction characteristics test of recycled asphalt mixtures. Comparing HNR and FNR at the same mixing temperature shows that FNR exhibits 11.6% higher dynamic stability, 13.4% higher bending strength, 13.3% higher maximum bending strain, 8.1% higher residual stability, and 8.5% higher freeze–thaw splitting strength ratio. These findings indicate that foam asphalt greatly enhances the temperature performance of the recycled mixture. This is attributed to the foaming technology effectively reducing the viscosity of the new asphalt, thereby improving the workability of the mixture and leading to better compaction of the recycled mixture. This results in a significant increase in the compactness of the recycled mixture. However, the improvement in water stability is relatively low. This can be attributed to the presence of residual trace water in foam asphalt, which may affect the adhesion at the aggregate–asphalt interface, allowing moisture to more easily penetrate the aggregate–foam-asphalt interface and lead to asphàlt stripping. [52].

3.3.2. Hamburg Wheel Tracking

Figure 5 shows that the rut depths of FN, FR, and FNR are all less than HNR, further confirming that FA can improve the high-temperature performance and water stability of HIR-AM. However, even for FNR, the rut depth is only reduced by about 0.7 mm compared to HNR, and the differences are relatively small. The rut depth curves also exhibit some similarities. Moreover, the stripping inflection points for these four specimens are very close, requiring nearly the same number of load cycles to achieve moisture damage. This is mainly because FA accounts for a relatively small proportion of the binder material in the recycled mixtures, with the majority of the binder material coming from the asphalt in the RAP. This limitation results in a relatively modest improvement in the performance of recycled mixtures due to FA.

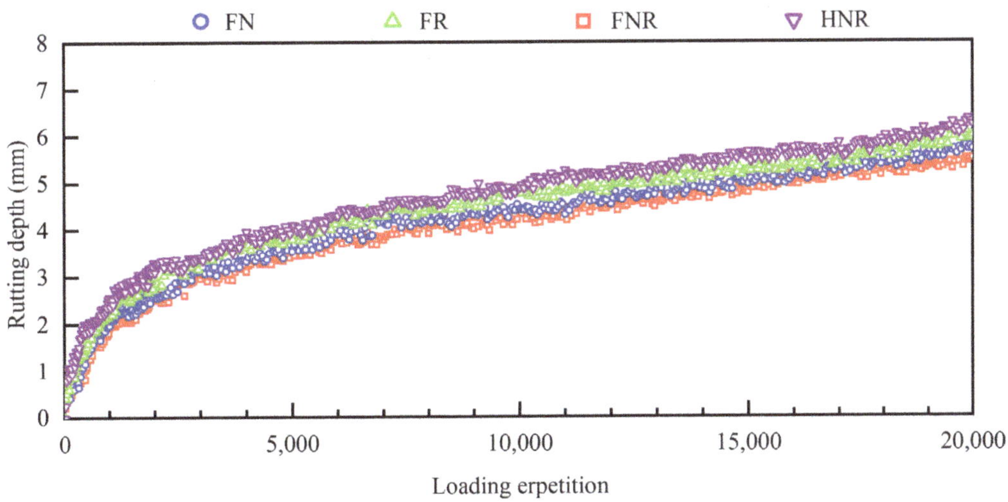

Figure 5. Hamburg wheel tracking test results.

3.3.3. Dynamic Modulus

The deformation resistance of asphalt mixtures is primarily affected by the asphalt binder and the mineral aggregate framework. In Figure 6, it is observed that under the same test conditions, HIR-FAM exhibits slightly higher dynamic modulus values compared to HIR-AM, suggesting that HIR-FAM has superior deformation resistance. This can be attributed to the increased adhesion between foam asphalt and the aggregate [53], as well as the denser skeletal structure of HIR-FAM [54]. The dynamic modulus of recycled mixtures decreases with increasing temperature. This is because the viscosity of the asphalt binder gradually decreases at high temperatures, leading to a reduction in the dynamic modulus. An increase in loading frequency enhances the viscosity of the asphalt binder, resulting in a higher dynamic modulus for the mixture. At low temperatures, the phase angle is primarily influenced by the viscosity of the asphalt binder [55]. The high fusion degree and good adhesion between foam asphalt and aged asphalt binder contribute to the relatively high phase angle of HIR-FAM compared to HIR-AM at low temperatures. However, as the temperature increases, the influence of the mineral aggregate framework surpasses the effect of asphalt viscosity, leading to a gradual reduction in the phase angle. Due to the denser framework, HIR-FAM exhibits lower phase angles. The phase angle is higher at low temperatures and lower at high temperatures at the same frequency. This indicates that such mixtures have a stronger resistance to low-frequency loads. Foam asphalt mixtures demonstrate better resistance to low-frequency loads compared to standard hot-mix asphalt mixtures.

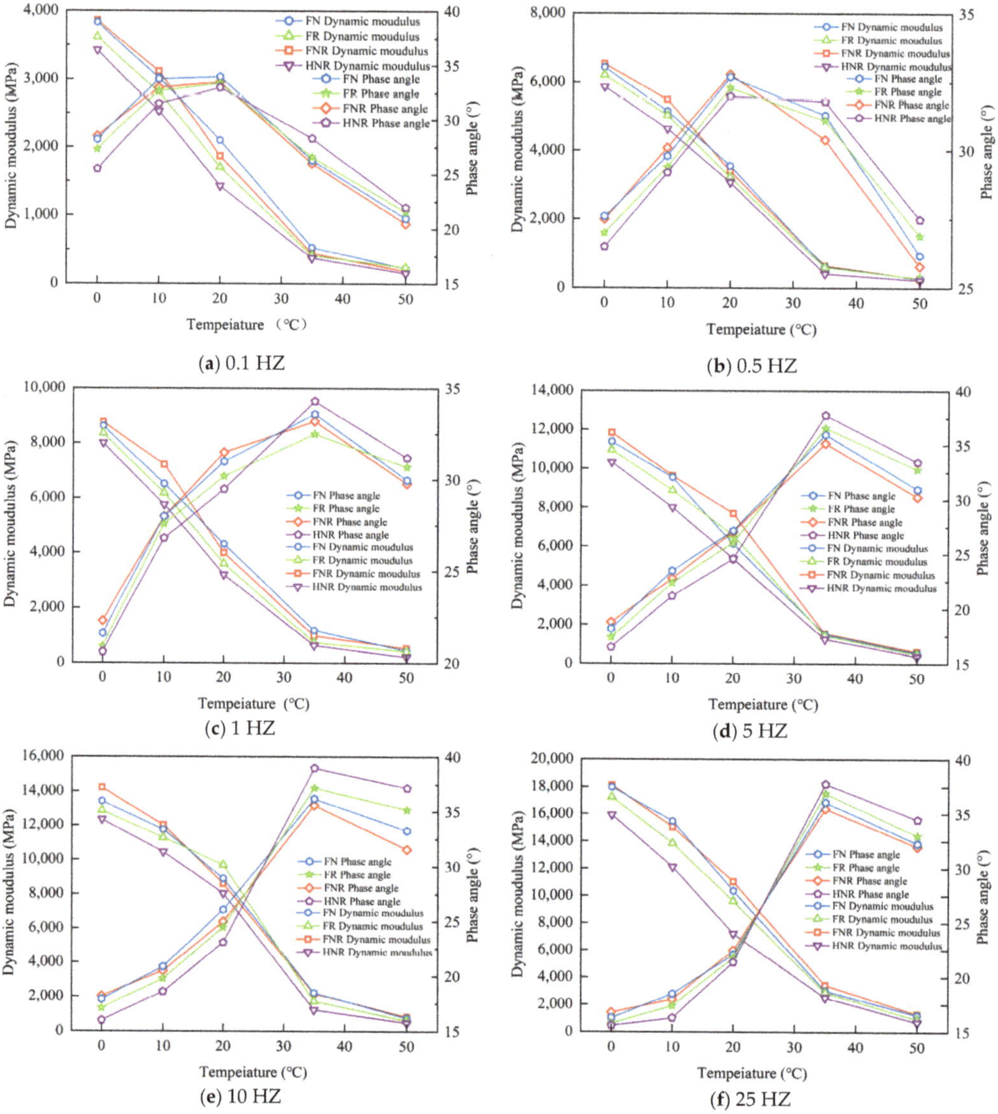

Figure 6. Effect of different factors on the dynamic modulus and phase angle of recycled mixes.

4. Assessment of the Application Effectiveness of HIR-FAM

4.1. On-Site Construction

We selected the first maintenance project of the major and medium-sized repair project of the section from Wudi Chengkou to the S236 intersection of the S511 Dacha Line to demonstrate the effect of FA on the compaction characteristics of HIR-AM. The engineering target mix ratio follows the mix ratio in Table 4, and based on the research results, the production mix ratio is determined after adjusting the mixing effect of the pre-mixed materials at the mixing station according to the research results for the target mix ratio, and the final synthesized gradation is shown in Figure 7. HIR-FAM Type Selection FNR, with the optimal asphalt content and rejuvenator content consistent with the target mix ratio.

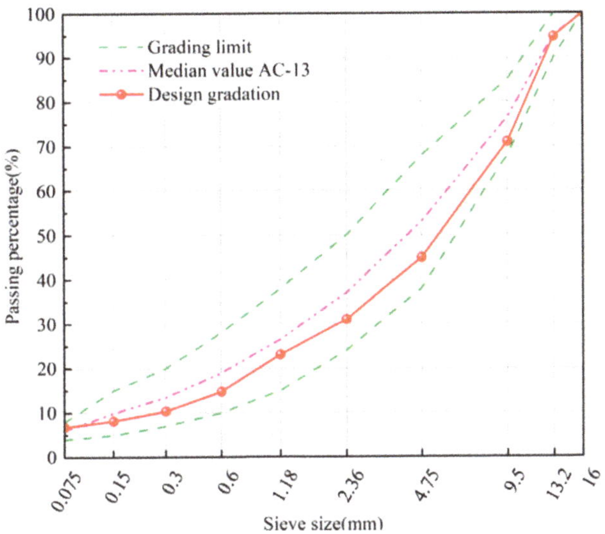

Figure 7. Gradation curve of the production mix ratio for AC-13.

The test was conducted in June 2023 and involved six main steps. (1) The selected road segment was pre-treated for HIR, which included road closure, removal of road markings, patching damaged road surfaces, and protecting bridge expansion joints Figure 8a. (2) The existing road surface was heated using three heating machines. The surface temperatures reached 150 °C, 180 °C, and 200 °C, fully softening the road surface Figure 8b. (3) An asphalt heating and loosening machine was used to loosen the heated original road surface. During this process, a rejuvenating agent and new FA were sprayed onto the road surface along the centerline, as shown in Figure 8c. The rejuvenating agent was heated to 85 °C, and the FA was heated to 160 °C. The measured expansion rate was 12, and the half life was 18 s. (4) A 4000-type asphalt mixing plant equipped with an asphalt foaming device was used to produce a new foam asphalt mixture. The aggregate heating temperature was set to 160 °C, the asphalt foaming temperature to 170 °C, and the foaming water content to 1.6%, considering asphalt foaming process parameters and construction conditions. The mixture was transported to the construction site by loading and unloading trucks (Figure 8d). (5) The new asphalt mixture was mixed with the reclaimed asphalt pavement (RAP) using a hot recycling machine. The hot recycling machine further heated the recycled mixture. The final temperature as it dropped into the paver's hopper was measured at 142 °C (Figure 8e). (6) The recycled mixture was paved and compacted using a paver and three rollers. The temperature of the recycled mixture was maintained at 135 °C during paving and decreased to 85 °C by the end of compaction (Figure 8f). The before-and-after construction states are illustrated in Figure 9, indicating an improvement in the road surface condition after HIR.

Figure 8. Construction process flow.

Figure 9. The test road before and after HIR.

4.2. Quality Inspection

To verify the improvement in compaction characteristics of FA on HIR-AM, a nuclear density gauge calibrated by the core drilling method was used to measure the bulk density of the test road for HIR-FAM and the HIR-AM, as shown in Figure 10a. For the HIR-AM test road, except for the addition of ordinary modified asphalt, all other construction parameters were the same as those for the HIR-FAM. To scientifically analyze the compaction status of the on-site measurement points, a compaction degree measurement point layout plan was designed, as shown in Figure 10b. The compaction degree of the recycled pavement was calculated according to Equation (3), and the test results are shown in Figure 11. The results indicate that the compaction degree of the road section using HIR-FAM is on average 0.79% higher than that of the road section using conventional HIR-AM, and it exhibits better uniformity, which is consistent with the conclusions of the indoor test.

$$K = \frac{\rho_d}{\rho_a} \times 100\% \tag{3}$$

where ρ_d is the measured density of the asphalt mixture (g/cm^3) and ρ_a is the maximum theoretical density of the asphalt mixture (g/cm^3).

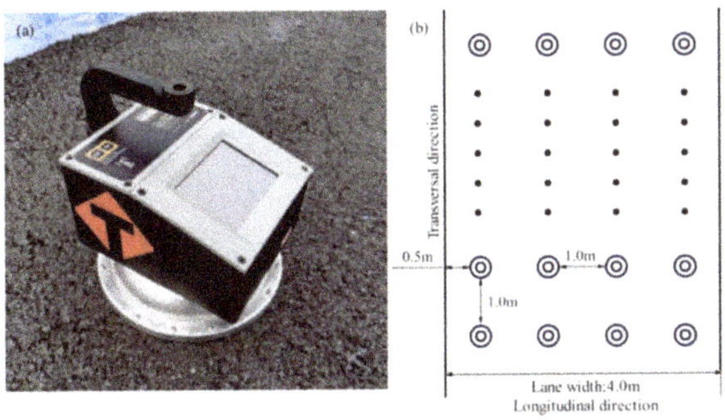

Figure 10. The compaction degree test: (**a**) testing equipment; (**b**) layout plan for measurement points.

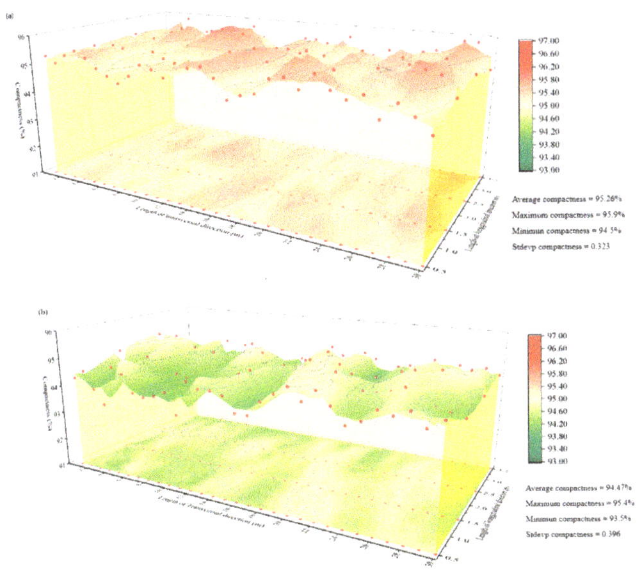

Figure 11. Results of compaction degree for different processes: (**a**) HIR-FAM; (**b**) HIR-AM.

5. Conclusions

To enhance the compaction efficiency of HIR-AM, this paper proposes an HIR technology based on FA. Focusing on the widely used SBS asphalt, the paper systematically investigates the influence of FA on the road performance and compaction quality of HIR-AM, including the FA foaming process, the preparation process of HIR-FAM, and the road performance of the recycled mixture. The findings are validated through practical engineering applications. The main conclusions are as follows.

(1) The FA foaming test indicated that parameters such as foaming temperature and water content significantly impact the foaming performance of SBS asphalt. There exists a good correlation between the expansion ratio and the half life of SBS asphalt. A foaming temperature of 170 °C achieves a relatively balanced expansion ratio and half life of the FA. Excessively high temperatures can reduce the elasticity of the asphalt film, leading to unstable asphalt foam. Additionally, at the same temperature, an increase in foaming water content results in a continuous increase in the expansion ratio and a gradual decrease in the half life, with the optimal foaming effect observed at 1.7%.

(2) The test results of percentage air voids for different recycled asphalt mixtures indicate that FA can effectively reduce the percent air voids of the mixture without increasing the mixing temperature. However, the compaction effects vary for different HIR-FAMs, with FNR>FN>FR. This difference may be attributed to a higher proportion of FA in the binder material of the recycled mixture, which leads to better viscosity-reduction effects.

(3) The performance testing results of recycled asphalt mixtures indicate that, at the same mixing temperature, FNR exhibits improvements in dynamic stability, flexural strength, maximum flexural strain, residual stability, and freeze–thaw splitting strength compared to HNR, with increases of 11.6%, 13.4%, 13.3%, 8.1%, and 8.5%, respectively. Additionally, it shows better resistance to water and heat stability, as well as low-frequency load resistance capabilities.

(4) Based on the research conducted above, an actual on-site project for HIR was carried out. The compaction quality of the pavement was evaluated through on-site testing using a nuclear density gauge. The results indicate that the compaction degree of the pavement section using HIR-FAM is improved by 0.79% compared to the HIR-AM pavement section, and it exhibits better uniformity. This demonstrates that FA effectively enhances the compaction quality of HIR-AM.

The following limitations exist in this study at present: (1) improvement and optimization of the foaming process, including the use of external modifiers to improve the foaming properties and stability of the foamed asphalt; (2) the effect of the regeneration process on the road performance of HIR-FAM; (3) more in-depth study of the material using microscopic characterization methods.

Author Contributions: Conceptualization, X.S.; Methodology, D.G.; Visualization, J.L.; Supervision, M.X.; Formal analysis, Z.L. and Q.H.; Investigation, Q.X.; Visualization, S.Y.; Writing—original draft preparation, X.S. and M.X.; funding acquisition, D.G. and M.X.; Resources, M.X. All authors have read and agreed to the published version of the manuscript.

Funding: This study is supported by the National Key R&D Program of China (2022YFB2601900), National Natural Science Foundation of China (NO.51208287 and U2233210), the Beijing Scholars Foundation (No. 067), and the Natural Science Foundation of Shandong Province (NO. ZR2016EEM42).

Data Availability Statement: The data presented in this study are available on request from the corresponding author.

Conflicts of Interest: The authors declare no conflicts of interest.

References

1. Research, G.; Institute, F. *Read Carbon Neutrality: China's Roadmap for Low-Carbon Development 2020–2050*; CITIC Press: Beijing, China, 2020; p. 316.
2. Liu, Z.; Sun, L.; Gu, X.; Wang, X.; Dong, Q.; Zhou, Z.; Tang, J. Characteristics, mechanisms, and environmental LCA of WMA containing sasobit: An analysis perspective combing viscosity-temperature regression and interface bonding strength. *J. Clean. Prod.* **2023**, *391*, 136255. [CrossRef]
3. Liu, Z.; Yang, Q.; Gu, X. Assessment of Pavement Structural Conditions and Remaining Life Combining Accelerated Pavement Testing and Ground-Penetrating Radar. *Remote Sens.* **2023**, *15*, 4620. [CrossRef]
4. Sha, A.M.; Liu, Z.Z.; Jiang, W.; Qi, L.; Hu, L.Q.; Jiao, W.X.; Barbieri, D. Advances and development trends in eco-friendly pavements. *J. Road Eng.* **2021**, *1*, 1–42. [CrossRef]
5. Chen, J.Q.; Dan, H.C.; Ding, Y.J.; Gao, Y.; Guo, M.; Guo, S.; Han, B.; Hong, B.; Hou, Y.; Hu, C.; et al. New innovations in pavement materials and engineering: A review on pavement engineering research 2021. *J. Trafic. Transp. Eng.* **2021**, *8*, 815–999. [CrossRef]
6. Cai, J.; Song, C.; Gong, X.B.; Zhang, J.P.; Pei, J.Z.; Chen, Z.W. Gradation of limestone-aggregate-based porous asphalt concrete under dynamic crushing test: Composition, fragmentation and stability. *Constr. Build. Mater.* **2022**, *323*, 126532. [CrossRef]
7. Pasetto, M.; Baloello, A.B.; Giacomello, G.; Pasquini, E. Towards very high RAP content asphalt mixes: A comprehensive performance-based study of rejuvenated binders. *J. Trafic. Transp. Eng.* **2021**, *8*, 1022–1035. [CrossRef]
8. Pan, Y.Y.; Yang, T.; Shang, Y.; Lin, M.; Liu, G.Q.; Xie, Y.C.; Zhao, Y.L. Field and laboratory evaluations of maintenance treatments for semi-rigid base asphalt pavement. *Constr. Build. Mater.* **2020**, *258*, 119726. [CrossRef]
9. Yao, Y.Q.; Yang, J.G.; Gao, J.; He, L.; Xu, J. Material composition volatility and control strategy for hot in-place recycling asphalt mixture. *Mater. Rep.* **2022**, *36*, 59–68.
10. Han, D.D.; Zhao, Y.L.; Pan, Y.Y.; Liu, G.Q.; Yang, T. Heating process monitoring and evaluation of hot in-place recycling of asphalt pavement using infrared thermal imaging. *Automat. Constr.* **2020**, *111*, 103005. [CrossRef]
11. Ding, X.H.; Chen, L.C.; Ma, T.; Ma, H.X.; Gu, L.H.; Chen, T.; Ma, Y. Laboratory investigation of the recycled asphalt concrete with stable crumb rubber asphalt binder. *Constr. Build. Mater.* **2019**, *203*, 552–557. [CrossRef]
12. Yu, B.; Wang, S.Y.; Gu, X.Y.; Ni, F.J.; Liu, Q. Environmental burden evaluation of hot in-place recycling of asphalt pavement based on discrete event simulation. *Transport. Res. D-Tr. E.* **2018**, *65*, 151–160. [CrossRef]
13. Li, J.; Ni, F.J.; Jin, J.; Zhou, Z. A comparison of rejuvenator and sryrene-butadiene rubber latex used in hot in-place recycling. *Road Mater. Pavement.* **2017**, *18*, 101–115. [CrossRef]
14. Yu, B.; Ni, F.J.; Zhang, L.D.; Gu, X.Y. Performance Evaluation of Aged Asphalt Mix for Hot In-Place Recycling. *J. Test. Eval.* **2015**, *44*, 770–780. [CrossRef]
15. Liu, Z.; Zhou, Z.; Gu, X.; Sun, L.; Wang, C. Laboratory evaluation of the performance of reclaimed asphalt mixed with composite crumb rubber-modified asphalt: Reconciling relatively high content of RAP and virgin asphalt. *Int. J. Pavement Eng.* **2023**, *24*, 2217320. [CrossRef]
16. Jiang, T.T.; Fan, Q.J.; Hou, M.Y.; Mi, S.Z.; Yan, X.H. Effects of Rejuvenator Dosage, Temperature, RAP Content and Rejuvenation Process on the Road Performance of Recycled Asphalt Mixture. *Sustainability* **2023**, *15*, 3539. [CrossRef]
17. Liu, Y.; Wang, H.N.; Tighe, S.L.; Zhao, G.Y.; You, Z.P. Effects of preheating conditions on performance and workability of hot in-place recycled asphalt mixtures. *Constr. Build. Mater.* **2019**, *226*, 288–298. [CrossRef]
18. Wang, T.; Jiang, W.; Xiao, J.J.; Guo, D.D.; Yuan, D.D.; Wu, W.J.; Wang, W.T. Study on the blending behavior of asphalt binder in mixing process of hot recycling. *Case Stud. Const. Mat.* **2022**, *17*, e01477. [CrossRef]
19. Yang, Q.; Liu, Z.; Sun, L.; Li, Z. Analysis of the viscosity and self-healing performance of warm-mix recycled asphalt: A method combining molecular dynamics simulations and laboratory tests. *Case Stud. Constr. Mat.* **2023**, *19*, e02593. [CrossRef]
20. Cheraghian, G.; Falchetto, A.C.; You, Z.P.; Chen, S.Y.; Kim, Y.S.; Westerhoff, J. Warm mix asphalt technology: An up to date review. *J. Clean. Prod.* **2020**, *268*, 122128. [CrossRef]
21. Mallick, R.B.; Kandhal, P.S.; Bradbury, R.L. Using Warm-Mix Asphalt Technology to Incorporate High Percentage of Reclaimed Asphalt Pavement Material in Asphalt Mixtures. *Transport. Res. Rec.* **2008**, *2051*, 71–79. [CrossRef]
22. Gkyrtis, K.; Plati, C.; Loizos, A. Structural Performance of Foamed Asphalt Base in a Full Depth Reclaimed and Sustainable Pavement. *Sustainability* **2023**, *15*, 3622. [CrossRef]
23. Ali, A.; Abbas, A.; Nazzal, M.; Alhasan, A.; Roy, A.; Powers, D. Effect of temperature reduction, foaming water content, and aggregate moisture content on performance of foamed warm mix asphalt. *Constr. Build. Mater.* **2013**, *48*, 1058–1066. [CrossRef]
24. Dong, F.Q.; Yu, X.; Wang, T.Y.; Yin, L.; Li, J.; Si, J.J.; Li, J. Influence of base asphalt aging levels on the foaming characteristics and rheological properties of foamed asphalt. *Constr. Build. Mater.* **2018**, *177*, 43–50. [CrossRef]
25. Hu, J.Y.; Ma, T.; Yin, T.; Zhou, Y. Foamed warm mix asphalt mixture containing crumb rubber: Foaming optimization and performance evaluation. *J. Clean. Prod.* **2022**, *333*, 130085. [CrossRef]
26. Alhasan, A.A.; Abbas, R.A.; Nazzal, M. Low-Temperature Characterization of Foamed Warm-Mix Asphalt Produced by Water Injection. *Transp. Res. Rec.* **2014**, *2445*, 1–11. [CrossRef]
27. Xu, S.Y.; Xiao, F.P.; Amirkhanian, S.; Singh, D. Moisture characteristics of mixtures with warm mix asphalt technologies-A review. *Constr. Build. Mater.* **2017**, *142*, 148–161. [CrossRef]
28. Monu, K.; Singh, S. Effect of long-term ageing on properties of RAP inclusive WMA mixes. *Constr. Build. Mater.* **2019**, *206*, 483–493. [CrossRef]

29. Tang, Y.C.; Wang, Y.F.; Wu, D.X. Exploring temperature-resilient recycled aggregate concrete with waste rubber: An experimental and multi-objective optimization analysis. *Rev. Adv. Mater. Sci.* **2023**, *62*, 20230347. [CrossRef]
30. Guo, M.; Liu, H.P.; Jiao, Y.B. Effect of WMA-RAP technology on pavement performance of asphalt mixture: A state-of-the-art review. *J. Clean. Prod.* **2020**, *266*, 121704. [CrossRef]
31. Rondón-Quintana, H.A.; Fernández-Gómez, W.; Zafra-Mejía, C.A. Behavior of a warm mix asphalt using a chemical additive to foam the asphalt binder. *Rev. Fac. Ing. Univ. Antioq.* **2016**, *78*, 129–138. [CrossRef]
32. Zheng, Z.G.; He, Y.W.; Wang, T. Study on the Performance of the Physical Foaming Warm-mix Recycled Asphalt Mixture. *E3S Web Conf.* **2021**, *261*, 02058. [CrossRef]
33. Abreu, L.P.F.; Oliveira, J.R.M.; Silva, H.M.R.D.; Palha, D.; Fonsece, P.V. Suitability of different foamed bitumens for warm mix asphalts with increasing recycling rates. *Constr. Build. Mater.* **2017**, *142*, 342–353. [CrossRef]
34. Zhao, S.; Huang, B.S.; Shu, X.; Jia, X.Y.; Woods, M.B. Laboratory Performance Evaluation of Warm-Mix Asphalt Containing High Percentages of Reclaimed Asphalt Pavement. *Transp. Res. Rec.* **2012**, *2294*, 98–105. [CrossRef]
35. Ashiqur, R.M.; Rouzbeh, G.; Musharraf, Z. Rutting and moisture-induced damage potential of foamed warm mix asphalt (WMA) containing RAP. *Innov. Infrastruct. So.* **2021**, *6*, 158. [CrossRef]
36. Kavussi, A.; Hashemian, L. Laboratory evaluation of moisture damage and rutting potential of WMA foam mixes. *Int. J. Pavement Eng.* **2012**, *13*, 415–423. [CrossRef]
37. Jeong, M.M.; Kim, S.G.; Shen, J.N. Laboratory evaluation of foamed asphalt mixtures with 100% RAP and rejuvenator. *Aust. J. Civ. Eng.* **2020**, *19*, 46–57. [CrossRef]
38. Transport, M.O. *Highway Performance Assessment Standards*; China Communications Press: Beijing, China, 2018.
39. Wang, D.; Liu, Z.; Gu, X.; Wu, W. Feature extraction and segmentation of pavement distress using an improved hybrid task cascade network. *Int. J. Pavement Eng.* **2023**, *24*, 2266098. [CrossRef]
40. Transport, M.O. *Standard Test Methods of Bitumen and Bituminous Mixtures for Highway Engineering*; China Communications Press: Beijing, China, 2011.
41. Transport, M.O. *Technical Specifications for Construction of Highway Asphalt Pavements*; China Communications Press: Beijing, China, 2004.
42. Liu, Z.; Gu, X.; Ren, H. Rutting prediction of asphalt pavement with semi-rigid base: Numerical modeling on laboratory to accelerated pavement testing. *Constr. Build. Mater.* **2023**, *375*, 130903. [CrossRef]
43. Liu, Z.; Gu, X.; Dong, Q. Permanent Deformation Evaluation and Instability Prediction of Semi-rigid Pavement Structure Using Accelerated Pavement Testing and Finite Element Method. *J. Test Eval.* **2023**, 52. [CrossRef]
44. Liu, Z.; Gu, X.; Dong, X.; Cui, B.; Hu, D. Mechanism and Performance of Graphene Modified Asphalt: An Experimental Approach Combined with Molecular Dynamic Simulations. *Case Stud. Constr. Mat.* **2022**, *18*, e01749. [CrossRef]
45. Transport, M.O. *Technical Specification for Highway Asphalt Pavement Recycling*; China Communications Press: Beijing, China, 2019.
46. Zhong, H.B.; Huang, W.D.; Yan, C.Q.; Zhang, Y.; Lv, Q.; Sun, L.J.; Liu, L.Y. Investigating binder aging during hot in-place recycling (HIR) of asphalt pavement. *Constr. Build. Mater.* **2021**, *276*, 122188. [CrossRef]
47. Tang, X.N. Study on Construction Quality Evaluation in Hot In-Place Recycling of Asphalt Pavement. Master's Thesis, Southeast University, Nanjing, China, 2020.
48. Liu, Z.; Gu, X.; Ren, H.; Li, S.; Dong, Q. Permanent Deformation Monitoring and Remaining Life Prediction of Asphalt Pavement Combining Full-Scale Accelerated Pavement Testing and FEM. *Struct. Control. Health Monit.* **2023**, *2023*, 6932621. [CrossRef]
49. Li, Y.S. Study on Application Performance of Foamed Asphalt in Seasonal Freezing Environment. Master's Thesis, Jilin University, Changchun, China, 2019.
50. He, Y.Z.; Wang, Q.G.; Zhang, J.P.; Li, Y.; He, H.Q.; Huang, G.J. Foamed crumb rubber asphalt binder: Preparation, rheological properties and adhesion characteristics. *J. Clean. Prod.* **2023**, *396*, 136516. [CrossRef]
51. Ali, A.; Abbas, A.; Nazzal, M.; Alhasan, A.; Roy, A.; Powers, D. Workability Evaluation of Foamed Warm-Mix Asphalt. *J. Mater. Civil Eng.* **2013**, *26*, 04014011. [CrossRef]
52. Zhao, H.; Ren, J.D.; Chen, Z.; Luan, H.; Yi, J.Y. Freeze and thaw field investigation of foamed asphalt cold recycling mixture in cold region. *Case Stud. Const. Mat.* **2021**, *15*, e00710. [CrossRef]
53. Wang, J.; Zheng, K.P.; Xu, J.; Liu, C.P. Influence of foaming water content on mechanical properties of foamed warm asphalt mixture. *J. Highw. Transp. Res. Deve.* **2023**, *40*, 1–8.
54. Meng, A.X.; Xu, H.N.; Fu, X.G.; Tan, Y.Q. Dynamic moduls and prediction model of asphalt mixture based on structural natural frequency. *China J. Highw. Transp.* **2019**, *32*, 31–38.
55. Ruan, L.; Luo, R.; Hu, X.D.; Pan, P. Effect of bell-shaped loading and haversine loading on the dynamic modulus and resilient modulus of asphalt mixtures. *Constr. Build. Mater.* **2018**, *161*, 124–131. [CrossRef]

buildings

MDPI

Article

Optimizing Rural Pavements with SBS-Modified Asphalt Binders and Petroleum Resin

Je-won Kim [1], Kyungnam Kim [2,*] and Tri Ho Minh Le [3]

[1] Department of Highway & Transportation Research, Korea Institute of Civil Engineering and Building Technology, 283 Goyangdae-ro, Ilsanseo-gu, Goyang-si 10223, Republic of Korea; jewonkim@kict.re.kr

[2] Korea Expressway Corporation Research Institute, Pavement Research Division, Dong-tansunhwan-daero 17-gil, Hwaseong-si 18489, Republic of Korea

[3] Faculty of Civil Engineering, Nguyen Tat Thanh University, 300A Nguyen Tat Thanh Street, District 4, Ho Chi Minh City 70000, Vietnam; lhmtri@ntt.edu.vn

* Correspondence: kkn248@ex.co.kr

Abstract: This study addresses the imperative for enhancing asphalt mixtures tailored for rural pavements, focusing on optimizing RAP mixtures with styrene–butadiene–styrene (SBS)-modified asphalt binders incorporating petroleum resin and oil. Through systematic investigation, the study examines the impact of varying RAP content (25% and 50%) and two SBS-modified asphalt binder types (Type 1 and Type 2) on mechanical properties and sustainability. Laboratory tests reveal that the mix of 25% RAP + 75% Type 1 exhibits exceptional flexibility, evidenced by a high ductility value of 880 mm at 25 °C, enhancing pavement resilience. Conversely, the 50% RAP + 50% Type 2 mixture displays vulnerability to fatigue cracking, while 25% RAP + 75% Type 1 demonstrates superior resistance, with a fatigue vulnerability value of 1524 kPa. The Hamburg Wheel Tracking test highlights the influence of RAP content on rut depth, with the mix of 50% RAP + 50% Type 1 achieving the lowest rutting at 3.9 mm. Overlay test results show the mix of 25% RAP + 75% Type 2's resilience, with the lowest load reduction at 64.5%, while the mix of 50% RAP + 50% Type 1 exhibits substantial load reduction at 82.1%. Field tests unveil differences in pavement bearing capacities, with the mix of 25% RAP + 75% Type 2 demonstrating a remarkable elastic modulus of 58.5 MPa, indicating heightened bearing capacity. The investigation underscores the significant role of SBS-modified asphalt binders with incorporated petroleum resin and oil in enhancing fatigue resistance for sustainable rural pavements.

Keywords: rural pavement; asphalt mixtures; extracted RAP binder; Hamburg wheel tracking; falling weight deflectometer

1. Introduction

The imperative for upgrading rural pavements stems from their pivotal role in the economic and agricultural landscape [1], with a focus on rice and fruit cultivation [2]. Reliable road networks are crucial for transporting perishable goods to urban markets [3], especially with evolving logistic demands such as increased container usage [4]. Rural pavements face unique challenges like adverse weather [5] and heavy agricultural machinery traffic, leading to accelerated deterioration and impacting safety and transportation efficiency [6]. Recognizing the economic importance of rural areas, systematic upgrades are essential to address degradation and yield profound economic benefits [7], fostering development and enhancing the quality of life for these communities [8]. Upgrading rural pavements is not just a technical necessity but a pivotal step towards sustainable development, ensuring the resilience and vitality of these transportation networks [9].

Recycled asphalt pavement (RAP) has gained widespread application in modern pavement construction as a sustainable and cost-effective alternative [10,11]. RAP, derived from reclaimed asphalt materials, is reintegrated into new pavement mixes, contributing to resource conservation and reducing environmental impact [12,13]. Its use aligns with the principles of circular economy, as it minimizes waste while capitalizing on the valuable properties of aged asphalt [14]. Incorporating RAP into pavement projects enhances sustainability by reducing the demand for virgin materials and lowering energy consumption during production [15]. The application of RAP not only addresses environmental concerns but also proves advantageous in improving pavement performance, including enhanced durability, flexibility, and rutting resistance [16,17]. This approach, combining environmental stewardship with engineering efficiency, underscores RAP's pivotal role in modern pavement practices [18–20].

Styrene–butadiene–styrene (SBS)-modified asphalt has become a focal point in pavement research and construction due to its exceptional performance-enhancing properties. SBS, a polymer additive, is incorporated into asphalt binder to improve its resistance to temperature fluctuations, fatigue cracking, and deformation under heavy loads [21,22]. This modification enhances the overall durability and longevity of asphalt pavements, making them better suited to withstanding the challenges of diverse climatic conditions and high traffic volumes [23]. The elastic and adhesive qualities of SBS-modified asphalt contribute to reduced rutting and cracking, providing significant advantages for both road safety and maintenance costs [24]. The current application of SBS asphalt in pavement research reflects a commitment to optimizing asphalt mixtures for superior performance and longevity, addressing the evolving demands of modern transportation infrastructure [25,26].

The application of RAP in rural areas faces certain limitations that warrant attention. In many cases, the utilization of RAP in these settings is hindered by the heterogeneous nature of reclaimed materials, varying significantly in composition and properties. Additionally, there is a lack of comprehensive research focus on the incorporation of SBS in RAP mixtures due to the increased stiffness issue [27]. This gap in research inhibits a nuanced understanding of the interaction between SBS-modified binders and high percentages of reclaimed materials, crucial for optimizing pavement performance. Furthermore, the behavior of the aging process on these binders remains insufficiently studied, impeding the development of robust pavement designs that account for the long-term effects of environmental exposure and traffic loading. Addressing these limitations is imperative to unlocking the full potential of RAP in rural pavement applications and to advance the utilization of SBS-modified asphalt in large RAP mixtures, contributing to sustainable and resilient infrastructure development.

This research introduces a pioneering approach to address the specific challenges of rural pavement development. Focused on upgrading rural infrastructure, the study investigates the application of RAP and SBS asphalt in asphalt mixtures. The research delves into the limitations of RAP in rural areas, the lack of research on large RAP mixtures (up to 50%), and the aging behavior of these binders. By evaluating sensitivity to fatigue cracking, and storage stability, and conducting field tests using a Light Falling Weight Deflectometer (LFWD), the study provides a comprehensive understanding of the performance of these asphalt mixtures in real-world rural conditions. This holistic and innovative approach positions the research at the forefront of advancing sustainable and resilient solutions for rural pavement challenges.

In this study, an extensive experimental program was conducted to assess the performance of asphalt mixtures designed for rural pavement applications. The materials utilized include RAP and SBS asphalt binder, with a particular emphasis on mixtures containing up to 50% RAP. This study investigates the influence of different RAP binder proportions, specifically 25% and 50% extracted RAP binders, combined with two types of modifiers (Type 1 and Type 2). The asphalt binder properties were characterized through Asphalt Binder Tests, Dynamic Shear Modulus (G) tests, and sensitivity to fatigue cracking assessments. Additionally, storage stability tests were conducted to evaluate the long-term structural integrity of rural pavement materials, including Control, Type 1, and Type 2 formulations. The performance of the asphalt mixtures was comprehensively evaluated through the Hamburg Wheel Tracking (HWT) test, Overlay Test (OT), and field tests employing the Light Falling Weight Deflectometer (LFWD). These diverse testing methods provide a robust foundation for analyzing the impact of RAP and SBS asphalt on rural pavement performance, setting the stage for a thorough exploration of sustainable and resilient solutions in this critical infrastructure domain. In general, this study innovatively evaluates the performance of asphalt mixtures for rural pavements by examining recycled asphalt pavement (RAP) with styrene–butadiene–styrene (SBS)-modified asphalt binders containing petroleum resin and oil, offering insights into the nuanced interplay between varying RAP content and modifier types on mechanical properties and sustainability. The overview of the research is presented in Figure 1.

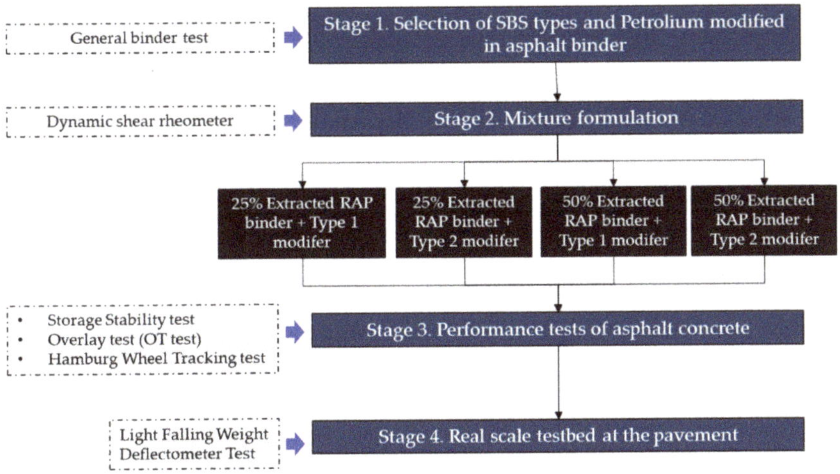

Figure 1. Research flowchart.

2. Materials and Methods

2.1. Materials

2.1.1. Recovery of RAP Binder

The selection of materials in this study is crucial for understanding the characteristics and performance of the asphalt binders. The following subsection provides a detailed overview of the materials employed in the research, encompassing the RAP binder, SBS binders (Type 1 and Type 2), and the specifications governing their use. The general properties of RAP and virgin binder used in this research are shown in Table 1.

Table 1. General properties of virgin and RAP binders.

Property	RAP	New Binder (PG64-22)
Penetration (1/10 mm) at 25 °C	82.5	64
Softening Point (°C)	68.7	50
Ductility (cm/min) at 5 °C	>100	>100

2.1.2. Recycled Asphalt Pavement Binder

The RAP binder used in this study was extracted from reclaimed asphalt pavement obtained from road resurfacing projects. The RAP material was carefully sampled and processed to isolate the binder content. The extraction process involved the use of a solvent, typically trichloroethylene, in accordance with established protocols such as ASTM D2172 [28]. The recovered RAP binder represents the aged asphalt binder present in the reclaimed pavement, which adds a unique dimension to the study by incorporating real-world, field-aged binder into the experimental matrix.

2.1.3. Styrene-Butadiene-Styrene Binders

Two distinct types of SBS binders, denoted Type 1 and Type 2, were employed in the investigation. These binders were selected based on their proven effectiveness in enhancing asphalt properties in previous research. The SBS polymers were sourced from reputable manufacturers, ensuring a consistent and high-quality supply. Both SBS binders were designed to modify and improve the rheological properties of asphalt binders, offering enhanced elasticity, durability, and resistance to aging.

2.1.4. Binder Blending Ratios

The binder combinations were formulated by blending the RAP binder with the SBS binders in different weight ratios. Two main combinations were investigated: 25% RAP binder with 75% SBS (Type 1, Type 2) and 50% RAP binder with 50% SBS (Type 1, Type 2). The rationale behind these ratios lies in achieving a balance between the rejuvenating effects of the RAP binder and the performance-enhancing properties of the SBS polymers. The selection of these ratios was informed by preliminary investigations and optimization studies, aiming to guarantee a thorough exploration of the binder space.

The recognition that the incorporation of polymer with RAP necessitates the addition of rejuvenating materials, preferably in liquid form, introduces a pivotal aspect of the discourse on asphalt binder formulation. This acknowledgment underscores the potential challenges associated with employing polymer-modified binders in the presence of RAP, emphasizing the need to address concerns related to stiffness and rejuvenation. In response to this valuable insight, the study strategically incorporates liquid rejuvenating agents, specifically petroleum resin (GX140) and oil, into the asphalt binder compositions. This purposeful integration seeks to improve the compatibility of the polymer with RAP, alleviate stiffness-related issues, and contribute to an overall enhancement of mechanical properties. Opting for liquid rejuvenators adheres to industry standards, providing a pragmatic solution to optimize asphalt mixtures for rural pavement applications, thereby ensuring improved performance and long-term durability.

2.1.5. SBS Molecular Structure Selection

In the realm of asphalt concrete materials, the judicious selection of SBS molecular structures emerges as a critical determinant for achieving optimal performance in rural pavement applications. Table 2 serves as a guide on this exploratory journey, categorizing SBS asphalt binder materials into two primary structural types while delineating their fundamental properties. The 3-Linear type, characterized by a softening point of 105.5 °C, an elevated ductility of 140 cm, and a penetration depth of 20.8 dm, epitomizes heightened flexibility and workability.

Table 2. Variations in Asphalt binder with SBS structures.

Property	3-Linear Type 1	3-Linear Type 2	3-Radial Type
Softening Point (°C)	105.5	92.4	111.3
Ductility (cm at 25 °C)	140	140	68
Penetration (dm at 25 °C, 100 g, 5 s)	20.8	23.4	13.6
Viscosity (cPs at 135 °C)	4800	5600	10,000
Elastic Recovery (% at 25 °C)	93	87	91

Conversely, the 3-Radial type, distinguished by a higher softening point of 111.3 °C, manifests with reduced ductility (68 cm) and a more restrained penetration (13.6 dm). A noteworthy contrast lies in the viscosity at 135 °C, where the 3-Linear type demonstrates a lower value of 4800 cPs compared to the 3-Radial type's 10,000 cPs. This structural differentiation significantly influences the elastic recovery rate, a pivotal attribute in asphaltic applications.

The 3-Linear type, boasting a superior elastic recovery rate of 93% at 25 °C, signifies resilience and the capacity to regain shape after deformation. In contrast, the 3-Radial type, with a marginally lower elastic recovery rate of 91% at the same temperature, introduces a nuanced aspect to the asphaltic composition. This nuanced exploration underscores the intricacies involved in the deliberate selection of SBS molecular structures for rural pavement, ultimately impacting performance metrics critical for the success of asphalt concrete materials.

Regarding the characterization testing process, the Softening Point, measured according to ASTM D36 [29], provides insights into the asphalt binder's temperature susceptibility. Penetration at 25 °C (100 g, 5 s), following ASTM D5 [29], offers a measure of the binder's consistency and hardness. Viscosity at 135 °C, conducted in triplicate according to ASTM D4402 [30], sheds light on the binder's flow behavior under elevated temperatures. Elastic Recovery at 25 °C, adhering to ASTM D113 [31], gauges the binder's ability to rebound after deformation. These additional tests contribute valuable information to the material characterization, ensuring a thorough understanding of the asphalt mixtures' properties and behaviors.

2.1.6. Selection of Enhanced Adhesion Agents

Ensuring the durability and integrity of the asphalt pavement while preventing damage and joint detachment requires a meticulous selection of adhesion agents that enhance the cohesion of materials.

In response to the Korean refined Company oil suggestion and trial test recommendation, the selection of enhanced adhesion agents for asphalt binder modification leans towards the application of GX140 petroleum resin.

GX140 is a thermoplastic resin derived from petroleum-based feedstocks. Its intricate composition involves the polymerization of hydrocarbons, resulting in a high-quality resin with distinctive properties. The development process focuses on achieving a fine balance between molecular weight, softening point, and compatibility with asphalt. The trial tests have highlighted its favorable characteristics, including a high softening point of 140.2 °C, making it ideal for demanding conditions requiring superior temperature resistance and asphalt adhesion. This strategic decision to utilize GX140 as the primary adhesion agent in the development of Type 1 and Type 2 modifiers is expected to enhance the overall performance and durability of the asphalt binders, aligning with the project's specific requirements. Table 3 provides a concise overview of the updated physical properties of the adhesion agent.

Table 3. Physical properties of enhanced adhesion agents.

Property	Petroleum Resin GX140
Softening Point (°C)	140.2
Stripping Resistance (kgf/cm at 90 °C)	2.15
Softening Point (°C)	140.2
Compatibility with Asphalt	Excellent

2.1.7. Asphalt Binder Formulation: Crafting Specialized Solutions

The asphalt pavement was meticulously developed, considering key components such as petroleum resin. Delving into the nuances of these formulations, Table 4 unveils the intricate compositions that lay the foundation for tailored filling asphalt materials, specifically engineered for pavement development. It should be noted that the inclusion of oil in the asphalt binder composition serves a multifaceted purpose. Its primary role is to replenish the aged, extracted asphalt binder from recycled asphalt pavement (RAP), contributing to the overall rejuvenation of the mixture. Furthermore, the strategic balance between asphaltene and oil content is crucial for achieving optimal binder properties.

Table 4. Composition of asphalt binder with modifiers for the research.

Division	Composition Ratio (%)	
	Type 1	Type 2
Asphalt (PG 64-22)	80	80
SBS	5	8
	5	2
Petroleum Resin (GX140)	8	7
Oil	2	3
Mineral	0	0

This nuanced composition strategy enables the creation of distinct filling asphalt materials, finely tuned to meet the specific demands and applications associated with pavement development. The goal is to contribute significantly to enhanced road infrastructure performance and longevity.

In-depth scrutiny of the quality test results highlights the success of the asphalt binder formulation. The meticulous balance achieved not only meets industry standards but also introduces substantial improvements in key physical properties. This ensures optimized performance across diverse environmental conditions, positioning the developed asphalt pavement as a robust and reliable solution for infrastructure longevity.

Table 5 outlines the exceptional performance of the initial rural pavement material (Type 1 and Type 2) designed to surpass ASTM D 6297 [32]. Both types exceed requirements for softening point, ductility, and adherence to the recommended application heating range, ensuring robust performance in diverse climates. The enhanced values for ductility, softening point, penetration, and viscosity reaffirm the material's suitability.

Table 5. Updated characteristics of binder.

ASTM D6297	ASTM Unit	Spec.	Type 1	Type 2
Softening Point D36	°C	83	108.0	103.0
Ductility (at 25 °C) D113	mm	>400	755	885
Recommended Application Heating	-	185–195	Satisfied	Satisfied
Penetration (at 25 °C, 100 g, 5 s) D5	dm	-	28.0	33.0
Elastic Recovery (at 25 °C)	%	-	77	94
Viscosity (at 135 °C) D446	cPs	-	6600	5700

Table 6 provides a detailed examination of the physical properties of the final rural pavement material (Type 1 and Type 2), meticulously evaluated against ASTM standards [32]. Both types exhibit outstanding performance, surpassing requirements for softening point, tensile adhesion, ductility, and cone penetration. Notably, the revised values for softening point, tensile adhesion, and ductility reinforce the material's exceptional suitability for diverse construction applications, promising longevity and reliability in various environmental conditions [33].

Table 6. Updated physical properties of final developed pavement material.

ASTM D6297	ASTM Unit	Spec.	Type 1	Type 2
Softening Point D36 [34]	°C	83	102.5	104.0
Tensile Adhesion D5329 [35]	%	>700	825	780
Ductility (at 25 °C) D113 [36]	mm	>400	765	560
Cone Penetration (at 25 °C, 100 g, 5 s) D5329 [35]	unit	<75	24.0	17.5
Cone Penetration (at −18 °C, 200 g, 60 s) D5329 [35]	unit	>10	11.0	10.6
Flow (at 60 °C, 5 h) D5329 [35]	mm	<3.0	0.55	0.51

Table 6. *Cont.*

ASTM D6297	ASTM Unit	Spec.	Type 1	Type 2
Resiliency (at 25 °C) D5329 [35]	%	40–70	66.1	61.1
Asphalt Compatibility D5329 [35]	-	Pass	Pass	Pass
Recommended Application Heating Temperature Range	-	182–199	Satisfied	Satisfied
Bond 3 Cycles (at −7 °C, 100% Elongation) D5329 [35]	-	Pass	Pass	Pass
Flexibility (at 23 °C)	-	Pass	Pass	Pass
Penetration (Max. at 25 °C, 100 g, 5 s) D5 [36]	dm	-	34.0	20.0
Elastic Recovery (at 25 °C)	%	-	77	100
Viscosity (at 135 °C) D446 [37]	cPs	-	3200	4500

2.1.8. Binder Blending

The binder blending process plays a pivotal role in formulating a high-performance asphalt pavement. This section outlines the meticulous steps undertaken for binder blending, emphasizing the careful selection of materials and their precise proportions to achieve the desired characteristics.

The primary constituents for binder blending include the extracted RAP binder and SBS polymer binders of Type 1 and Type 2. The blending ratios were systematically varied to explore different binder combinations, specifically 25% extracted RAP binder with 75% SBS (Type 1, Type 2) and 50% extracted RAP binder with 50% SBS (Type 1, Type 2). These combinations were chosen to strike a balance between the rejuvenating properties of the RAP binder and the enhanced performance attributes imparted by the SBS polymer binders.

Following the binder blending, the resultant mixtures underwent an aging process through the Pressure Aging Vessel (PAV) method (see Figure 2a). This controlled aging procedure simulates the long-term oxidative aging that asphalt binders experience during service. The PAV method allows for the assessment of binder performance under realistic aging conditions based on AASHTO R28 [38], providing valuable insights into the long-term durability and stability of the blended binders. After aging, the blended binders were subjected to a battery of performance tests to evaluate crucial properties such as stiffness, elasticity, and resistance to high-temperature distress.

Figure 2. Illustration of the laboratory tests. (**a**) PAV curing of specimens; (**b**) dynamic shear rheometer; (**c**) HWT test; (**d**) HWT specimens; (**e**) OT test.

2.1.9. Mixing of Aggregate to the Proposed Binders

In the meticulous mixing phase, the designated binders, enriched with petroleum resin and oil, are expertly combined with the chosen aggregate, utilizing a Nominal Maximum Aggregate Size (NMAS) of 13 mm. This intricate process occurs at controlled mixing temperatures, specifically set at 160 °C for optimal blending and interaction between the binders and aggregates. The mixture undergoes precise compaction using a gyratory compactor, where the compaction temperature is carefully maintained at 150 °C. These temperature-controlled procedures are foundational in shaping the subsequent testing and evaluation phases, providing crucial insights into the performance and sustainability aspects of the asphalt mixtures tailored for rural pavement applications. The aggregate gradation used in the research is shown in the following Table 7.

Table 7. Aggregate gradation of proposed mixtures.

Percent Passing (%)	20 mm	13 mm	10 mm	5 mm	2.5 mm	0.6 mm	0.3 mm	0.15 mm	0.075 mm
13 mm PMA	100	83.9	62.1	43.5	32.6	18.3	12.1	7.2	3.8

2.2. Testing Methods

In the comprehensive evaluation of asphalt pavement to ascertain its enhanced performance, a meticulously devised and quantitative testing protocol was rigorously implemented. This section delineates the detailed methodologies employed to assess the physical and mechanical properties of the blended binders and the resulting asphalt mixtures, adhering strictly to industry-standard procedures and specifications.

2.2.1. Binder Performance Testing

Dynamic Shear Rheometer (DSR) Test: The DSR test, precisely conforming to the stringent ASTM D7552 standards [39], was used to meticulously assess the viscoelastic properties of the binders (see Figure 2b). The specimens, measuring 25 mm in diameter and 1.5 mm in thickness, underwent a comprehensive testing regime. The frequency sweep test spanned from 0.1 to 10 Hz, and the temperature ramp test ranged from −16 to 76 °C. Each binder specimen underwent triplicate testing, ensuring a robust characterization of rheological properties, and contributing to the comprehensiveness and precision of the performance analysis.

2.2.2. Mixture Performance Testing

Hamburg Wheel Tracking Test (HWT): The HWT, adhering strictly to AASHTO T324 [40], was used to conduct a meticulous examination of the asphalt mixtures' resistance to rutting (see Figure 2c,d). The specimens, 150 mm in diameter and 63.5 mm in height, underwent precisely 20,000 loading cycles, allowing for an in-depth assessment of the mixtures' rutting resistance. The triple replication of this test guaranteed the consistency and replicability of the observed data, providing a reliable basis for the ensuing performance evaluations and analysis.

Overlay Test (OT): The crack resistance of asphalt mixes was systematically assessed through a laboratory test known as the Overlay Test (OT). This evaluation method, aligned with the TxDOT Tex-248-F standard [41], serves as a crucial measure for potential crack reflection within the asphalt mixtures. The specimens undergo assessment at a controlled temperature of 25 °C, employing the OT method to determine the mix's resistance to cracking. The evaluation is based on the number of cycles required to induce failure, with a focus on the number of OT loops needed to achieve a 93% reduction in the initial load during the first cycle. The experiment is designed to cease if the specified reduction is not observed within a maximum set number of iterations, ensuring a comprehensive understanding of the mix's crack resistance. In this study, a total of 1000 loading cycles were applied, or until the experiment was stopped, providing valuable insights into the asphalt mixtures' durability and performance under overlay conditions.

This comprehensive and meticulously detailed testing methodology, crafted in strict adherence to standardized protocols, incorporated replicates and precise specimen sizes. This meticulous approach ensured a thorough and reliable assessment of the asphalt-polymer performance, underlining the credibility and accuracy of the ensuing analysis and evaluation.

2.3. Field Application

The empirical validation of the proposed methodology was meticulously conducted through a field testbed within the GS Caltex Complex, an industrially significant petrochemical entity situated in Yeosu City, South Korea. Recognizing the natural degradation of asphalt pavement subjected to vehicular loads over time, a periodic repair protocol was instituted. The test construction meticulously involved sequential processes such as transportation, laying, and compaction. The dimensions of the strategically positioned testbed within the GS Caltex Complex were 50 m in length and 10 m in width.

Throughout the service duration, the compaction dynamics of the asphalt pavement were diligently scrutinized via advanced non-destructive testing instruments, encompassing the Light Falling Weight Deflectometer (LFWD). The strategic selection of testing points for these field experiments is depicted in Figure 3.

(a) (b)

Figure 3. On-site use of Light Falling Weight Deflectometer on the field test bed: (**a**) pavement at the field, (**b**) LFWD test.

Light Falling Weight Deflectometer Test

The Light Falling Weight Deflectometer (LFWD) is employed in this study to directly measure the elastic modulus of asphalt mixtures in the field. By assessing pavement deflection caused by a falling weight, the LFWD provides a non-destructive and efficient method to gauge material stiffness and deformation under load. This in situ approach offers valuable insights into real-world performance, aiding in the optimization of pavement design for enhanced durability.

To quantitatively assess the post-construction pavement performance, the deflection magnitude and modulus of elasticity were systematically evaluated, leveraging the LFWD and following the stringent guidelines outlined in ASTM 2835 [42]. This compact version of the Falling Weight Deflectometer (FWD) was used to effectively measure deflection induced by the impact load of a falling weight. The LFWD test facilitated a rapid and robust assessment of the elastic settlement of the ground, with the measured elastic settlement promptly transmuted into an elastic modulus through the integration of a data logger. The LFWD field test conditions, detailed in Table 8, incorporated a load plate diameter of 150 mm, a load weight of 15 kg, a load height of 27 inches, and a Poisson's ratio of 0.4.

Table 8. LFWD field test conditions.

LFWD Test Conditions	
The diameter of the load plate	150 mm
The weight of the load	15 kg
Load height	27 inches
Poisson's ratio	0.4″

3. Results and Discussions

3.1. Binder Test Results

3.1.1. General Binder Test

This section delves into the specific findings of the Asphalt Binder Test, a critical phase in evaluating the performance and characteristics of various asphalt mixtures for rural pavements. The focus is on understanding the nuanced behaviors of asphalt binders under different compositions, shedding light on their influence on the overall performance of rural pavement.

The asphalt binder properties depicted in the table showcase distinct characteristics among various mixture compositions, providing insights into their thermal and mechanical behaviors. As presented in Table 9, the softening point, a key parameter reflecting the binder's vulnerability to high temperatures, exhibits a range of values. The mix of 50% RAP + 50% Type 2 demonstrates the highest softening point at 108.3 °C, indicating its superior resistance to deformation under elevated temperatures. In contrast, the mix of 25% RAP + 75% Type 2 has the lowest softening point at 102.5 °C, suggesting a binder more susceptible to temperature-induced changes. Ductility, representing the binder's ability to deform without breaking, is most pronounced in the mix of 25% RAP + 75% Type 1, boasting a value of 880 mm at 25 °C. This indicates excellent flexibility and resilience against deformation. On the other hand, penetration, a measure of binder consistency and hardness, is highest in the mix of 25% RAP + 75% Type 2, with a value of 34.4 dm at 25 °C, suggesting a softer binder. These nuanced variations emphasize the intricate interplay between RAP content and modifier type, influencing the thermal and mechanical properties of the binder. Such detailed insights are vital for tailoring asphalt mixtures to specific performance requirements, guiding the optimization of pavement design for enhanced durability and longevity in real-world applications.

Table 9. General asphalt binder test results.

Mixture Composition	Softening Point (°C)	Ductility (mm at 25 °C)	Penetration (dm at 25 °C)
25% RAP + 75% Type 1	103.4	880	33.9
25% RAP + 75% Type 2	102.5	760	34.4
50% RAP + 50% Type 1	104.5	760	27.8
50% RAP + 50% Type 2	108.3	570	20.4

The outcomes of this study are in line with previous research on asphalt mixtures for rural pavements, underscoring the intricate relationship between recycled asphalt pavement (RAP) content and modifier type. The observed patterns, such as the association between higher RAP content and elevated softening points, resonate with the established literature [43]. This alignment substantiates the reliability and relevance of the results, emphasizing their contribution to optimizing asphalt mixtures for enhanced durability and longevity in rural pavement applications.

3.1.2. Dynamic Shear Modulus (G) Test Results

In the Dynamic Shear Modulus (G) Test, the asphalt binder mixtures exhibit distinctive mechanical responses, particularly noteworthy in the cases of the 25% RAP + 75% Type 2 and 50% RAP + 50% Type 1 compositions. As shown in Figure 4, the former stands out as the softest among the tested mixtures at low temperatures, a characteristic attributed to its combination of low RAP content and the presence of the Type 2 modifier. With a limited RAP content, this mixture showcases enhanced flexibility, and the Type 2 modifier further contributes to its softer behavior, rendering it more susceptible to deformation under lower temperature conditions. Conversely, the latter composition, featuring 50% RAP + 50% Type 1, emerges as the stiffest among the tested mixtures at low temperatures. The pronounced stiffness can be primarily ascribed to the higher RAP content in this mixture, underscoring the dominant influence of RAP content on the mechanical response of the binder. This composition exhibits notable resistance to deformation at lower temperatures, highlighting the crucial role of RAP concentration in shaping the mixture's rheological properties.

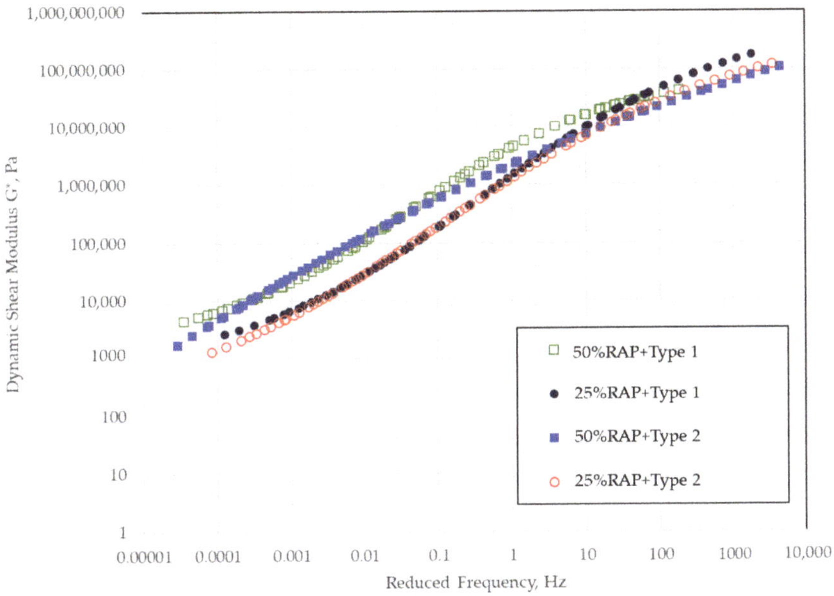

Figure 4. Dynamic shear modulus (G*) test results.

These results in the Dynamic Shear Modulus Test echo established research [44], underscoring the pivotal role of RAP content in asphalt binder performance. In line with the existing literature, our study reaffirms that lower RAP content, as in the 25% RAP + 75% Type 2 mixture, promotes flexibility and softness. Conversely, higher RAP concentration, exemplified by the 50% RAP + 50% Type 1 composition, results in increased stiffness and heightened resistance to deformation at lower temperatures. This concordance with prior findings enhances the credibility of our study and contributes to the collective understanding of the interplay between binder composition and rheological characteristics.

3.1.3. Sensitivity to Fatigue Cracking

The fatigue characteristics of the asphalt binder mixes were systematically evaluated across varying temperature ranges, ranging from 34 to 90 °C, as illustrated in the provided Figure 5. The susceptibility to fatigue crack initiation, as measured by the fatigue values (G*sin δ), is a critical parameter in assessing the thermal cracking resistance of asphalt binders. The Figure further elucidates the results of the Sensitivity to Fatigue Cracking Test after Pressure Aging Vessel (PAV) conditioning.

The Sensitivity to Fatigue Cracking Test after PAV conditioning provides valuable insights into the thermal cracking resistance of asphalt binder mixes. The fatigue characteristics, measured in terms of fatigue values (kPa), are indicative of the susceptibility of each mixture to fatigue cracking, especially under low-temperature conditions. Examining the results, the mixture composition with 50% RAP + 50% Type 2 modifier stands out with the highest fatigue value at 23,874 kPa, suggesting a notable vulnerability to fatigue cracking. On the contrary, the composition of 25% RAP + 75% Type 1 exhibits the lowest fatigue value at 1524 kPa, indicating superior resistance to thermal cracking. This observed trend aligns with the general understanding that higher RAP content and specific modifier types contribute to enhanced resistance against fatigue-induced thermal cracking. The intricate mechanism at play involves the dynamic interaction between RAP and the modifier, influencing the binder properties and, subsequently, the overall performance of the asphalt mixture under diverse temperature conditions. These results underscore the nuanced relationship between binder composition and fatigue resistance, emphasizing the need for careful consideration in designing asphalt mixtures for optimal performance and longevity.

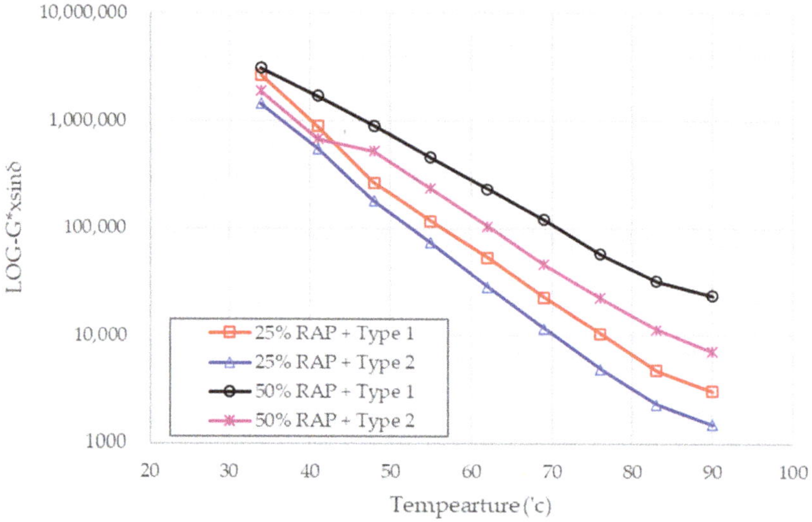

Figure 5. Sensitivity to Fatigue Cracking test results.

This sensitivity to fatigue cracking result aligns with previous studies, reinforcing the correlation between higher RAP content, specific modifier types, and enhanced resistance against fatigue-induced thermal cracking [45]. The observed trends substantiate the established understanding in the literature. This concordance underscores the reliability and consistency of our findings with the broader body of research in asphalt mixture behavior and design.

3.1.4. Analysis of Softening Point Variation in Storage Stability Test

The following Figure 6 outlines findings from the Storage Stability Test, specifically focusing on softening point variations in Type 1 and Type 2 asphalt binder samples. In their unaged states, Type 1 shows a softening point of 64.5 °C, slightly lower than Type 2's 65.2 °C. Post the storage stability test, both types are evaluated in their upper and lower sections.

Remarkably, Type 1 consistently maintains the highest softening point, recording 64.9 °C on top and 64.8 °C at the bottom. Conversely, Type 2 displays lower values, with 64.1 °C on top and 64.2 °C at the bottom. The softening point difference, indicating the change between the top and bottom sections post-test, is minimal for Type 1 at 0.3 °C, highlighting exceptional uniformity and stability. In contrast, Type 2 exhibits a slightly higher difference of 0.6 °C.

These results emphasize that Type 1 exhibits superior storage stability, with less variation between its top and bottom sections compared to Type 2. This is crucial for potential applications in expansion joints, ensuring improved long-term performance and resistance to temperature-induced changes.

This corroborates findings from other studies [46], reinforcing the importance of binder selection, for applications requiring enhanced long-term stability, such as in expansion joints. Our research substantiates these trends and contributes to the collective understanding of asphalt binder performance, providing valuable insights for optimizing asphalt mixtures in real-world pavement applications.

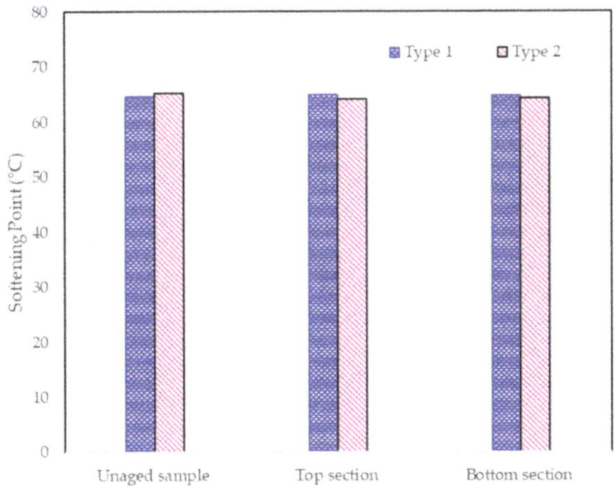

Figure 6. Storage stability test results are based on the difference in the softening point.

3.2. Asphalt Concrete Test Results

3.2.1. Hamburg Wheel Tracking (HWT) Test Results

The Hamburg Wheel Tracking (HWT) test is a crucial evaluation of asphalt concrete mixtures' rutting resistance, simulating the effects of repeated wheel loading and moisture infiltration. In this study, four distinct asphalt concrete mixtures were examined: 25% RAP + 75% Type 1, 25% RAP + 75% Type 2, 50% RAP + 50% Type 1, and 50% RAP + 50% Type 2.

Figure 7 details the HWT test results, offering a comprehensive view of rut depth variations across distinct asphalt concrete mixtures. Notably, the mixture composition featuring 25% RAP + 75% Type 1 modifier exhibits the highest rut depth at 5.4 mm. This seemingly counterintuitive result can be ascribed to the relatively lower RAP content, contributing to a softer mix that is more prone to rutting. In contrast, the composition with 50% RAP + 50% Type 1 modifier showcases the lowest rut depth of 3.9 mm, indicating a stiffer mix with heightened resistance to rutting. Here, the RAP content plays a pivotal role in mitigating rutting effects, underscoring the intricate interplay between RAP content and mixture performance. The mixtures with 25% RAP + 75% Type 2 and 50% RAP + 50% Type 2 fall between the extremes, portraying a balanced performance in rutting resistance. These findings illuminate the nuanced impact of RAP content and modifier type on rut depth, emphasizing the need for a holistic evaluation strategy that considers various mixture components for optimal performance across diverse conditions in real-world applications.

These HWT test results illuminate the complex interplay between RAP content and asphalt binder properties, providing valuable information for optimizing asphalt concrete mixtures to meet specific performance requirements in real-world pavement applications.

In general, this study's HWT test results are consistent with existing literature [47], emphasizing the intricate relationship between RAP content and modifier type in asphalt concrete mixtures. The higher rut depth in the 25% RAP + 75% Type 1 mixture aligns with prior research due to the softer characteristic, highlighting the nuanced impact of RAP proportions on asphalt stiffness. Concurrently, the superior rutting resistance in the 50% RAP + 50% Type 1 mixture corresponds to the recognized stiff benefits of increased RAP content. These findings reinforce the significance of considering RAP proportions and modifier types for optimizing asphalt mixtures, contributing valuable insights to pavement design and durability.

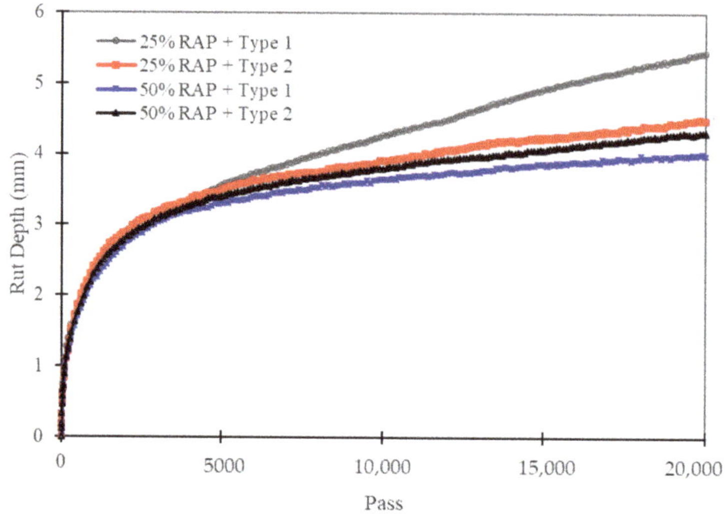

Figure 7. Hamburg Wheel Tracking Test results.

3.2.2. Overlay Test (OT) Results

The OT results reveal distinct patterns in the performance of asphalt concrete mixtures under simulated overlay conditions. The findings align with the overall trend observed in all asphalt concrete performance tests, highlighting specific characteristics of each mixture composition.

The outcomes depicted in Figure 8, derived from the OT, unfold a comprehensive narrative regarding the intricate interplay between varying asphalt concrete mixtures and their performance under simulated overlay conditions. The load reduction after 1000 cycles serves as a critical indicator, with higher percentages suggesting diminished resistance to overlay-induced distress. Surprisingly, the composition featuring 25% RAP + 75% Type 2 modifier exhibits remarkable resilience, showcasing the lowest load reduction at 64.5%. In contrast, the mixture with 25% RAP + 75% Type 1 modifier registers a slightly higher load reduction of 69.8%. The most intriguing revelation comes from the composition with 50% RAP + 50% Type 1 modifier, displaying a substantial load reduction of 82.1%. This unexpected outcome prompts a reevaluation of assumptions, indicating that higher RAP content contributes to a stiffer mixture, resulting in heightened vulnerability to overlay-induced distress. This insight challenges conventional expectations and underscores the need for a nuanced understanding of the relationship between RAP content, modifier type, and mixture stiffness in the context of overlay conditions. A deeper exploration of these findings is imperative for guiding informed decisions for optimizing asphalt mixtures, ensuring enhanced pavement durability in the face of overlay challenges.

A higher percentage of RAP, as seen in the 50% RAP + 50% Type 1 modifier composition, induces a stiffer mixture, amplifying its susceptibility to distress under overlay conditions. The stiffness of the asphalt concrete, in this case, hinders its ability to flex and accommodate dynamic forces, leading to increased susceptibility to cracking and rutting. On the contrary, the mixture with 25% RAP + 75% Type 2 modifier, displaying superior resilience with the lowest load reduction, suggests a more balanced blend of flexibility and stiffness. The interaction between RAP content and modifier type unfolds a complex dynamic, highlighting the need for a nuanced approach in mixture design. These findings underscore the importance of deciphering the intricate mechanisms governing asphalt mixture behavior under overlay conditions, paving the way for tailored solutions that prioritize durability and performance.

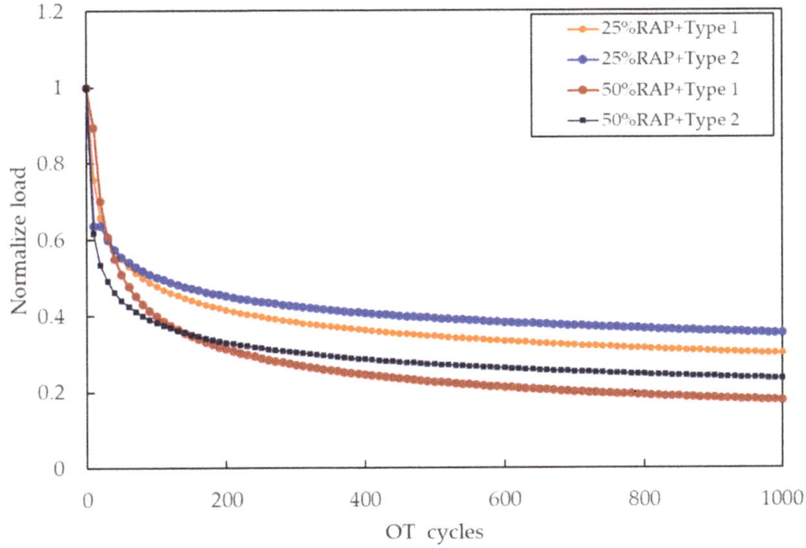

Figure 8. OT test results.

Overall, the OT test results align with existing research [12], providing insights into the intricate relationship between asphalt concrete mixtures and their performance under simulated overlay conditions. The observed higher load reduction in the 50% RAP + 50% Type 1 modifier composition is consistent with previous studies, emphasizing that higher RAP content contributes to a stiffer mixture, rendering it more vulnerable to overlay-induced distress. Surprisingly, the composition with 25% RAP + 75% Type 2 modifier exhibits remarkable resilience, challenging conventional expectations. These findings highlight the complex interplay of RAP content and modifier type, urging a nuanced understanding of optimized asphalt mixtures and enhanced pavement durability in overlay scenarios.

3.3. Field Test Results

The detailed examination of field test outcomes, with a specific focus on elastic modulus (MPa), unveils distinctive characteristics in the pavement-bearing capacities of various mixtures. As shown in Table 10, particularly noteworthy is the 25% RAP + 75% Type 2 composition, showcasing a remarkable elastic modulus of 58.5 MPa, indicating heightened bearing capacity in the pavement structure. This aligns with expectations that the specific combination of a higher proportion of the Type 2 modifier and a lower percentage of RAP contributes to enhanced rigidity and, consequently, increased load-bearing capacity. In contrast, the 50% RAP + 50% Type 1 mixture exhibits a relatively lower elastic modulus of 42.3 MPa, suggesting a pavement with reduced bearing capacity. The intricate interplay between RAP content and the type of modifier becomes apparent, influencing the overall bearing capacities of the asphalt mixtures. While the maximum load-bearing capacities of these mixtures are comparable, the nuanced variations in elastic modulus shed light on the complex dynamics governing the performance of these asphalt compositions in real-world field conditions. The test results reveal that the nuanced variations in elastic modulus, as observed between the 25% RAP + 75% Type 2 and 50% RAP + 50% Type 1 mixtures, provide valuable insights for strategic decisions in binder selection and asphalt composition to optimize pavement behavior and achieve superior bearing capacity. The higher elastic modulus in the 25% RAP + 75% Type 2 composition signifies enhanced rigidity and bearing capacity, attributed to the specific combination of a higher proportion of Type 2 modifier and a lower percentage of RAP. Conversely, the lower elastic modulus in the 50% RAP + 50% Type 1 mixture indicates reduced bearing capacity. These findings underscore the importance of carefully balancing RAP content and modifier type in asphalt mixtures, emphasizing the need for tailored solutions in real-world field conditions. Strategic decisions in binder selection should consider the desired level of rigidity and deformation resistance based on pavement performance requirements, ensuring optimal bearing capacity and long-term durability.

Table 10. Summary of LFWD Test results for different asphalt mixtures.

Mixture Composition	Maximum Load (N)	Maximum Displacement (mm)	Elastic Modulus (MPa)
25% RAP + 75% Type 1	4572	0.092	51
25% RAP + 75% Type 2	4590	0.08	58.5
50% RAP + 50% Type 1	4466	0.111	42.3
50% RAP + 50% Type 2	4577	0.117	44.2

In summary, the field test results reveal that the asphalt mixtures, while exhibiting comparable load-bearing capacities, demonstrate nuanced differences in terms of deformation resistance and stiffness. The findings underscore the significance of selecting appropriate binder types and proportions to tailor asphalt mixtures for optimal performance in real-world pavement applications. In general, these field test results, particularly in elastic modulus (MPa), echo trends observed in related research. The 25% RAP + 75% Type 2 composition showcases a notably high elastic modulus of 58.5 MPa, aligning with expectations of enhanced bearing capacity due to a higher proportion of the Type 2 modifier and lower RAP content. Conversely, the 50% RAP + 50% Type 1 mixture exhibits a comparatively lower elastic modulus of 42.3 MPa, indicating reduced bearing capacity. This nuanced interplay between RAP content and modifier type in influencing bearing capacities aligns with broader research trends, emphasizing the critical role of intentional binder selection for optimal pavement performance in real-world conditions.

In general, these research findings align with Zheng et al. in corroborating the impact of varying content [48], specifically RAP-SBS, on asphalt performance. The observed influence of RAP content and loading frequency on viscoelastic properties in Zheng's finding resonates with our study's focus on the nuanced effects of RAP content and SBS-modified binders on mechanical properties. Furthermore, our research confirms the insights from Ahmed et al. [49], emphasizing the importance of additives like aromatic oil and SBS copolymers for enhanced performance. The strategic inclusion of petroleum resin and oil in our asphalt binder compositions corresponds to improved compatibility with RAP, addressing stiffness concerns and ultimately enhancing overall mechanical properties. These comparisons reinforce the relevance of our study in the context of existing research, contributing to the collective understanding of optimal asphalt mixture formulations.

4. Conclusions

This study assesses the performance of asphalt mixtures for rural pavements, specifically examining RAP with SBS-modified asphalt binders containing petroleum resin and oil. The focus is on the impact of varying RAP content (25% and 50%) and two types of SBS-modified asphalt binders (Type 1 and Type 2) on mechanical properties and sustainability. A comprehensive set of laboratory tests, including Asphalt Binder Tests, Sensitivity to Fatigue Cracking assessments, HWT test, OT test, and field tests using the LFWD, is conducted. Key conclusions are presented as follows:

- Regarding asphalt binder properties, the mix of 25% RAP + 75% Type 1 exhibits outstanding flexibility and deformation resistance, evident in its high ductility value of 880 mm at 25 °C, potentially enhancing pavement resilience.

- The mixture with 50% RAP + 50% Type 2 modifier displayed the highest fatigue value at 23,874 kPa, indicating notable vulnerability to fatigue cracking. Conversely, the composition of 25% RAP + 75% Type 1 exhibited the lowest fatigue value at 1524 kPa, showcasing superior resistance to thermal cracking. These findings align with expectations that higher RAP content and specific modifier types enhance resistance against fatigue-induced thermal cracking.
- The storage stability tests reveal that Type 1 asphalt binder consistently maintains higher softening points with minimal variation (0.3 °C) between top and bottom sections, showcasing superior uniformity and stability compared to Type 2
- The HWT test emphasized the nuanced impact of RAP content and modifier type on rut depth. The mixture with 25% RAP + 75% Type 1 exhibited the highest rut depth at 5.4 mm, attributed to its relatively lower RAP content and softer nature. In contrast, the composition with 50% RAP + 50% Type 1 showcased the lowest rut depth of 3.9 mm, indicating heightened resistance to rutting.
- The OT results challenged conventional expectations, revealing unexpected outcomes. The mixture with 25% RAP + 75% Type 2 modifier displayed remarkable resilience, with the lowest load reduction at 64.5%. Surprisingly, the composition with 50% RAP + 50% Type 1 modifier exhibited substantial load reduction at 82.1%, challenging assumptions about higher RAP content contributing to a stiffer mixture.
- The field test results highlighted nuanced differences in pavement bearing capacities among asphalt mixtures. The 25% RAP + 75% Type 2 composition demonstrated a remarkable elastic modulus of 58.5 MPa, indicating heightened bearing capacity. In contrast, the 50% RAP + 50% Type 1 mixture exhibited a relatively lower elastic modulus of 42.3 MPa.
- The study highlights the significant impact of SBS modifiers with petroleum resin and oil on Type 1 and Type 2 asphalt binders, shaping various asphalt mixture properties. Type 2, with 50% RAP, demonstrates balanced flexibility and stiffness, performing well in overlay conditions and field tests. These insights underscore the crucial role of SBS modifiers in customizing asphalt mixture properties for sustainable and resilient pavement design. Limitations include the focus on specific compositions, and future research should explore a broader spectrum of variations for comprehensive pavement design.

Author Contributions: Conceptualization, J.-w.K., K.K. and T.H.M.L.; methodology, T.H.M.L.; validation J.-w.K. and K.K.; formal analysis, J.-w.K. and T.H.M.L.; investigation, J.-w.K. and K.K.; resources, J.-w.K. and T.H.M.L.; writing—original draft preparation, J.-w.K., K.K. and T.H.M.L.; writing—review and editing, J.-w.K. and T.H.M.L.; visualization, J.-w.K. and K.K.; supervision, J.-w.K.; project administration, J.-w.K. and K.K.; funding acquisition, J.-w.K. and K.K. All authors have read and agreed to the published version of the manuscript.

Funding: Research for this paper was carried out under the KICT Research Program (Development of innovative trenching and pavement restoration technology based on Smart QSE (3rd year)) funded by the Ministry of Science and ICT.

Data Availability Statement: The data presented in this study are available on request from the corresponding author.

Conflicts of Interest: Author Kyungnam Kim was employed by the company Korea Expressway Corporation. The remaining authors declare that the research was conducted in the absence of any commercial or financial relationships that could be construed as a potential conflict of interest.

References

1. Saniga, A.M.; Rajish, A.; Shankar, S.; Vishnu, R. Optimal Pavement Maintenance and Management Strategies for Flood Affected Low Volume Rural Roads. *Mater. Today Proc.* **2023**. [CrossRef]
2. Tang, Y.; Xiao, J.; Liu, Q.; Xia, B.; Singh, A.; Lv, Z.; Song, W. Natural Gravel-Recycled Aggregate Concrete Applied in Rural Highway Pavement: Material Properties and Life Cycle Assessment. *J. Clean. Prod.* **2022**, *334*, 130219. [CrossRef]

3. Wood, J.; Donnell, E.T. Empirical Bayes Before-after Evaluation of Horizontal Curve Warning Pavement Markings on Two-Lane Rural Highways in Pennsylvania. *Accid. Anal. Prev.* **2020**, *146*, 105734. [CrossRef] [PubMed]
4. Tawalare, A.; Vasudeva Raju, K. Pavement Performance Index for Indian Rural Roads. *Perspect. Sci.* **2016**, *8*, 447–451. [CrossRef]
5. Hoy, M.; Samrandee, V.; Samrandee, W.; Suddeepong, A.; Phummiphan, I.; Horpibulsuk, S.; Buritatum, A.; Arulrajah, A.; Yeanyong, C. Evaluation of Asphalt Pavement Maintenance Using Recycled Asphalt Pavement with Asphalt Binders. *Constr. Build. Mater.* **2023**, *406*, 133425. [CrossRef]
6. Rout, M.D.; Biswas, S.; Shubham, K.; Sinha, A.K. A Systematic Review on Performance of Reclaimed Asphalt Pavement (RAP) as Sustainable Material in Rigid Pavement Construction: Current Status to Future Perspective. *J. Build. Eng.* **2023**, *76*, 107253. [CrossRef]
7. Peduzzi, A.; Franco, A.; De Luca, G.; Coppola, O.; Bonati, A. Economical Assessment of Recycled Asphalt Pavement (RAP) Aggregate for Structural Concrete Production in Italy. *Buildings* **2023**, *13*, 2191. [CrossRef]
8. Zhang, S.; Luo, C.; Huang, Z.; Li, J. Study on Thixotropy of Mastic Asphalt Binder and Asphalt Mastic. *Buildings* **2023**, *13*, 2380. [CrossRef]
9. Zhao, W.; Yang, Q. Design and Performance Evaluation of a New Green Pavement: 100% Recycled Asphalt Pavement and 100% Industrial Solid Waste. *J. Clean. Prod.* **2023**, *421*, 138483. [CrossRef]
10. Lee, S.Y.; Le, T.H.M. Evaluating Pavement Performance in Bus Rapid Transit Systems: Lessons from Seoul, South Korea. *Case Stud. Constr. Mater.* **2023**, *18*, e02065. [CrossRef]
11. Dinh, B.H.; Park, D.W.; Le, T.H.M. Effect of Rejuvenators on the Crack Healing Performance of Recycled Asphalt Pavement by Induction Heating. *Constr. Build. Mater.* **2018**, *164*, 246–254. [CrossRef]
12. Lee, S.Y.; Kim, Y.M.; Le, T.H.M. Laboratory and Field Testbed Evaluation of the Performance of Recycled Asphalt Mixture Using High-Penetration Asphalt. *Buildings* **2023**, *13*, 529. [CrossRef]
13. Lee, S.Y.; Ho Minh Le, T.; Kim, Y.M. Full-Scale and Laboratory Investigations on the Performance of Asphalt Mixture Containing Recycled Aggregate with Low Viscosity Binder. *Constr. Build. Mater.* **2023**, *367*, 130283. [CrossRef]
14. Newcomb, D.; Ray Brown, E.; Epps, J.A. *Designing HMA Mixtures with High RAP Content: A Practical Guide*; National Asphalt Pavement Association: Greenbelt, Maryland, 2007; p. 41.
15. Qian, Y.; Guo, F.; Leng, Z.; Zhang, Y.; Yu, H. Simulation of the Field Aging of Asphalt Binders in Different Reclaimed Asphalt Pavement (RAP) Materials in Hong Kong through Laboratory Tests. *Constr. Build. Mater.* **2020**, *265*, 120651. [CrossRef]
16. Li, W.; Yao, H.; Yang, D.; Peng, C.; Wang, H.; Chen, Z.; Zhao, Y. Study on Pavement Performance of Recycled Asphalt Mixture Modified by Carbon Nanotubes and Waste Engine Oil. *Appl. Sci.* **2023**, *13*, 10287. [CrossRef]
17. Wang, P.; Chen, J.; Wang, J.; Li, J.; Ning, H.; Liang, C.; Ge, X.; Wang, X. Evaluation of the Refined Decomposition Effect of Reclaimed Asphalt Pavement Materials. *Buildings* **2023**, *13*, 2240. [CrossRef]
18. Yun, J.; Na, I.H.; Choi, P.; Ji, B.; Kim, H. Laboratory Evaluation of High-Temperature Properties of Recycled PMA Binders. *Sustainability* **2023**, *15*, 12744. [CrossRef]
19. Sun, M.; Wang, J.; Sun, H.; Hong, B. Feasibility Analysis of Polyurethane-Prepolymer-Modified Bitumen Used for Fully Reclaimed Asphalt Pavement (FRAP). *Materials* **2023**, *16*, 5686. [CrossRef] [PubMed]
20. Chen, Z.; Liu, B.; Feng, D.; Li, G. Analysis of Factors Influencing the Modulus of Hot-Recycled Asphalt Mixture with High RAP. *Materials* **2023**, *16*, 5280. [CrossRef]
21. Kim, K.N.; Le, T.H.M. Durability of Polymer-Modified Asphalt Mixture with Wasted Tire Powder and Epoxy Resin under Tropical Climate Curing Conditions. *Polymers* **2023**, *15*, 2504. [CrossRef]
22. Pazzini, M.; Tarsi, G.; Tataranni, P.; Lantieri, C.; Dondi, G. Mechanical Characterization of Thin Asphalt Overlay Mixtures with 100% Recycled Aggregates. *Materials* **2023**, *16*, 188. [CrossRef] [PubMed]
23. Wang, H.; Huang, Y.; Jin, K.; Zhou, Z. Properties and Mechanism of SBS/Crumb Rubber Composite High Viscosity Modified Asphalt. *J. Clean. Prod.* **2022**, *378*, 134534. [CrossRef]
24. Porto, M.; Caputo, P.; Loise, V.; Abe, A.A.; Tarsi, G.; Sangiorgi, C.; Gallo, F.; Rossi, C.O. Preliminary Study on New Alternative Binders through Re-Refined Engine Oil Bottoms (Reobs) and Industrial by-Product Additives. *Molecules* **2021**, *26*, 7269. [CrossRef] [PubMed]
25. Hamid, A.; Baaj, H.; El-Hakim, M. Rutting Behaviour of Geopolymer and Styrene Butadiene Styrene-Modified Asphalt Binder. *Polymers* **2022**, *14*, 2780. [CrossRef] [PubMed]
26. Zhu, H.; Xu, G.; Gong, M.; Yang, J. Recycling Long-Term-Aged Asphalts Using Bio-Binder/Plasticizer-Based Rejuvenator. *Constr. Build. Mater.* **2017**, *147*, 117–129. [CrossRef]
27. Zhang, L.; Wang, H.; Zhang, C.; Wang, S. Laboratory Testing and Field Application of Devulcanized Rubber/SBS Composite Modified Asphalt. *Case Stud. Constr. Mater.* **2023**, *19*, e02285. [CrossRef]
28. *ASTM D2172-17*; Standard Test Methods for Quantitative Extraction of Asphalt Binder from Asphalt Mixtures. American Society for Testing and Materials: West Conshohocken, PA, USA, 2017; pp. 1–10.
29. *ASTM D36-16*; Standard Test Method for Softening Point of Bitumen (Ring-and-Ball Apparatus). ASTM International: West Conshohocken, PA, USA, 2016.
30. *ASTM D4402*; Standard Test Method for Viscosity Determination of Asphalt at Elevated Temperatures Using a Rotational Viscometer. American Society for Testing and Materials: West Conshohocken, PA, USA, 2012.

31. *ASTM D113-17*; Standard Test Method for Ductility of Asphalt Materials. American Society for Testing and Materials: West Conshohocken, PA, USA, 2008.
32. *ASTM D6927*; Standard Test Method for Marshall Stability and Flow of Asphalt Mixtures. ASTM International: West Conshohocken, PA, USA, 2015.
33. Li, J.; Fan, Y.; Dai, L.; Liu, J. Fundamental Performance Investigation on Reactive Liquid Asphalt. *J. Clean. Prod.* **2019**, *225*, 315–323. [CrossRef]
34. *ASTM D36-06*; Standard Test Method for Softening Point of Bitumen (Ring-and-Ball Apparatus). American Society for Testing and Materials: West Conshohocken, PA, USA, 2014.
35. *ASTM D 5329*; Standard Test Methods for Sealants and Fillers, Hot-Applied, for Joints and Cracks in Asphalt Pavements and Portland Cement Concrete Pavements. American Society for Testing and Materials: West Conshohocken, PA, USA, 2020.
36. *ASTM D5*; Standard Test Method for Penetration of Bituminous Materials. ASTM International: West Conshohocken, PA, USA, 2019.
37. *ASTM D446-97*; Standard Specifications and Operating Instructions for Glass Capillary Kinematic Viscometers. ASTM International: West Conshohocken, PA, USA, 2007; Volume 5, pp. 1–26.
38. *AASHTO R 28-09*; Accelerated Aging of Asphalt Binder Using a Pressurized Aging Vessel (PAV). ASTM International: West Conshohocken, PA, USA, 2007; pp. 1–8.
39. *ASTM D7552-22*; Standard Test Method for Determining the Complex Shear Modulus (G *) Of Bituminous Mixtures Using Dynamic Shear Rheometer. American Society for Testing and Materials: West Conshohocken, PA, USA, 2016; pp. 1–11.
40. *AASHTO T324*; Standard Method of Test for Hamburg Wheel-Track Testing of Compacted Hot-Mix Asphalt (HMA). AASHTO: Washington, DC, USA, 2014.
41. Tex-248-F; Test Procedure for Overlay Test. Texas Department of Transportation, 2014; pp. 1–8. Available online: https://rosap.ntl.bts.gov/view/dot/25533 (accessed on 11 November 2023).
42. *ASTM E2835-07*; Standard Test Method for Measuring Deflections with a Light Weight Deflectometer (LWD). ASTM International: West Conshohocken, PA, USA, 2015; Volume 7, pp. 5–7.
43. Xing, C.; Li, M.; Liu, L.; Lu, R.; Liu, N.; Wu, W.; Yuan, D. A Comprehensive Review on the Blending Condition between Virgin and RAP Asphalt Binders in Hot Recycled Asphalt Mixtures: Mechanisms, Evaluation Methods, and Influencing Factors. *J. Clean. Prod.* **2023**, *398*, 136515. [CrossRef]
44. Sabouri, M. Evaluation of Performance-Based Mix Design for Asphalt Mixtures Containing Reclaimed Asphalt Pavement (RAP). *Constr. Build. Mater.* **2020**, *235*, 117545. [CrossRef]
45. Xu, H.; Sun, Y.; Chen, J.; Li, J.; Yu, B.; Qiu, G.; Zhang, Y.; Xu, B. Investigation into Rheological Behavior of Warm-Mix Recycled Asphalt Binders with High Percentages of RAP Binder. *Materials* **2023**, *16*, 1599. [CrossRef] [PubMed]
46. Zalghout, A.; Castro, S.; Karam, J.; Kaloush, K. Laboratory and Field Evaluation of Plant Produced Asphalt Mixtures Containing RAP in Hot Climate: A Case Study from Phoenix, Arizona. *Constr. Build. Mater.* **2022**, *351*, 128322. [CrossRef]
47. Rafiq, W.; Napiah, M.B.; Sutanto, M.H.; Alaloul, W.S.; Zabri, Z.N.B.; Khan, M.I.; Musarat, M.A. Investigation on Hamburg Wheel-Tracking Device Stripping Performance Properties of Recycled Hot-Mix Asphalt Mixtures. *Materials* **2020**, *13*, 4704. [CrossRef] [PubMed]
48. Zheng, K.; Xu, J.; Wang, J. Viscoelasticity of Recycled Asphalt Mixtures with High Content Reclaimed SBS Modified Asphalt Pavement. *Sustainability* **2023**, *15*, 2515. [CrossRef]
49. Eltwati, A.; Al-Saffar, Z.; Mohamed, A.; Rosli Hainin, M.; Elnihum, A.; Enieb, M. Synergistic Effect of SBS Copolymers and Aromatic Oil on the Characteristics of Asphalt Binders and Mixtures Containing Reclaimed Asphalt Pavement. *Constr. Build. Mater.* **2022**, *327*, 127026. [CrossRef]

buildings

MDPI

Article

Evaluation of the Effect of C_9 Petroleum Resin on Rheological Behavior, Microstructure, and Chemical Properties of Styrene–Butadiene–Styrene Modified Asphalt

Chaoqun Yan [1], Taoli Zhang [2,*], Kui Hu [2], Syed Tafheem Abbas Gillani [2] and Wengang Zhang [3]

1 Yangling Vocational & Technical College, Xianyang 712100, China; ycq@ylvtc.edu.cn
2 College of Civil Engineering, Henan University of Technology, Zhengzhou 450001, China; mailhukui@haut.edu.cn (K.H.); gillani@stu.haut.edu.cn (S.T.A.G.)
3 School of Civil and Architectural Engineering, Shandong University of Technology, Zibo 255000, China; ziwuzizwg@sdut.edu.cn
* Correspondence: taolizhang@stu.haut.edu.cn

Abstract: Understanding the modification mechanism of C_9 petroleum resin (C_9PR) on styrene–butadiene–styrene (SBS) polymer modified asphalt properties is of significant importance. In this paper, dynamic shear rheometer (DSR), storage stability, fluorescence morphology (FM), scanning electron microscope (SEM), Fourier transform-infrared (FTIR) spectrometer, and molecular dynamic (MD) simulation were adopted to evaluate the rheological, chemical, and microstructure molecular motion state of C_9PR and SBS composite modified asphalt at different aging states. The DSR storage results indicate that the addition of C_9PR could improve the high-temperature property, storage stability, and temperature susceptibility. FM and SEM results indicate that the network microstructure was enhanced and the roughness between polymer resins and virgin asphalt was improved at the microscopic scale. The MD results indicate that the heterogeneity between C_9PR and SBS modified asphalt was demonstrated, and the bonding energies were enhanced with the addition of C_9PR. Moreover, the FTIR results indicate that new function groups were generated in addition to C_9PR. In general, the addition of C_9PR is a good approach to promote high-quality polymer modified asphalt (PMA) for pavement engineering.

Keywords: C_9 petroleum resin; SBS modified asphalt; rheological properties; microstructure; chemical properties

1. Introduction

As a byproduct from petrochemical processes, C_9 petroleum resin (C_9PR) has garnered extensive attention for its utilization within the realm of material science, attributable to its cost efficiency and superior chemical attributes [1–3]. Notably, within the domain of polymer modified asphalt (PMA) research and applications, it has been authenticated that C_9PR markedly modulates the compositional distribution and microstructural state of asphalt [4,5]. Such modulation not only augments asphalt's performance by enhancing its abrasion resistance and anti-aging capabilities, but also exhibits significant potential in cost reduction for production [6,7]. Consequently, the exploration of C_9PR not only bears significant implications for the enhancement of current asphalt material performance, but also paves new pathways for the development of novel high-performance materials for road construction endeavors.

C_9PR, derived from the polymerization of light aromatic hydrocarbons during petroleum cracking, exhibits excellent chemical stability, thermoplastic adhesion, and compatibility as a thermoplastic resin [8]. These characteristics enable C_9PR to play a crucial role in the production of adhesives, inks, coatings, and rubber products in multiple industrial fields [9]. Specifically, the chemical composition of C_9PR enables it to play a unique role in modified asphalt systems. Due to its ability to interact with various chemical components in asphalt through the presence of polycyclic aromatic hydrocarbons, C_9PR can effectively increase the viscosity and elastic modulus of asphalt, while reducing its temperature sensitivity [10–12]. This characteristic makes C_9PR an ideal modifier for improving the road performance of styrene–butadiene–styrene (SBS) modified asphalt. SBS is a commonly used high-performance polymer modifier that can significantly improve the elasticity and durability of asphalt [13,14]. However, the cost of SBS modified asphalt is relatively high, and there is still room for improvement in certain performance aspects. The introduction of C_9PR provides a new path for optimizing SBS modified asphalt [15]. The research shows that C_9PR can improve the anti-aging performance of modified asphalt, reduce the damage of ultraviolet ray and oxidation to asphalt, and enhance its performance stability under extreme weather conditions [16,17]. In addition, the low cost of C_9PR also helps to reduce the production cost of overall modified asphalt materials and improve their market competitiveness [18]. In summary, C_9PR, with its unique chemical and physical properties, not only has theoretical significance in improving and optimizing the road performance of SBS modified asphalt, but also provides strong support for practical applications, significantly promoting the development of materials science and road engineering.

In recent years, C_9PR has been extensively studied as an effective modifier to optimize the performance of polymer modified asphalt (PMA) [19]. The introduction of C_9PR has been found to significantly enhance the overall performance of modified asphalt, including improved high-temperature rheological properties, enhanced low-temperature cracking resistance, and increased resistance to water damage [20]. These performance enhancements are primarily attributed to the good compatibility and synergistic effects between C_9PR and the asphalt matrix, as well as SBS and other polymer modifiers, thus optimizing the microstructure and macro performance of the modified asphalt [21]. Previous academic research has shown that C_9PR can effectively adjust the polymer network structure in SBS modified asphalt, increasing the uniform distribution of the polymer phase [22]. This structural improvement contributes to an increase in the viscoelasticity of SBS modified asphalt, thereby providing better deformation resistance at high temperatures, while maintaining sufficient flexibility at low temperatures to reduce the occurrence of cracks [23–25]. Furthermore, the addition of C_9PR has been found to improve the aging resistance of SBS asphalt mixtures, due to the specific chemical composition of C_9PR that can slow down the oxidation reaction rate in asphalt, thus prolonging its service life [26–28]. Overall, C_9PR has shown a wide range of application potential in enhancing the performance of SBS modified asphalt.

It is noteworthy that previous scholars have conducted a series of experiments and analyses, such as dynamic shear rheological tests, infrared spectroscopy analysis, and microscopic structure observations, to deeply explore the specific mechanism by which C_9PR improves PMA performance [29–31]. These studies reveal how C_9PR affects the chemical and physical structure of modified asphalt, further proving its potential as an efficient modifier [32]. Therefore, based on the results of these scientific studies, the application of C_9PR in improving PMA performance has shown great application prospects and practical value [33,34]. With the rapid development of polymer modified asphalt (PMA) technology, research methods are also constantly advancing [35]. Among these, molecular dynamics (MD) simulation has become an important tool, capable of simulating the interaction between C_9PR and the asphalt matrix at the molecular level, providing a new perspective for understanding and predicting material performance. MD simulation, by calculating the forces and movements between molecules, can reveal the dispersion state of the modifier in the asphalt, compatibility, and the mechanism of its impact on asphalt performance, details that are difficult to directly observe with traditional experimental methods [36]. Moreover, by combining experimental data with MD simulation results, researchers can optimize the formula of modified asphalt and predict material performance under different conditions. This approach not only improves research efficiency and reduces experimental costs, but also provides strong theoretical support for the design of high-performance modified asphalt materials.

In this study, five modified bitumen with different contents of C_9PR and SBS were prepared. According to the previous studies, the optimal dosage of SBS in composite modified bitumen was 5 wt%, and C_9PR at dosage of 3 wt%, 6 wt%, 9 wt%, and 12 wt% was selected. The rheological properties were tested with a dynamic shear rheometer (DSR) in temperature sweep, the storage stability was characterized through difference in softening point, and the functional group changes were represented with a Fourier transform-infrared (FTIR) spectrometer. The microscopic phase structure and interface morphology were observed with a fluorescence microscope (FM) and scanning electron microscope (SEM), respectively. The diffusion and aggregation behaviors of SBS and four components were carried out by MD simulation in terms of mean square displacement (MSD), radial distribution function (RDF), and relative concentration distribution (RCD).

2. Materials and Methods

2.1. Raw Materials

The 70# bitumen utilized in this study was provided by Sinopec Maoming Petrochemical Company (Maoming, China), and its physical properties are listed in Table 1. The SBS modifier for the liner was purchased from Sinopec Paling Petrochemical Co., Ltd. (Yueyang, China), with a butadiene/styrene ratio of 30/70. The C_9PR was produced by Zibo Luhua Hongjin New Material Group Co., Ltd. (Zibo, China). The specific model used for production was PRR1-120, and its technical indices are shown in Table 2.

Table 1. Physical properties of bitumen.

Item	Unit	Test Results
Penetration (25 °C, 100 g, 5 s)	0.1 mm	65.2
Softening point	°C	47.5
Ductility (15 °C, 1 cm/min)	cm	145.0
Viscosity (135 °C)	Pa·s	0.593

Table 2. Technical indices of C_9PR.

Item	Unit	Test Results
Softening point	°C	118.4
Gardner color scale	#	8
Acid value	mgKOH/g	0.15

2.2. Preparation of C_9PR/SBS Modified Bitumen

To evaluate the effect of C_9PR on the rheological, microstructure, and chemical characteristics of SBS modified bitumen, SBS modified bitumen and C_9PR/SBS modified bitumen were prepared in this study. Firstly, virgin bitumen was put into the oven (160 °C, 1.5 h) until it was molten. And the virgin bitumen was poured into the beaker and put in the oil bath. Subsequently, 5 wt% SBS was added into the bitumen and sheared (5000 r/min, 45 min). After that, C_9PR (0 wt%, 3 wt%, 6 wt%, 9 wt% and 12 wt%) was mixed into SBS modified bitumen and sheared (5000 r/min, 45 min). Finally, C_9PR/SBS modified bitumen was put the oven to swell (160 °C, 2 h). Meanwhile, the preparation procedures were replicated for SBS modified bitumen, as depicted in Figure 1.

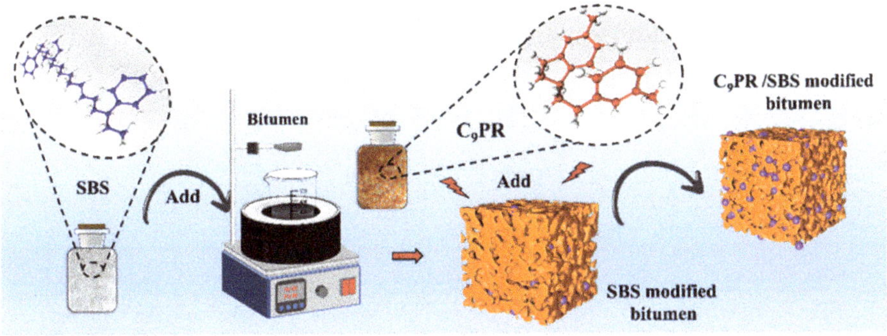

Figure 1. Preparation procedure of C9PR/SBS modified bitumen samples.

2.3. Test Methods

2.3.1. Dynamic Shear Rheometer (DSR)

To assess the rheological properties of the modified bitumen, temperature sweep tests were conducted using the MCR 702 DSR from Anton Paar Inc. (Graz, Austria). The complex modulus (G^*) and phase angles (δ) were measured in accordance with ASTM D7175 [37], and the rutting resistance was calculating the rutting factor ($G^*/sin\delta$). The range of temperature sweep test was set from 46 °C to 88 °C with an increase of 6 °C, and the test frequency was set to 10 rad/s. The diameter of the rotating shaft plate was set to 25 mm, and the spacing between the rotating shaft plate and the bottom plate was set to 1 mm.

2.3.2. Scanning Electron Microscope (SEM)

The interface morphology of SBS modified asphalt and C9PR/SBS modified asphalt was studied using a scanning electron microscope (JSM 5800LV) produced by Japan Electron Optics Laboratory Co., Ltd. (Akishima, Japan). During the SEM testing, the samples were gold-coated for 100 s in a vacuum using the JFC-1600 precision coating machine (JEOL, Akishima, Japan) to minimize charging effects. The operating temperature of the SEM equipment is typically around 20 °C to ensure the normal operation of the device and to avoid image distortion caused by temperature fluctuations.

2.3.3. Fluorescence Microscope Test

The microstructure and swelling of the polymers and bitumen were determined using FM, MATLAB R2018a, and Image J software. The modified bitumen samples were heated in an oven to a fluid state and then dropped onto slides using a glass rod. Subsequently, a coverslip was pressed onto the slide to form a film of bitumen, ensuring that it can be observed under a fluorescence microscope. Finally, the fluorescence images were binarized using MATLAB software, and the particle sizes of polymer were measured using Image J software to quantitatively analyze the effect of C_9PR on SBS modified asphalt.

2.3.4. Storage Stability

The storage stability is used to assess the stability and quality changes in bitumen during storage process. In this study, the compatibility of SBS modifier and bitumen was determined by the softening point test. After the modified bitumen was heated to a flow state, the glass segregation tubes with a diameter of about 25 mm and a length of about 140 mm were vertically placed on a support in an oven at a temperature of 163 °C for 48 h. Subsequently, the softening points measured at the top and bottom one-third of the sample were used to evaluate the storage stability of the modified bitumen. According to the literature, if the difference in softening points between the two parts is less than 3 °C, the sample is considered to have good high-temperature storage stability [38].

2.3.5. Fourier Transform Infrared Spectroscopy (FTIR)

FTIR was adopted to investigate the change in function group of SBS modified with (0 wt%, 3 wt%, 6 wt%, 9 wt% and 12 wt%) C_9PR contents. And the instrument used in this study was the Nicolet iS10 Fourier transform infrared spectrometer. The FTIR spectra were recorded in Attenuated Total Reflection (ATR) mode with the wavelength range from 4000 cm^{-1}~500 cm^{-1} at a resolution of 4 cm^{-1} and 64 scans, and the measurements were usually conducted at room temperature.

2.4. Simulation Methods

2.4.1. Model Construction

Bitumen is a complex mixture composed of various organic compounds, including hydrocarbons, oxides, and nitrides. In addition, trace metal elements exist in the form of inorganic salts in bitumen, such as vanadium, nickel, iron, sodium, calcium, and copper [39]. Currently, Greenfield proposed the AAA-1 12-component bitumen model, which is closest to real bitumen and widely used for molecular dynamics simulations. It mainly includes asphaltenes (Phenol, Pyrrole, Thiophene), resins (Quinolinohopane, Thioisorenieratane, Benzobisbenzothiophene, Pyridinohopane, Trimethybenzene), saturates (Trimethybenzene, Hopane), and aromatics (PHPN, DOCHN) [40]. The molecular model is shown in Figure 2.

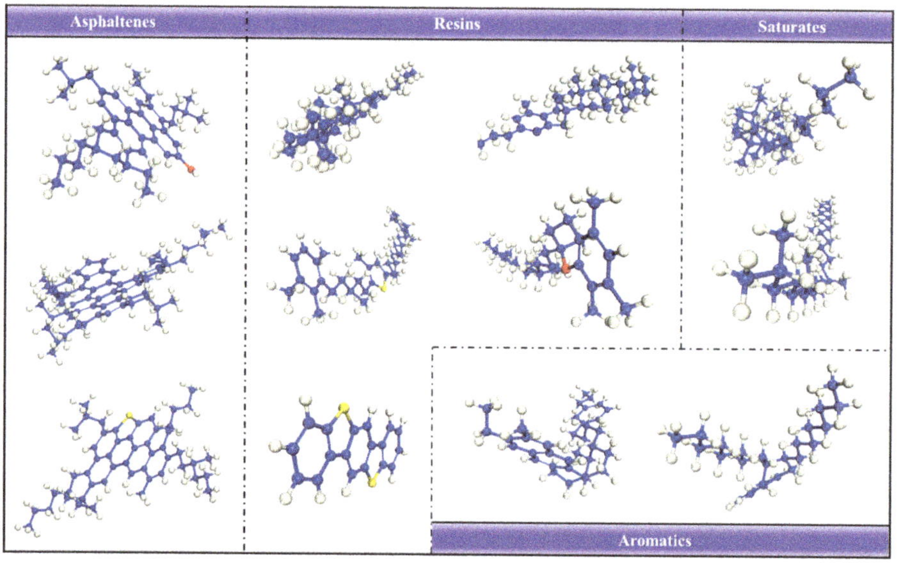

Figure 2. Twelve-component models of virgin bitumen.

Molecular dynamic (MD) is a computational modeling technique that reveals the macroscopic properties and microscopic mechanisms of materials by simulating the motion trajectories and interaction forces of molecules [41]. It is based on the principles of Newtonian mechanics and primarily calculates various characteristics of molecules from a molecular perspective [42]. As an extension of the COMPASS force field, COMPASS II offers higher precision, better parameter methods, and higher computational efficiency, making it particularly suitable for studying the properties and behaviors of organic compounds, inorganic small molecules, polymers, and metallic materials [43]. Therefore, the dynamics simulation in this study employed the COMPASS II force field.

The model of virgin bitumen in this study was constructed using the Amorphous Cell module in Materials Studio 2018 software. The specific simulation process is as follows: (1) Twelve molecules of bitumen were added to the cubic in a certain proportion, and the construction of the initial structure was completed with a density of 0.1 g/cm^3. (2) The geometric optimization in the Forcite module was performed to eliminate unfavorable contacts of molecules and achieve energy minimization of the initial structure. The iteration was performed for 100,000 times, and the COMPASS II force field was used. (3) Annealing of the bitumen model was carried out for 5 cycles in the range of 298 K~498 K using the NVE system to eliminate the local high-energy spots in the model. (4) Dynamic simulations were performed at 500 ps under the NPT system, which resulted in full relaxation of the model to reach equilibrium, and the density gradually approached the real value at 1 atm. The models of SBS modified bitumen and C_9PR/SBS modified bitumen were duplicated for the construction of virgin bitumen. The parameters are listed in Table 3 and the models are shown in Figure 3.

Table 3. Molecular parameters of the model.

SARA	Molecules	Molecular Formula	Molar Mass (g/mol)	Number	Mass Ratio (%)
Asphaltenes	Phenol	$C_{42}H_{54}O$	575	3	16.6
	Pyrrole	$C_{66}H_{81}N$	888.5	2	
	Thiophene	$C_{51}H_{62}S$	707.2	3	
Resins	Quinolinohopane	$C_{40}H_{59}N$	554.0	4	41.9
	Thioisoremieratane	$C_{40}H_{60}S$	573.1	4	
	Trimethybenzeneoxane	$C_{29}H_{50}O$	414.8	15	
	Pyridinohopane	$C_{36}H_{57}N$	530.9	5	
	Benzobisbenzothiophene	$C_{18}H_{10}S_2$	290.4	4	
Aromatics	PHPN	$C_{35}H_{44}$	464.8	11	30.8
	DOCHN	$C_{30}H_{46}$	406.8	13	
Saturates	Squalane	$C_{30}H_{62}$	422.9	4	10.7
	Hopane	$C_{35}H_{62}$	483.0	4	
SBS	-	$C_{180}H_{192}$	2355.52	1	5
C_9PR	-	$C_{190}H_{202}$	2485.71	1	9

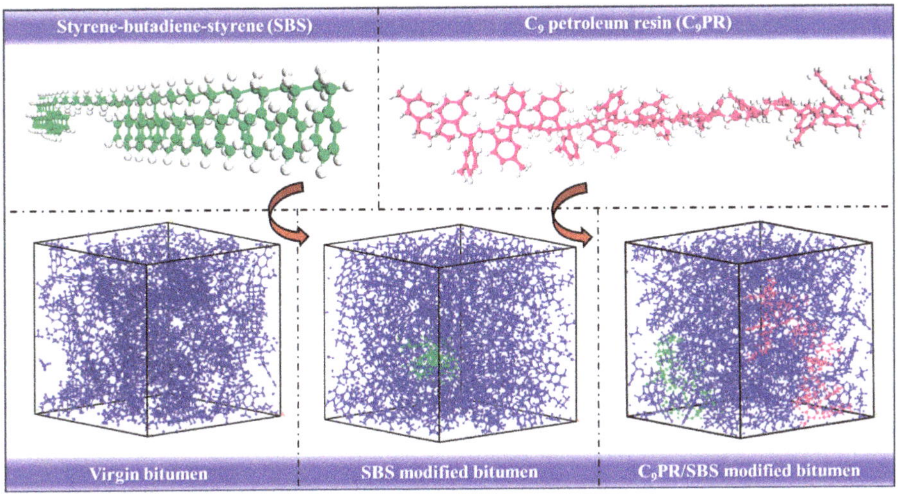

Figure 3. Construction of amorphous cell models for bitumen.

2.4.2. Model Validation

Density. In this study, density was utilized to validate the models for virgin bitumen, SBS modified asphalt, and C_9PR/SBS modified asphalt. The simulation results were used to generate density curves under various conditions, as illustrated in Figure 4. The figure indicates that the initial density of virgin bitumen is 0.1 g/cm³, which gradually approaches equilibrium starting from 200 ps, ultimately reaching 0.987 g/cm³. This value is close to the actual density range of virgin bitumen, from 1.01 to 1.04 g/cm³ [44], thus confirming the model's validity. Simulations for SBS modified asphalt showed that its initial density also starts at 0.1 g/cm³ and stabilizes at 1.005 g/cm³ after approximately 250 ps. For C_9PR /SBS modified asphalt, the initial density begins at 0.1 g/cm³ and reaches equilibrium at 1.002 g/cm³ after 300 ps. These results demonstrate the accuracy and practicality of the models in depicting the actual density characteristics of modified asphalts, further supporting the reliability of the simulation methods and models used in this research.

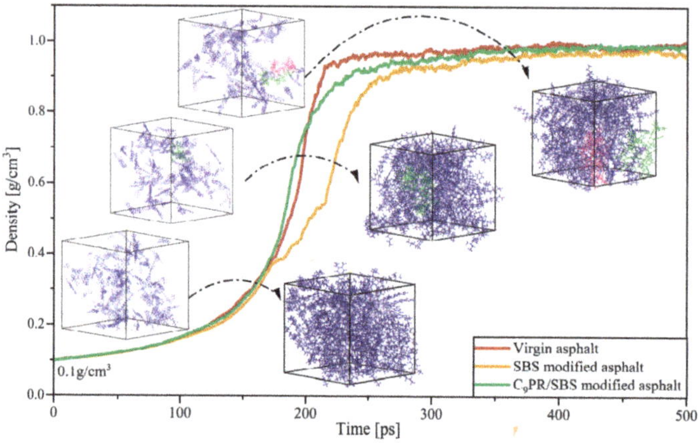

Figure 4. Density curve of bitumen models.

3. Results and Discussion

3.1. Evaluation of Macroscopic Performance of C_9PR in SBS Modified Bitumen

Rheological properties analysis at high temperature. The rheological properties of SBS modified bitumen and C_9PR/SBS modified bitumen were evaluated using the temperature scanning method, and the complex shear modulus (G^*) and phase angle (δ) were obtained at temperatures ranging from 46 °C to 82 °C. G^* is a measure of the total resistance of the material to repeated shearing, with a higher G^* indicating increased stiffness and improved resistance to flow deformation of the bitumen. For δ, as an indicator of the viscoelastic proportion of the bitumen, a smaller phase angle indicates a stronger resistance to high-temperature deformation of the bitumen. The rutting factor ($G^*/sin\delta$) is employed as an evaluation index for the high-temperature rutting resistance of the bitumen, and a larger $G^*/sin\delta$ indicates a stronger resistance to rutting at high temperatures [45].

As illustrated in Figure 5a,b, G^* and $G^*/sin\delta$ of SBS modified bitumen exhibit a sharp decrease as the temperature increases. This indicates that higher temperatures result in increased flowability and reduced high-temperature rutting resistance of SBS modified bitumen. However, the addition of C_9PR significantly enhances the high-temperature rutting resistance of G^* and $G^*/sin\delta$ of the modified bitumen, as evidenced by the significant increase in G^* and $G^*/sin\delta$. When the C_9PR content reaches 12 wt%, G^* and $G^*/sin\delta$ are increased by 60% and 57.3% at 46 °C, respectively, further enhancing the high-temperature rutting resistance. On the one hand, C_9PR promotes the swelling of SBS in the bitumen, reducing the light components in the bitumen and increasing the viscosity of the bitumen as a solvent. On the other hand, C_9PR, being a commonly used plasticizer, can increase the elasticity and ductility of the bitumen, which makes the instantaneous elastic recovery of C_9PR/SBS modified bitumen increase, and thus shows that the rutting factor becomes greater with the increase in C_9PR content. As can be seen from Figure 5c, the δ curve shows a gradual increase with the increase in temperature. The addition of C_9PR makes the δ of

SBS modified bitumen decrease, which results in a change in the ratio of elastic and viscous components within the bitumen. In summary, C_9PR can further improve the rheological properties of SBS modified bitumen and enhance its stability and durability under different temperature conditions.

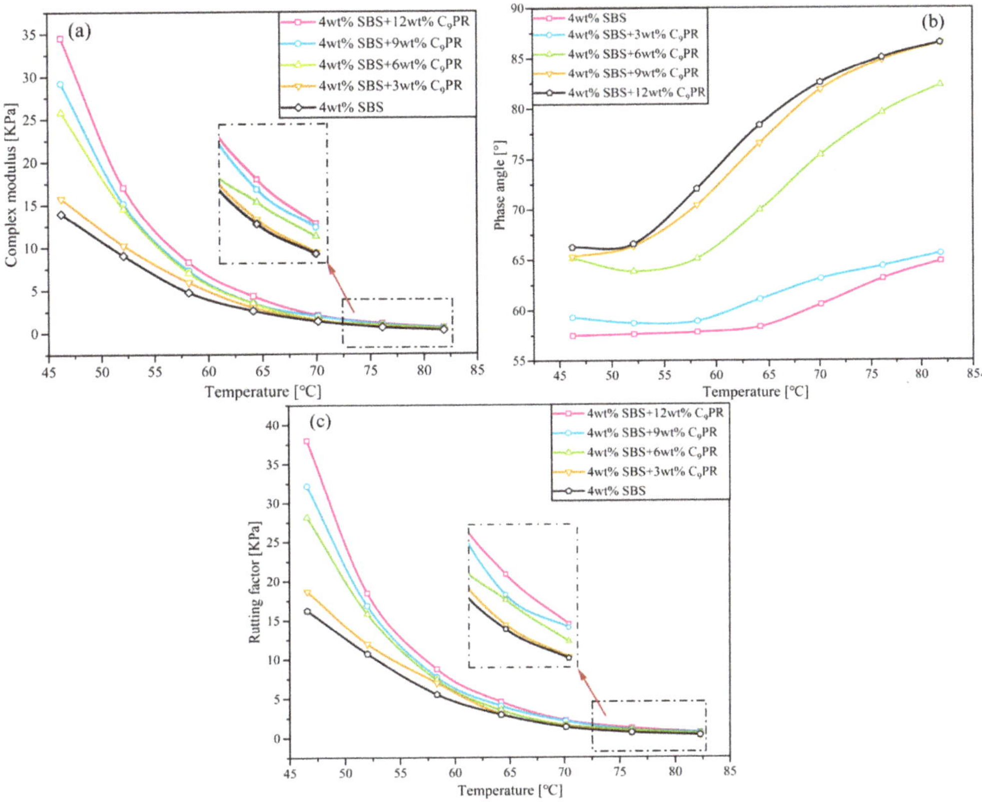

Figure 5. Rheological properties of modified bitumen: (**a**) complex modulus, (**b**) phase angle and (**c**) rutting factor.

Storage Stability Analysis. To investigate the effect of C_9PR content on the storage stability of C_9PR/SBS modified bitumen, five groups of samples were selected for high-temperature storage tests, as shown in Figure 6. Observing the SBS modified bitumen and C_9PR/SBS modified bitumen (C_9PR with 3 wt% content) in Figure 6, the softening point difference showed unstable fluctuations, which may be attributed to the fact that the C_9PR/SBS modified bitumen was not sufficiently sheared and miscible during the preparation process, and failed to achieve the role of promoting compatibility. With the increase in C_9PR content from 3 wt% to 12 wt%, the difference in softening point appeared to be significantly reduced, which also indicates that the addition of C_9PR gradually improved the compatibility of SBS and bitumen. This is due to the fact that C_9PR, as a product of petroleum distillation, has excellent compatibility with bitumen, which is beneficial for the preparation of modified bitumen with good storage stability. In addition, petroleum resin as a thermoplastic resin has good solubility and plasticity, while SBS as an elastomer has good elasticity and flexibility. Since their chemical structures are similar, C_9PR can interact with SBS, showing excellent compatibility.

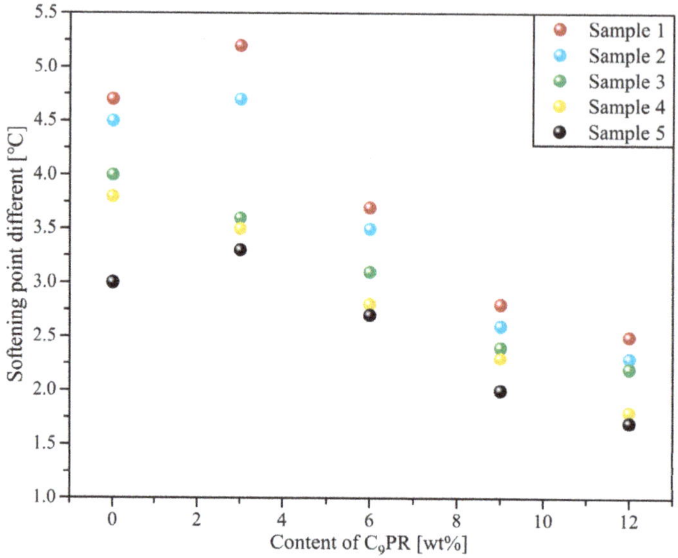

Figure 6. Storage stability results of modified bitumen samples.

Based on the results of the five samples in Figure 6, the average value of the difference in softening point of C_9PR/SBS modified bitumen with different contents was calculated, as shown in Table 4. The data in Table 4 show that the difference in softening point is less than 3 only when the content of C_9PR is 9 wt% and 12 wt%, which meets the requirement of storage stability for modified bitumen. This indicates that, although C_9PR has the potential to improve the compatibility between SBS and bitumen, the storage stability of C_9PR/SBS modified bitumen is sufficiently good when the C_9PR content reaches 9 wt% and 12 wt%, especially at 12 wt%.

Table 4. Average of softening point differences in modified bitumen samples.

C_9PR Content	0 wt%	3 wt%	6 wt%	9 wt%	12 wt%
Average values	4	4.06	3.16	2.42	2.1

3.2. Evaluation of Microscopic Morphologies of C_9PR on SBS Modified Bitumen

Functional groups characterization. FTIR spectroscopy was used to analyze the changes in bitumen functional groups after the addition of SBS and different contents of C_9PR (0 wt%, 3 wt%, 6 wt%, 9 wt%, 12 wt%). From the spectra, it can be observed that the SBS modified bitumen spectrum exhibits a strong absorption peak at 2924 cm^{-1}, which is attributed to the anti-symmetric stretching vibration of the methyl ethyl C=H bond. A peak at 2863 cm^{-1} corresponds to the symmetric stretching vibration of methylene groups, while a peak at 1603 cm^{-1} represents the stretching vibration of aromatic rings. Peaks at 1458 cm^{-1} and 1375 cm^{-1} correspond to the in-plane and out-of-plane bending vibrations of saturated C=H bonds (Figure 7).

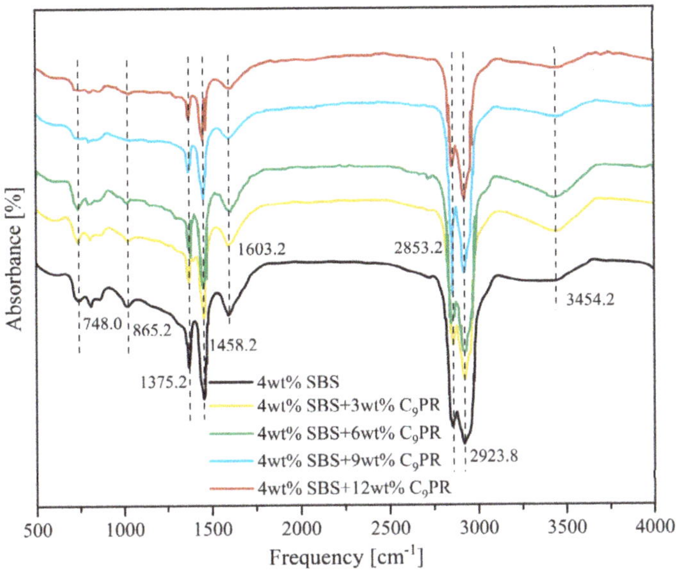

Figure 7. FTIR spectroscopy of modified bitumen.

Comparing the infrared spectra of SBS modified bitumen and C_9PR/SBS modified bitumen, it is found that the positions of the absorption peaks remain relatively unchanged. However, a relatively weak absorption peak appears at 1157 cm^{-1}, which is attributed to the stretching vibration of the C–O bond. The peak at 1458 cm^{-1} increases with the increase in C_9PR content, confirming that C_9PR can effectively improve the compatibility of bitumen. Additionally, the introduction of different contents of C_9PR (0 wt%, 3 wt%, 6 wt%, 9 wt%, 12 wt%) alters the intensity of multiple absorption peaks at 1375 cm^{-1}, 1458 cm^{-1}, and 2923 cm^{-1}, indicating the superposition or weakening effect of C_9PR on SBS. Therefore, it can be inferred that the modification of SBS modified bitumen with C_9PR is a physical modification.

Interfacial morphologies characterization. The interfacial morphologies of SBS and bitumen before and after C_9PR modification were observed using SEM, and the comparison images are shown in Figure 8. From Figure 8a, it can be seen that there are large agglomerates of SBS polymer particles, and bitumen in SBS modified bitumen, and the number of agglomerates is high. The surface between SBS and bitumen is also clearly observed, indicating that SBS is not uniformly dispersed in the bitumen, with poor compatibility, and prone to phase separation. When the interface is subjected to force and water damage, the adhesion and mechanical properties will sharply decrease. On the contrary, the incorporation of C_9PR changed the morphology of the interface between SBS and bitumen, as shown in Figure 8b. Comparing with Figure 8a, under the dispersing effect of C_9PR, SBS particles are more uniformly dispersed with minimal agglomeration. This is manifested by a smoother and flatter bitumen surface, indicating an improvement in the compatibility between SBS and bitumen and an enhancement of the adhesion between the interfaces. In summary, a stable and uniform interface relationship can improve the adhesion and compatibility between SBS and bitumen.

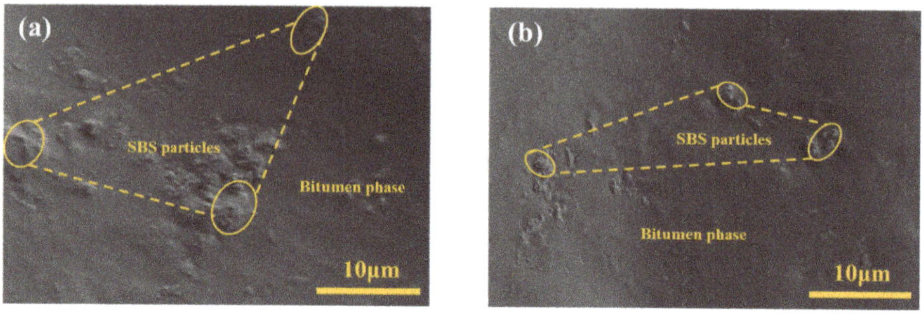

Figure 8. Interfacial morphologies of SBS and bitumen: (**a**) SBS modified asphalt and (**b**) C_9PR/SBS modified asphalt.

Dispersed phase structure characterization. Fluorescence microscopy was used to observe the phase distribution of SBS with bitumen at different contents of C_9PR (0 wt%, 3 wt%, 6 wt%, 9 wt%, 12 wt%), as shown in Figure 9. Figure 9a–e shows that SBS is distributed in the bitumen phase in a micro-filamentary form. When the C_9PR content is 0 wt%, the distribution of SBS in the bitumen phase is minimal and the interface is obvious. With the addition of C_9PR, the distribution of SBS gradually increases, and the interface between SBS and bitumen gradually tends to be blurred. This phenomenon promotes the dispersion of SBS in bitumen more uniformly, and gradually forms a three-dimensional network structure.

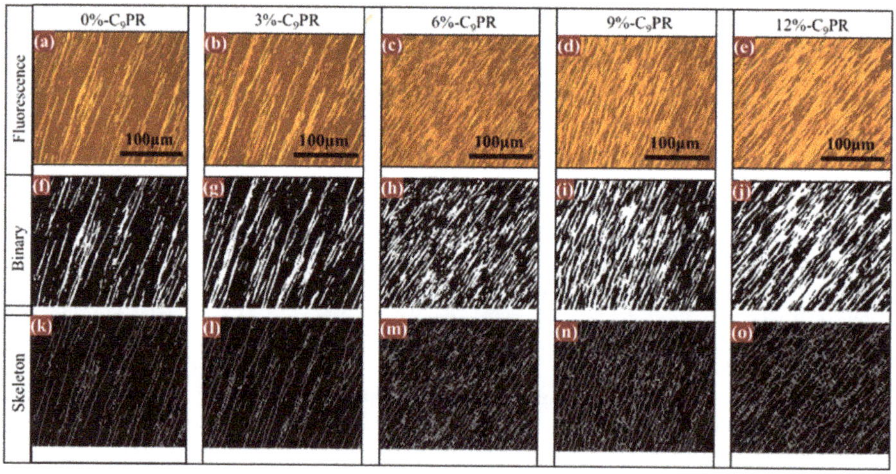

Figure 9. Microscopic morphologies of SBS and modified bitumen: (**a–e**) fluorescence, (**f–j**) binary and (**k–o**) skeleton.

To quantitatively analyze the effect of C_9PR on SBS modified bitumen, the fluorescence microscope images were binarized using MATLAB software and their skeleton images were obtained. Then, the Image J software was used for measurements and calculations, and the results of image processing and measurement are shown in Figure 9 and Table 5. When C_9PR content was 0 wt%, SBS dissolved in bitumen showed short and discontinuous molecular chains, which was the reason for the uplift and segregation of SBS during storage and the poor modification effect. When the C_9PR content was 3 wt%, SBS continued to swell and form partially continuous molecular chains; however, there was no obvious entanglement between the molecular chains. When the content reached 6 wt%, the entanglement phenomenon between SBS molecular chains was extremely obvious. As the content of C_9PR continued to increase, SBS molecular chains appeared in numerous short-branched chains, and the degree of swelling increased significantly. This is attributed to the high polarity and active groups of C_9PR, which can undergo cross-linking reactions with the active groups in SBS to form a network structure and improve the compatibility.

Table 5. Area and skeleton length of SBS.

C_9PR Content	0 wt%	3 wt%	6 wt%	9 wt%	12 wt%
Area of SBS	40,398	41,845	45,861	48,298	49,954
Skeleton length of SBS	295	316	457	674	717

For the same SBS content, the difference in the state after swelling is very large due to the different particle sizes after shearing. As far as the SBS area is concerned, the maximum value (C_9PR content of 12 wt%) increased by 23.65% compared to the minimum value (C_9PR content of 0 wt%). Meanwhile, the maximum SBS skeleton length (C_9PR content of 12 wt%) is 2.43 times that of the minimum value (C_9PR content of 0 wt%). The results show that, the larger the C_9PR content, the longer the SBS skeleton length, and the easier it is to cross-link and form a network structure. When the C_9PR content is less than 9 wt%, the increase in SBS area and skeleton length is larger. On the contrary, when the C_9PR content is higher than 9 wt%, the increase tends to level off. Therefore, it can be considered that C_9PR doping at 9 wt% is optimal.

3.3. Analysis of Interaction Behavior between C_9PR and SBS Modified Bitumen

Diffusion behavior analysis. Mean square displacement is an average and unique indicator that describes the movement of molecules within a certain period, reflecting the changes in the motion properties and kinetic behavior of molecules in bitumen [46]. The MSD curve of bitumen's four components and SBS molecules is obtained by calculating the average of the unique squares within different time intervals using Equation (1), as shown in Figure 10 [47]. The slope of the MSD curve can reflect the diffusion of molecules, with a larger slope indicating faster molecular diffusion, as shown in Figure 11.

$$MSD(t) = \left\langle |r_i(t) - r_i(0)|^2 \right\rangle \tag{1}$$

$$D = \frac{1}{6}\lim_{t\to\infty}\frac{d}{dt}\sum_{i=1}^{N}\left\langle |r_i(t) - r_i(0)|^2 \right\rangle_i \tag{2}$$

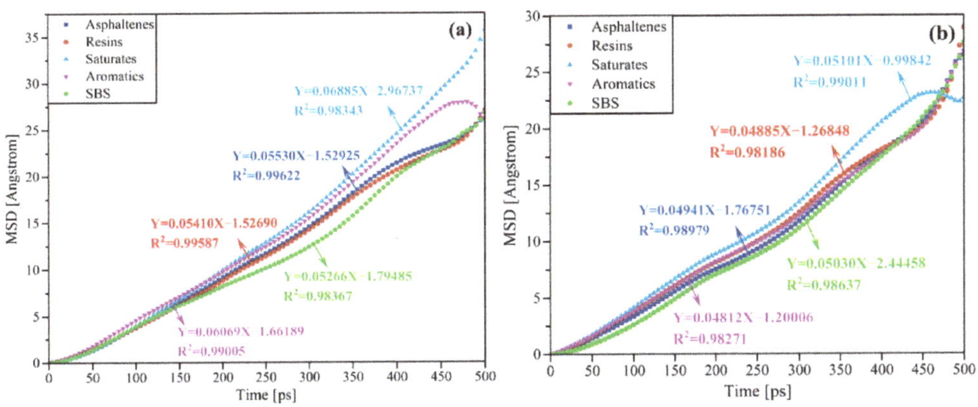

Figure 10. MSD and linear fitting curves of SBS and four components: (**a**) SBS modified asphalt and (**b**) C9PR/SBS modified asphalt.

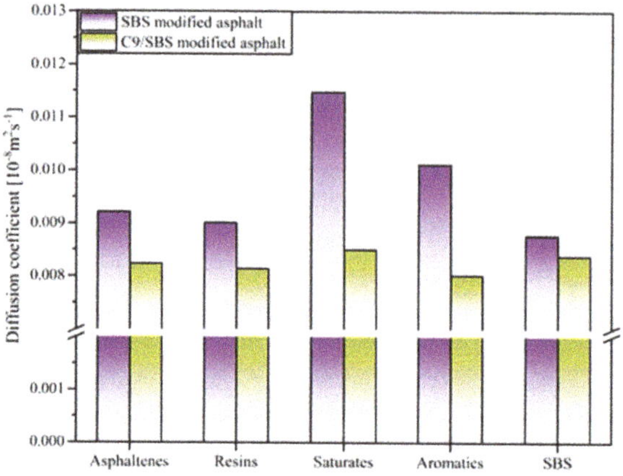

Figure 11. Diffusion coefficient of SBS and four components.

In Equation (1), ⟨⟩ denotes the average value, $r_i(t)$ denotes the position vector of particle the at time t, N and D denote the number of particles in the system and the molecular diffusion coefficient, respectively, while d/dt denotes the differentiation of the equation, $MSD(t)$ represents the mean square displacement curve, and D is the diffusion coefficient.

The figures show the MSD values of bitumen, colloid, aromatic fraction, and saturated fraction in SBS modified bitumen and C_9PR/SBS modified bitumen systems, respectively. The curves exhibit a rapid increase in the initial stage of simulation, and the MSD values in the C_9PR/SBS modified bitumen are greater than those in the SBS modified bitumen. In the second stage of simulation, the diffusion rate gradually stabilizes, and the MSD values in the SBS modified bitumen are greater than those in the C_9PR/SBS modified bitumen. Initially, the four components of bitumen and SBS molecules rapidly come into contact, filling the pores. Subsequently, under the influence of van der Waals forces and electric field forces, the molecules continue to diffuse, reaching a state of equilibrium in molecular interactions. The changes in the MSD curves indicate that the addition of C_9PR makes the SBS modified bitumen system more stable.

According to the fitting results of the MSD curves in Figure 10, the correlation coefficients (R^2) of the four components in the SBS modified bitumen and C9PR/SBS modified bitumen systems are all greater than 0.99, indicating a high linear correlation of the MSD curves and that the slope can satisfy the accuracy of diffusion coefficients. Comparing the diffusion results of the four components in the SBS modified bitumen and C_9PR/SBS modified bitumen, it is found that the addition of C_9PR reduces the diffusion coefficients of the four components, indicating faster diffusion of the four component molecules in the SBS modified bitumen. However, SBS molecules exhibit the opposite diffusion trend, with a diffusion coefficient of 2.194×10^{-8} $m^2 \cdot s^{-1}$ in SBS modified bitumen and 0.237×10^{-8} $m^2 \cdot s^{-1}$ in C_9PR/SBS modified bitumen, reaching the maximum diffusion coefficient in the C_9PR/SBS modified bitumen. The four components and SBS molecules interact with each other during the diffusion process, and the addition of C_9PR enhances the interaction strength between SBS and the four components of bitumen, thereby restraining the diffusion motion of the four components. Meanwhile, SBS molecules move closer to each component of bitumen during the diffusion process. This also verified the results according to which the elastic components of bitumen decreased and the viscous components gradually became dominant in the macroscopic experiment.

Aggregation behavior analysis. According to the RDF curve of SBS modified bitumen in Figure 12, both SBS and bitumen components exhibit peaks between 1.03 Å and 1.19 Å. The peak for SBS is at (1.09 Å, 139.65%), the peak for asphaltene is at (1.11 Å, 59.58%), the peak for resin is at (1.11 Å, 18.31%), the peak for aromatic fraction is at (1.11 Å, 30.18%), and the peak for saturates is at (1.09 Å, 77.44%). Figure 12 shows the RDF curve of

SBS and bitumen components after the addition of C_9PR, and the peaks remain within the range from 1.03 Å to 1.19 Å. The peak for SBS is at (1.09 Å, 154.40%), the peak for asphaltene is at (1.11 Å, 61.32%), the peak for resin is at (1.11 Å, 30.49%), the peak for aromatic fraction is at (1.09 Å, 30.34%), and the peak for saturates is at (1.09 Å, 91.62%). Compared to SBS modified bitumen, the peaks of SBS and bitumen components are enlarged, indicating an enhancement in molecular order. This result is attributed to the cross-linking structure formed by C_9PR and SBS, which maintains stability even at high temperatures and alleviates disorderly movement. From a macroscopic perspective, the increased order of SBS and bitumen components significantly improves the temperature sensitivity of bitumen, allowing it to remain stable at high temperatures while maintaining flexibility at low temperatures. Additionally, the peaks of SBS and bitumen components do not shift, indicating that C_9PR does not affect the aggregation of SBS modified bitumen. Comparing the changes in peak values, the aromatic fraction increases by 0.16% and the saturates increase by 14.18%, which also suggests that the addition of C_9PR effectively supplements the light components in bitumen, increasing their proportion and enhancing system stability.

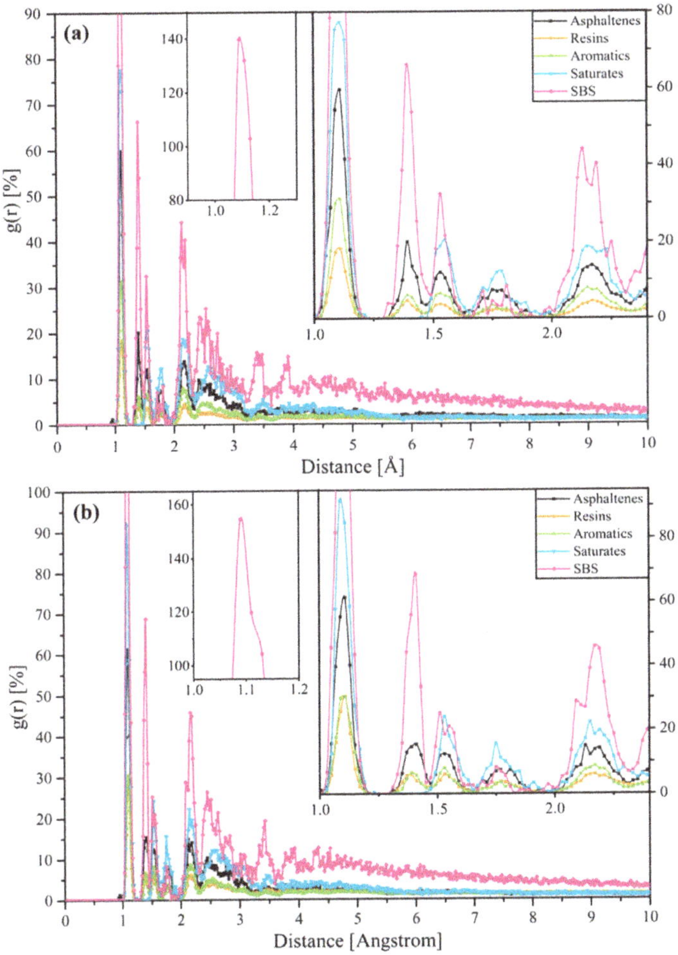

Figure 12. RDF of SBS and four components: (**a**) SBS modified bitumen and (**b**) C_9PR/SBS modified bitumen.

Concentration distribution analysis. The relative concentration distribution can deeply analyze the spatial distribution of molecules in the system. Figure 13 shows the distribution of four components of SBS modified bitumen and C_9PR/SBS modified bitumen system, as well as the distribution of SBS molecules. In the SBS modified bitumen system, asphaltene, colloid, and aromatic components exhibit relatively uniform distribution, while saturated components and SBS show varying degrees of aggregation in different directions (X, Y, Z). Among them, saturated components and SBS molecules always maintain adjacent aggregation states. The aggregation peak of saturated components appears near the coordinates (11, 6, 23), and SBS shows strong aggregation near (23 ± 7, 14 ± 5, 28 ± 5), which is the adsorption effect of SBS on saturated components. Similarly, when saturated components and aromatic components show strong aggregation behavior, the aggregation behavior of asphaltene weakens. When asphaltene shows a strong aggregation peak, the aggregation peaks of saturated components and aromatic components are smaller, which also demonstrates the competition between SBS and asphaltene for light components.

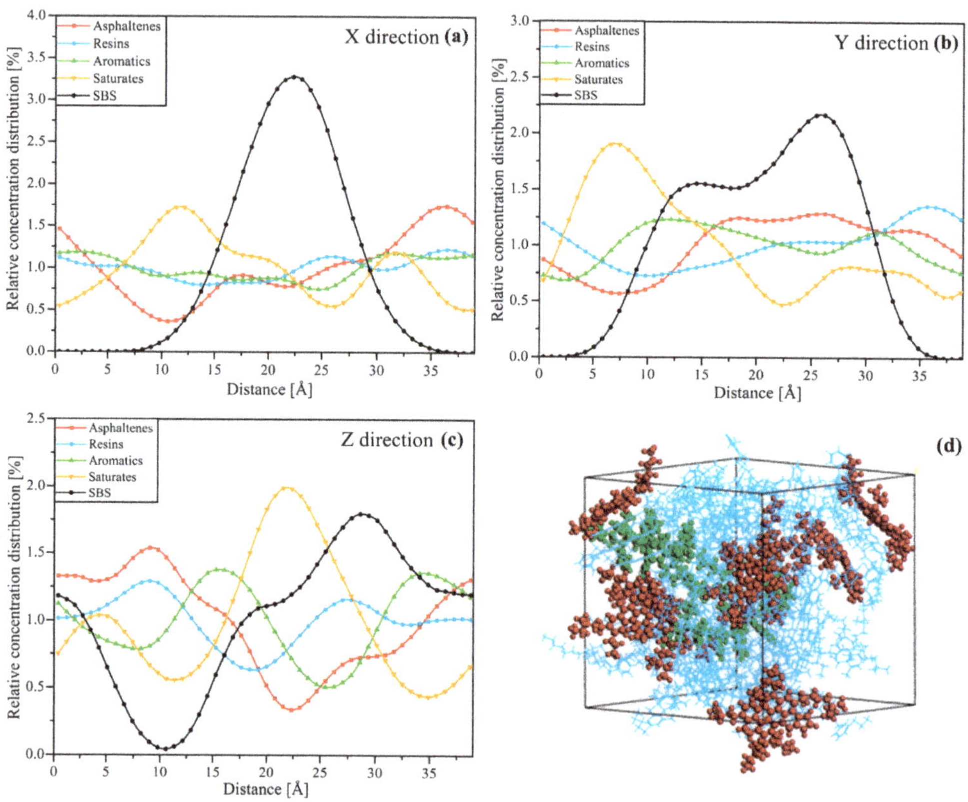

Figure 13. Relative concentration distribution: (**a**) X direction, (**b**) Y direction, (**c**) X direction and molecular morphology (**d**) of SBS modified bitumen.

By comparing the concentration distribution of the four components and SBS in the C_9PR/SBS modified bitumen system with Figure 14, significant changes can be observed, especially the significant fluctuations in the colloid curve, indicating strong aggregation behavior. SBS and saturated components still maintain strong adjacent aggregation behavior, and the peak values of asphaltene are enhanced in the X and Z directions. Comparing the average upper and lower limits of the relative concentrations of the four components in the C_9PR/SBS modified bitumen system in the three directions, the average values of asphaltene, colloid, aromatic components, and saturated components are 1.05%, 1.02%, 1.03%, and 1.17%, respectively. However, for SBS modified bitumen, these values are 0.97%, 0.95%, 0.96%, and 1.17%, respectively. The addition of C_9PR increases the peak values of the four components of bitumen, especially the peak of heavy components. At the same time, compared with SBS modified bitumen, the average peak value decreases by 0.12%, which is due to the further promotion of SBS by C_9PR in competing for light components and enhancing adsorption capacity. This result is consistent with the conclusion of RDF analysis.

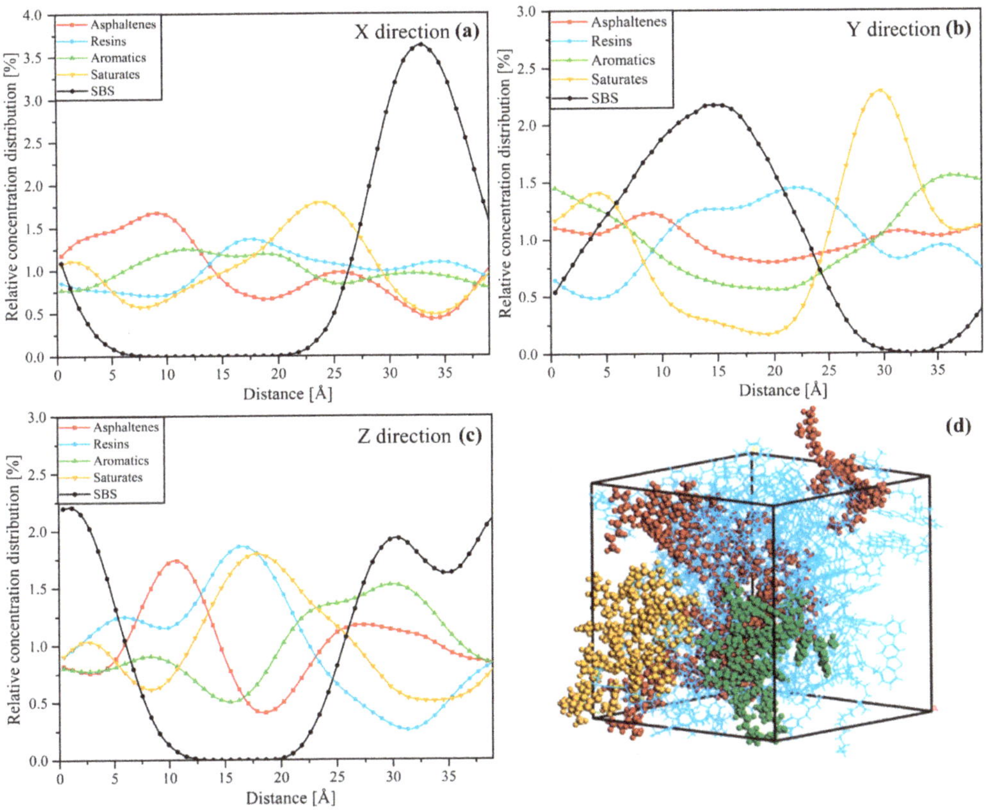

Figure 14. Relative concentration distribution: (**a**) X direction, (**b**) Y direction, (**c**) X direction and molecular morphology (**d**) of C_9PR/SBS modified bitumen.

4. Conclusions

This study has substantially enhanced our understanding of the influence of C_9PR on SBS modified asphalt, elucidating its effects on rheological, microstructural, and chemical properties through a detailed examination using DSR, storage stability tests, fluorescence microscopy, SEM, FTIR, and molecular dynamic simulations. The results highlight C_9PR's significant improvements in the performance of SBS modified asphalt, crucial for developing advanced pavement materials.

(1) A key finding from this research is C_9PR's role in optimizing the balance between the elastic and viscous components of the asphalt. This optimization enhances the material's stability and resistance to deformation under varying temperatures, crucial for the durability of pavement surfaces. Notably, a C_9PR content of 12 wt% was found to be optimal for maximizing storage stability. This concentration enhances the

compatibility of SBS and bitumen without causing significant chemical reactions, as confirmed by FTIR analysis.
(2) Microscopic evaluations revealed that C_9PR positively affects the distribution and adhesion of SBS within the bitumen matrix. Adding C_9PR leads to a more uniform dispersion of SBS, smoothing the interface between SBS and bitumen and enhancing material cohesion and adhesion. This improvement marks significant progress in developing asphalt composites with superior mechanical properties and longevity.
(3) MD*-simulations offered deeper insights into the molecular interactions facilitating these enhancements. The inclusion of C_9PR limits the diffusion of bitumen molecules, suggesting a stronger intermolecular force. This mechanism indicates that C_9PR improves stability by promoting a competitive absorption process, aiding the integration of SBS into the bitumen and enhancing the overall performance of the composite.
(4) This research conclusively shows that integrating C_9PR into SBS modified asphalt is an effective strategy to improve the material's performance across various metrics, including low-temperature properties, storage stability, and temperature susceptibility. The combination of empirical testing and molecular simulation has validated the benefits of C_9PR and laid a strong foundation for its future use in pavement engineering. The outcomes of this study are expected to guide the development of more advanced asphalt materials, contributing to the creation of more durable, sustainable, and high-performing road infrastructures.

Author Contributions: Conceptualization: C.Y. and T.Z.; methodology, C.Y.; software, T.Z.; validation, S.T.A.G.; formal analysis, S.T.A.G.; investigation, K.H. and T.Z.; data curation, S.T.A.G.; writing—original draft preparation, C.Y. and T.Z.; writing—review and editing, K.H.; visualization, W.Z.; supervision, W.Z.; project administration, K.H.; funding acquisition, K.H. All authors have read and agreed to the published version of the manuscript.

Funding: This research was funded by Fundamental Research Fund of Natural Science Foundation of China [52208434], the Fundamental Research Funds for the Central Universities, CHD [300102213507], Yangling Vocational & Technical College 2023 Science and Technology Innovation Key Projects [ZKZ23-04], Natural Science Foundation of Henan Province [222300420142], Postdoctoral Re-search Grant in Henan Province [202103107], Science and Technology Project of Henan Province [192102310229], China Postdoctoral Science Foundation [2022M711079], Young backbone teachers plan of Henan University of Technology [21420156], Open Fund of National Engineering Research Center of Highway Maintenance Technology (Changsha University of Science and Technology) [kfj230206], Tianshan leading talents in scientific and technological innovation [2022TSYCLJ0045].

Data Availability Statement: The original contributions presented in the study are included in the article, further inquiries can be directed to the corresponding author.

Conflicts of Interest: The authors declare no conflicts of interest.

References

1. Joohari, I.B.; Giustozzi, F. Chemical and high-temperature rheological properties of recycled plastics-polymer modified hybrid bitumen. *J. Clean. Prod.* **2020**, *276*, 123064. [CrossRef]
2. Xu, J.; Fan, Z.; Lin, J.; Liu, P.; Wang, D.; Oeser, M. Study on the effects of reversible aging on the low temperature performance of asphalt binders. *Constr. Build. Mater.* **2021**, *295*, 123604. [CrossRef]
3. Yao, H.; Liu, J.; Xu, M.; Ji, J.; Dai, Q.; You, Z. Discussion on molecular dynamics (MD) simulations of the materials. *Adv. Colloid Interface Sci.* **2022**, *299*, 102565. [CrossRef] [PubMed]
4. Hassanpour-Kasanagh, S.; Ahmedzade, P.; Fainleib, A.M.; Behnood, A. Rheological properties of asphalt binders modified with recycled materials: A comparison with Styrene-Butadiene-Styrene (SBS). *Constr. Build. Mater.* **2020**, *230*, 117047. [CrossRef]
5. Yin, W.; Ye, F.; Lu, H. Establishment and experimental verification of stability evaluation model for SBS modified asphalt: Based on quantitative analysis of microstructure. *Constr. Build. Mater.* **2021**, *131*, 291–302. [CrossRef]
6. Xu, X.; Sreeram, A.; Leng, Z.; Yu, J.; Li, R.; Peng, C. Challenges and opportunities in the high-quality rejuvenation of unmodified and SBS modified asphalt mixtures: State of the art. *J. Clean. Prod.* **2022**, *378*, 134634. [CrossRef]
7. Azizi, H.; Faghihi, J. Effect of nanoclay on mechanical and thermal properties of triblock SBS (styrene–butadiene–styrene) and its binary blend nanocomposites: Comparison between polyethylene and polystyrene. *Plast. Rubber Compos.* **2018**, *47*, 429–437. [CrossRef]

8. Jiang, Z.; Easa, S.M.; Hu, C.; Zheng, X. Evaluation of new aspect of styrene-butadiene-styrene modified bitumens: Damping property and mechanism. *Constr. Build. Mater.* **2020**, *242*, 118185. [CrossRef]
9. Xu, X.; Chen, G.; Wu, Q.; Leng, Z.; Chen, X.; Zhai, Y.; Tu, Y.; Peng, C. Chemical upcycling of waste PET into sustainable asphalt pavement containing recycled concrete aggregates: Insight into moisture-induced damage. *Constr. Build. Mater.* **2022**, *360*, 129632. [CrossRef]
10. Liu, Z.; Gu, X.; Dong, X.; Cui, B.; Hu, D. Mechanism and performance of graphene modified asphalt: An experimental approach combined with molecular dynamic simulations. *Case. Studi. Constr. Mater.* **2023**, *18*, e01749. [CrossRef]
11. Huang, T.; Cao, Z.; Lv, S.; Yu, H.; Peng, X.; Jiang, H.; Wu, L. Road behavior and micro characteristics of high-performance asphalt based on compound modification approach. *Constr. Build. Mater.* **2023**, *371*, 130622. [CrossRef]
12. Wei, Y.; Hu, C.; Muhammad, Y.; Chen, L.; Zhou, D.; Wang, S.; Li, J.; Chen, Q. Fabrication and performance evaluation of aminopropyl triethoxysilane-dopamine-MoS2 incorporated SBS modified asphalt. *Constr. Build. Mater.* **2020**, *265*, 120346. [CrossRef]
13. Zhang, W.; Qiu, L.; Liu, J.; Hu, K.; Zou, L.; Chen, Y.; Yang, C.; Wang, F.; Zang, J. Modification mechanism of C9 petroleum resin and its influence on SBS modified asphalt. *Constr. Build. Mater.* **2021**, *306*, 124740. [CrossRef]
14. Nie, X.; Hou, T.; Yao, H.; Li, Z.; Zhou, X.; Li, C. Effect of C9 petroleum resins on improvement in compatibility and properties of SBS-modified asphalt. *Petrol. Sci. Technol.* **2019**, *306*, 124740. [CrossRef]
15. Yaphary, Y.L.; Leng, Z.; Wang, H.; Ren, S.; Lu, G. Characterization of nanoscale cracking at the interface between virgin and aged asphalt binders based on molecular dynamics simulations. *Constr. Build. Mater.* **2022**, *335*, 127475. [CrossRef]
16. Yao, L.; Leng, Z.; Lan, J.; Chen, R.; Jiang, J. Environmental and economic assessment of collective recycling waste plastic and reclaimed asphalt pavement into pavement construction: A case study in Hong Kong. *J. Clean. Prod.* **2022**, *336*, 130405. [CrossRef]
17. Ma, F.; Li, C.; Fu, Z.; Huang, Y.; Dai, J.; Feng, Q. Evaluation of high temperature rheological performance of polyphosphoric acid-SBS and polyphosphoric acid-crumb rubber modified asphalt. *Constr. Build. Mater.* **2021**, *306*, 124926. [CrossRef]
18. Xu, G.; Yao, Y.; Ma, T.; Hao, S.; Ni, B. Experimental study and molecular simulation on regeneration feasibility of high-content waste tire crumb rubber modified asphalt. *Constr. Build. Mater.* **2023**, *369*, 130570. [CrossRef]
19. Tang, Y.; Fu, Z.; Ma, F.; Liu, J.; Sun, Q.; Li, C. Carbon nanotubes for improving rheological and chemical properties of styrene–butadiene–styrene modified asphalt binder. *Int. J. Pavement Eng.* **2023**, *24*, 2211212. [CrossRef]
20. Wang, P.; Zhai, F.; Dong, Z.-J.; Wang, L.-Z.; Liao, J.-P.; Li, G.-R. Micromorphology of asphalt modified by polymer and carbon nanotubes through molecular dynamics simulation and experiments: Role of Strengthened Interfacial Interactions, Energy. *Fuels* **2018**, *32*, 1179–1187. [CrossRef]
21. Zhao, P.; Song, X.; Dong, M.; Sun, H.; Wu, W.; Zhang, R.; Sun, M.; Zhao, X. Preparation and characterization of CQDs/SBS composites and its application performance as asphalt modifier. *Constr. Build. Mater.* **2022**, *320*, 126312. [CrossRef]
22. Liu, S.; Zhou, S.B.; Peng, A.; Xuan, W.; Li, W. Analysis of the performance and mechanism of desulfurized rubber and low-density polyethylene compound-modified asphalt. *J. Appl. Polym. Sci.* **2019**, *136*, 48194. [CrossRef]
23. Zhang, W.; Zou, L.; Wang, Y.; Liu, J.; Yang, C.; Di, J.; Hu, H.; Yang, Z. Influence of high viscosity petroleum resin (HV-PR) on the intermediate and high temperature performances of Styrene–Butadiene–Styrene block copolymer (SBS) modified bitumen. *Arab. J. Sci. Eng.* **2022**, *47*, 12521–12533. [CrossRef]
24. Zhang, F.; Kaloush, K.; Underwood, S.; Hu, C. Preparation and performances of SBS compound modified asphalt mixture by acidification and vulcanization. *Constr. Build. Mater.* **2023**, *296*, 123693. [CrossRef]
25. Zhang, J.; Huang, W.; Zhang, Y.; Yan, C.; Lv, Q.; Guan, W. Evaluation of the terminal blend crumb rubber/SBS composite modified asphalt. *Constr. Build. Mater.* **2021**, *278*, 122377. [CrossRef]
26. Yang, Q.; Yu, C. Deterioration effects of oxidative aging on graphene-asphalt nanocomposite interfaces: Multiscale Modeling. *Langmuir* **2023**, *39*, 8339–8353. [CrossRef] [PubMed]
27. Roja, K.L.; Rehman, A.; Ouederni, M.; Krishnamoorthy, S.K.; Abdala, A.; Masad, E. Influence of polymer structure and amount on microstructure and properties of polyethylene-modified asphalt binders. *Mater. Struct.* **2021**, *54*, 2. [CrossRef]
28. Xu, J.; Sun, L.; Pei, J.; Xue, B.; Liu, T.; Li, R. Microstructural, chemical and rheological evaluation on oxidative aging effect of SBS polymer modified asphalt. *Road Constr. Build. Mater.* **2021**, *267*, 121028. [CrossRef]
29. Li, J.; Jiang, H.; Han, F.; Lin, Z.; Zhao, Z.; Jin, X.; Liu, Y. Investigation of the effects of chemical modification and oxidative aging on the properties and compatibility of rubber asphalt based on thermodynamic principles. *J. Clean. Prod.* **2023**, *428*, 139070. [CrossRef]
30. Zhou, Y.; Peng, Z.-C.; Huang, J.-L.; Ma, T.; Huang, X.-M.; Miao, C.-W. A molecular dynamics study of calcium silicate hydrates-aggregate interfacial interactions and influence of moisture. *Appl. Surf. Sci.* **2021**, *28*, 16–28. [CrossRef]
31. Cui, B.; Wang, H. Molecular modeling of asphalt-aggregate debonding potential under moisture environment and interface defect. *Appl. Surf. Sci.* **2022**, *606*, 154858. [CrossRef]
32. Fan, J.; Zhu, Y.; Ma, T.; Xu, G.; Ding, X. Multiscale study on aging process and mechanism of high viscosity Asphalt: Macro-micro experiments and quantum chemical molecular simulation. *Constr. Build. Mater.* **2023**, *402*, 133059. [CrossRef]
33. Yu, C.; Yang, Q. Investigation of the interfacial interaction of carbon nanomaterials with asphalt matrix: Insights from molecular simulations. *Mol. Simulat.* **2022**, *49*, 208–222. [CrossRef]
34. Wang, Y.; Lin, D.; Xiang, M.; Cui, M.; Liu, N. Experimental study on aging performance of polyethylene gas pipelines. *IOP Conf. Ser. Earth Environ. Sci.* **2021**, *631*, 012066. [CrossRef]

35. Li, M.; Min, Z.; Wang, Q.; Huang, W.; Shi, Z. Effect of epoxy resin content and conversion rate on the compatibility and component distribution of epoxy asphalt: A MD simulation study. *Constr. Build. Mater.* **2021**, *319*, 126050. [CrossRef]
36. Shi, K.; Fu, Z.; Liu, J.; Song, R.; Ma, F.; Dai, J.; Li, C.; Wen, Y. Multiscale investigation of waste soybean oil rejuvenated asphalt binder utilising experimental methodologies and molecular dynamics simulations. *Int. J. Pavement Eng.* **2023**, *24*, 2181961. [CrossRef]
37. *ASTM D7175*; Standard Test Method for Determining the Rheological Properties of Asphalt Binder Using a Dynamic Shear Rheometer. ASTM: West Conshohocken, PA, USA, 2005.
38. Qian, C.; Fan, W.; Yang, G.; Han, L.; Xing, B.; Lv, X. Influence of crumb rubber particle size and SBS structure on properties of CR/SBS composite modified asphalt. *Constr. Build. Mater.* **2020**, *235*, 117517. [CrossRef]
39. Du, Z.; Zhu, X.; Zhang, Y. Diffusive dynamics and structural organization of moisture in asphaltic materials based on molecular dynamics simulation. *J. Mater. Civil. Eng.* **2021**, *33*, 04020403. [CrossRef]
40. Li, D.D.; Greenfield, M.L. Chemical compositions of improved model asphalt systems for molecular simulations. *Fuel* **2014**, *115*, 347–356. [CrossRef]
41. Jiao, B.; Pan, B.; Che, T. Evaluating impacts of desulfurization and depolymerization on thermodynamics properties of crumb rubber modified asphalt through molecular dynamics simulation. *Constr. Build. Mater.* **2022**, *323*, 126360. [CrossRef]
42. Cui, W.; Huang, W.; Hassan, H.M.Z.; Cai, X.; Wu, K. Study on the interfacial contact behavior of carbon nanotubes and asphalt binders and adhesion energy of modified asphalt on aggregate surface by using molecular dynamics simulation. *Constr. Build. Mater.* **2022**, *316*, 125849. [CrossRef]
43. Cui, B.; Wang, H. Molecular interaction of Asphalt-Aggregate interface modified by silane coupling agents at dry and wet conditions. *Appl. Surf. Sci.* **2022**, *572*, 151365. [CrossRef]
44. Long, Z.; You, L.; Tang, X.; Ma, W.; Ding, Y.; Xu, F. Analysis of interfacial adhesion properties of nano-silica modified asphalt mixtures using molecular dynamics simulation. *Constr. Build. Mater.* **2020**, *255*, 119354. [CrossRef]
45. Yu, C.; Hu, K.; Yang, Q.; Chen, Y. Multi–scale observation of oxidative aging on the enhancement of high–temperature property of SBS–modified asphalt. *Constr. Build. Mater.* **2021**, *313*, 125478. [CrossRef]
46. Yu, J.; Zheng, H.; Hou, D.; Zhang, J.; Xu, W. Silane coupling agent modification treatment to improve the properties of rubber–cement composites. *ACS Sustain. Chem. Eng.* **2021**, *9*, 12899–12911. [CrossRef]
47. Luo, L.; Chu, L.; Fwa, T.F. Molecular dynamics analysis of oxidative aging effects on thermodynamic and interfacial bonding properties of asphalt mixtures. *Constr. Build. Mater.* **2021**, *269*, 121299. [CrossRef]

buildings

MDPI

Article

Modeling and Assessment of Temperature and Thermal Stress Field of Asphalt Pavement on the Tibetan Plateau

Bin Li [1], Yadong Xie [1], Yanqiu Bi [1], Xiaoling Zou [1,*], Fafu Tian [1] and Zhimin Cong [2]

[1] School of Civil Engineering, Chongqing Jiaotong University, Chongqing 400074, China; bin2020@cqjtu.edu.cn (B.L.); 622210111002@mails.cqjtu.edu.cn (Y.X.); biyanqiu@cqjtu.edu.cn (Y.B.); 622190951050@mails.cqjtu.edu.cn (F.T.)

[2] Shanghai Municipal Engineering Design Institute (Group) Co., Ltd., Shanghai 200092, China; congzhimin@smedi.com

* Correspondence: zouxiaoling@cqjtu.edu.cn

Abstract: The Qinghai–Tibet Plateau (QTP) is the highest altitude plateau in the world, characterized by strong solar radiation and large diurnal temperature differences and so on, which brings a great negative impact on the temperature and thermal stress field of asphalt pavement. The purpose of this study is to analyze the temperature field and thermal stress status of asphalt pavement in the QTP to provide a reference for pavement design and maintenance in high-altitude areas. The finite element method was applied to establish the temperature field model to study the distribution and variation of pavement temperature. On this basis, the influence of cooling amplitude on pavement thermal stress was studied during cold waves. In addition to this, the key internal factors affecting the thermal stress of pavement, such as surface thickness, surface temperature shrinkage coefficient, surface modulus, and base modulus, were analyzed by an orthogonal test. It was found that temperature and solar radiation have a significant effect on the pavement temperature field. When the cold wave came, the cooling rate had a considerable impact on the thermal stress of the pavement, that is, every 5 °C increase in cooling rate would increase the thermal stress by more than 50%. The temperature shrinkage coefficient and surface modulus of the surface layer material had the greatest influence on the pavement thermal stress. The thermal stress could be reduced by more than 0.4 Mpa for every $5 \times 10^{-6}/°C$ reduction in the surface temperature shrinkage coefficient or every 1000 Mpa reduction in the surface modulus. This study can provide a reference for improving the temperature field and thermal stress field of asphalt pavement in the plateau area.

Keywords: pavement engineering; plateau areas; low temperature; cold wave; temperature field; thermal stress

1. Introduction

The QTP, the youngest and highest plateau on Earth, is known as the "third pole" of the planet, which exerts influence on global climate change and is rich in mineral resources [1–3]. It appears to have long been a focus of attention for researchers in domestic and overseas. Apparently, the construction of the QTP road is of crucial interest to scientific research and resource development. Nevertheless, its average altitude of 4500 m indicates its harsh climatic conditions, such as low temperature, strong solar radiation, and large diurnal temperature difference. The mean temperature in the cold season is below -10 °C and the daily temperature variation is above 15 °C [4,5]. The performance of asphalt pavements is highly susceptible to environmental influences, notably the spheric temperature. More to the point, the thermal stress of asphalt pavement is dramatically increased under the combined effect of low temperature and cold waves [6–8]. Asphalt pavements are significantly affected by solar radiation and external atmospheric convective exchanges, which can lead to the formation of different temperature gradients within the pavement, causing different degrees of temperature strain and thus different degrees of thermal stress. And when the temperature is lower, the tensile strength of the asphalt mixture decreases. The coupling of temperature and vehicle loading makes the stress greater than the tensile strength of the asphalt mixture, which leads to thermal cracks in the asphalt pavement. Water entering the pavement structure along the cracks will cause erosion, spalling, loosening, and other diseases. This contributes to the frequent occurrence of cracks on the QTP [9–11] (Figure 1).

Figure 1. The thermal cracks.

Numerous studies demonstrated that the characteristics of asphalt can be critically affected by temperature [12–14]. As a consequence, the process of heat transfer in asphalt pavements is of great significance and it affects the properties of asphalt pavements [15–18]. There were extensive studies on the temperature and thermal stress fields of asphalt pavements. First of all, Barber et al. [19] researched the association of pavement temperature with environmental conditions such as solar radiation, wind velocity, and atmospheric temperature, and introduced the first theoretical analysis approach for predicting pavement temperature. Based on the finite difference method, Christison et al. [20] developed software that can predict the temperature field of pavement using continuous temperature records from several pavement systems in Western Canada and analyzed the impact of low temperatures on asphalt pavement. Roque et al. [21] evaluated the interaction of coupled low temperatures and loads on asphalt pavements through experiments. Lytton et al. [22] proposed an integrated model that reflects the climate effects on pavements. These studies clarified the impact of low temperatures on asphalt pavement, which mainly focuses on theoretical analysis. In the 1990s, SHRP, C-SHRP, LTPP, and other projects in the United States and Canada gathered a great amount of pavement temperature and meteorological data to identify the extreme temperature conditions that asphalt pavements can withstand in service and established a database to provide a database basis for statistical analysis. However, it was discovered by researchers that the applicability of predictive models using statistical analysis methods is limited and fails to reveal the impacts of structures and materials [23].

Subsequently, with the advancement of the IT industry, simulation software such as ABAQUS and ANSYS were broadly utilized in the research of pavement temperature fields [24–26]. Hall et al. [27] developed a simplified one-dimensional (1D) heat flow modeling tool to study the influence of different pavement thermophysical properties on the thermal response of pavement by using a finite difference solution method. Kang et al. [27] employed a finite element model to evaluate the stress reaction of pavement layers and subgrade under continuous variable temperature conditions during the day. The foregoing studies indicated that climatic conditions such as low temperatures, large temperature differences, and cold waves have obvious consequences on the thermal stresses in asphalt pavements. The influence of temperature field and thermal stress on asphalt pavement has been widely studied, but the research on temperature field and thermal stress for high altitude climate is still in the preliminary stage, and no systematic research has been conducted. And there are fewer studies on the temperature field for cold waves. In this paper, a finite element model for calculating the pavement temperature field was developed, which features the simulation of three processes: solar radiation, convective heat transfer, and heat conduction. The pavement temperature field can be calculated as long as meteorological data are available, and the calculation process is simple and intuitive.

It is noticeable that during the cold season in the QTP, low temperatures, large temperature differences, and cold wave climate, which are very detrimental to asphalt pavement will appear at the same time, which can cause a drastic growth in thermal stress and have a highly serious impact on the properties and lifespan of asphalt pavement.

Therefore, this study focuses on the stress response of asphalt pavement during the cold wave in the cold season. By using the finite element software ABAQUS, the temperature field changes in asphalt pavement during the cold season were investigated and the thermal stresses in different cooling ranges during cold spells were studied. The most important factors affecting the surface thermal stresses were determined by the sensitivity coefficient analysis, which provides a reference for the pavement crack resistance design. The research route is shown in Figure 2.

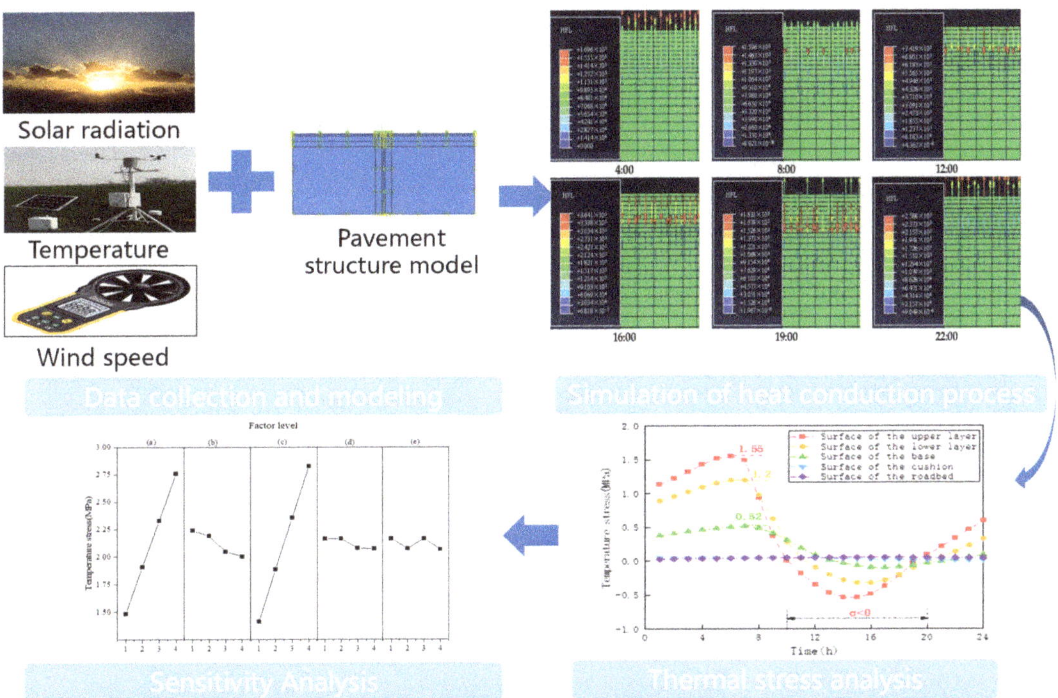

Figure 2. Research route for the temperature and thermal stress field of asphalt pavement on the Qinghai–Tibetan Plateau.

2. Climatic Characteristics of QTP

2.1. Temperature

According to the statistical data of China Meteorological Data Network from 1981 to 2010, the data of five observation stations at different altitudes of Dangxiong, Jiangzi, Naqu, Zedang, and Linzhi were selected to analyze the annual temperature situation (Figure 3). Their geographical location is shown in Table 1.

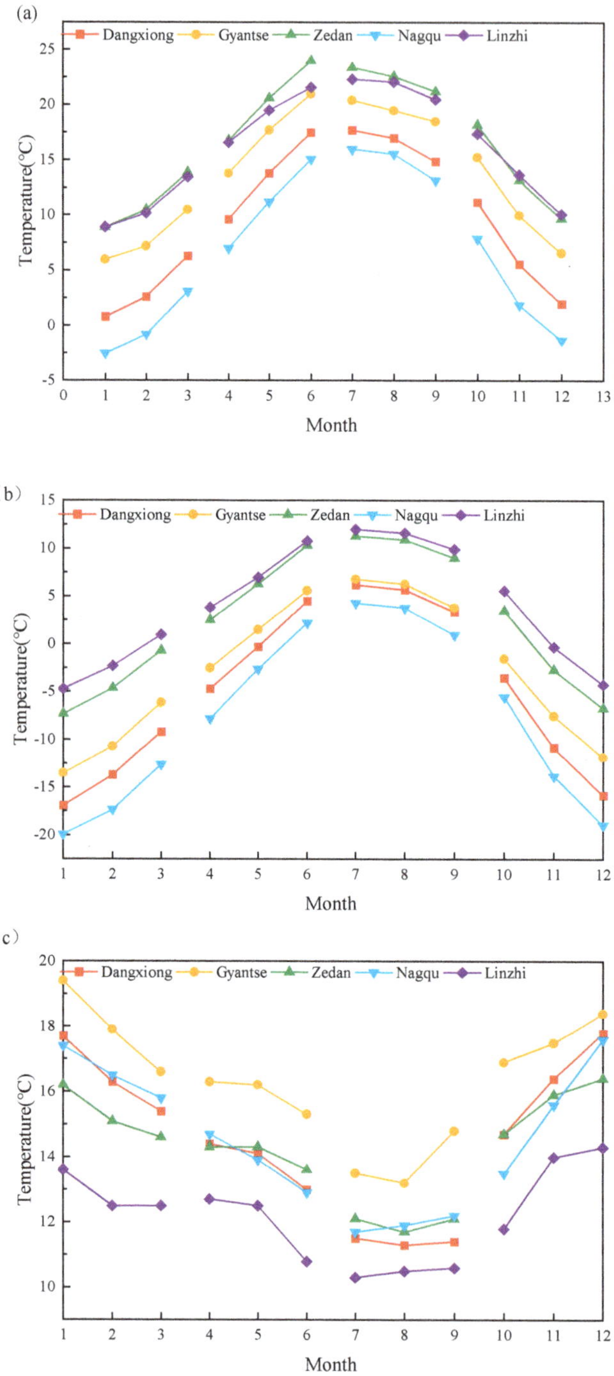

Figure 3. The cumulative average monthly temperature in QTP (**a**) monthly average maximum temperature, (**b**) monthly average minimum temperature, and (**c**) monthly average daily temperature range.

Table 1. The geographical location of the five observation stations.

Region	Longitude	Dimensional	Elevation (m)
Dangxiong	90°45′–91°31′ E	29°31′–31°04′ N	4300
Jiangzi	89°06′–90°12′ E	28°30′–29°18′ N	4000
Naqu	83°55′–95°05′ E	29°55′–36°30′ N	4450
Zedang	91°48′ E	29°14′ N	3580
Linzhi	92°09′–98°47′ E	26°52′–30°40′ N	3100

The low temperatures months in QTP were mainly in January, February, and December. The mean of the minimum monthly temperature and the mean of the maximum monthly temperature spread both emerged in January. The mean coldest monthly temperature in Nagqu reached −19.9 °C in January and the highest temperature difference in Gyantse was 19.4 °C. In conclusion, the annual temperature in QTP is relatively low, with temperatures below 10 °C most of the time. Thus, asphalt pavements in QTP are very susceptible to shrinkage cracks caused by low temperatures. The extreme temperature difference in winter is large and can even reach more than 30 °C in some areas.

2.2. Solar Radiation

It is apparent from Figures 4 and 5 that QTP is a region of the world with strong solar radiation. The radiation in QTP is strongest in spring and summer and reaches up to 25 MJ/m^2. Even in winter, the total solar radiation can reach about 15 MJ/m^2 (Figure 6). The negative climatic conditions in winter can lead to substantial temperature cracks in asphalt pavements in the QTP.

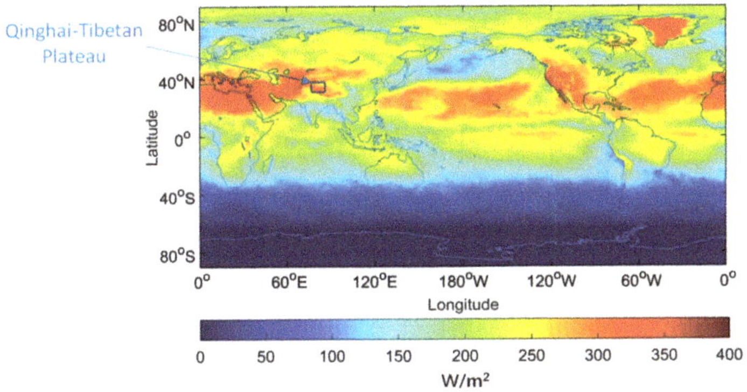

Figure 4. Global annual total solar radiation cloud map.

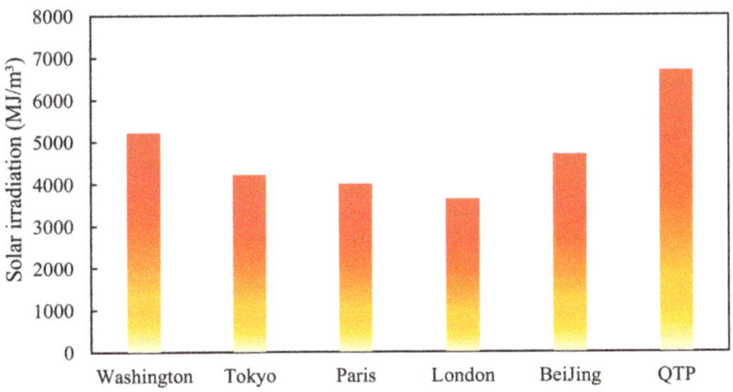

Figure 5. Total annual radiation in major cities of the world.

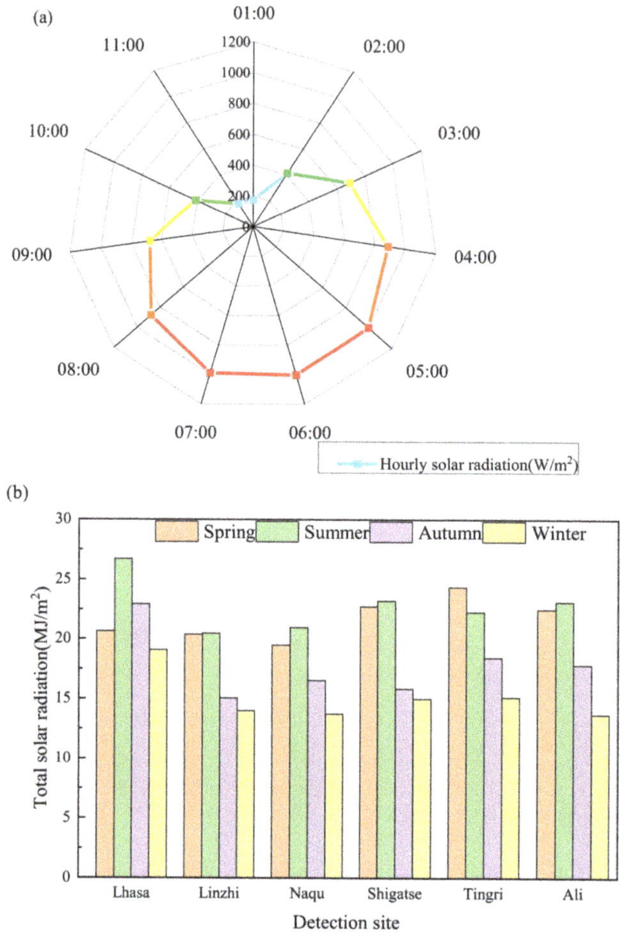

Figure 6. Variation of solar radiation in the QTP (**a**) hourly variation of annual total solar radiation in Lhasa, and (**b**) seasonal average daily total solar radiation exposure at six stations.

3. Methodology

3.1. Theoretical Equations for Temperature Fields

3.1.1. Differential Equations of Heat Conduction

When the object is a homogeneous isotropic material, and the physical parameters such as material density, specific heat, and thermal conductivity are constants and internal heat sources, the differential equation of heat conduction can be simplified as:

$$\frac{\partial T}{\partial t} = \alpha\left(\frac{\partial^2 T}{\partial x^2} + \frac{\partial^2 T}{\partial y^2} + \frac{\partial^2 T}{\partial z^2}\right) \tag{1}$$

where $\alpha = \lambda / c\rho$ is the heat diffusivity; λ is the thermal conductivity, J/(m·h·°C); c is the specific heat, J/(kg·°C); ρ is the density of the material, kg/m^3; and T is the temperature in the pavement.

3.1.2. Boundary Condition

When analyzing the temperature field of the pavement structure, only the temperature change along the thickness direction was considered. The exterior transfer of heat is mostly at the road surface, which transfers heat mainly through thermal radiation and thermal convection.

The heat transferred to the road surface by air convection can be calculated by:

$$q_c = h_c(T_a - T_1|_{Z=0}) \tag{2}$$

The heat transferred to the road surface by radiation can be calculated by:

$$q_r = \alpha_s \cdot q - q_F \tag{3}$$

where T_a is the atmospheric temperature, °C; q is the heat transferred to road surface per unit area by solar radiation per unit time; α_s is the radiation absorption rate of pavement surface; and q_F is the effective radiation of pavement surface.

(1) Solar radiation

Solar radiation consists of direct radiation and atmospheric scattering, so solar radiation is impacted by cloud and air transparency. Baber and Middel [28,29] investigated the diurnal variation curve of solar radiation. The diurnal variation curve of solar radiation can be analogous to a sinusoidal half-wave curve. The function of the fitting curve is:

$$q_t = \begin{cases} 0 & 0 \le t < 12 - \frac{c}{2} \\ q_0 \cos m\omega(t-12) & 12 - \frac{c}{2} \le t \le 12 + \frac{c}{2} \\ 0 & 12 + \frac{c}{2} < t \le 24 \end{cases} \tag{4}$$

where q_0 is the maximum radiation in a day, which can be calculated by $q_0 = 0.131mQ$, $m = 12/c$; Q is the total solar daily radiation, J/m^2; c is the actual effective sunshine time, h; and ω is the angular frequency, which can be calculated by $\omega = 2\pi/24\text{rad}$.

To avoid jumping discontinuity points in the calculation of the temperature fields, the above equations can be extended to trigonometric form by means of a Fourier series as follows:

$$q(t) = \frac{\alpha_0}{2} + \sum_{k=1}^{\infty} \alpha_k \cos \frac{k\pi(t-12)}{12} \tag{5}$$

where $\alpha_0 = \frac{2q_0}{m\pi}$; $\alpha_k = \begin{cases} \frac{q_0}{\pi}\left[\frac{1}{m+k}\sin(m+k)\frac{\pi}{2m} + \frac{\pi}{2m}\right] & k = m \\ \frac{q_0}{\pi}\left[\frac{1}{m+k}\sin(m+k)\frac{\pi}{2m} + \frac{1}{m-k}\sin(m-k)\frac{\pi}{2m}\right] & k \ne m \end{cases}$;

The intensity of solar radiation absorbed by the road surface can be calculated by:

$$q_s = \alpha_s \cdot q \tag{6}$$

where α_s is the solar radiation intensity absorption rate of road surface, which was set to 0.90 in this study.

(2) The change in air temperature

There is a disparity between the time of rise and fall of temperature during one day, with the cooling time being about 4 h longer than the heating time. By using a linear combination of two sinusoidal functions, the practical variation of the temperature process can be well simulated:

$$T_a = \overline{T_a} + T_m[0.96 \sin \omega(t-t_0) + 0.14 \sin 2\omega(t-t_0)] \tag{7}$$

where $\overline{T_a}$ is the mean daily air temperature, which can be calculated by $\overline{T_a} = 1/2(T_a^{max} + T_a^{min})$, °C; T_m is the daily temperature variation range, which can be calculated by $T_m = 1/2(T_a^{max} - T_a^{min})$, °C; ω is the angular frequency, which can be calculated by $\omega = 2\pi/24\text{rad}$, rad; t_0 is the primary phase.

The coefficient of heat interaction between the pavement and the atmosphere is primarily governed by the wind speed, which can be expressed by the following equation:

$$\text{h}_c = 3.7v_\text{W} + 9.4 \quad (8)$$

where h_c is the heat exchange coefficient, $W/(m^2 \cdot °\text{C})$; v_w is the daily mean wind speed, m/s.

(3) Effective radiation of the pavement

The effective radiation of pavement is the difference between the emission of radiation from the pavement and the absorption of inverse atmospheric radiation. The effective radiation of the pavement can be impacted by the internal temperature of the structure and the external meteorological conditions:

$$q_F = \varepsilon\sigma[(T_1|_{Z=0} - T_Z)^4 - (T_a - T_Z)^4] \quad (9)$$

where q_F is the effective radiation of road surface, $W/(m^2 \cdot °\text{C})$; ε is the road surface emissivity, which can be set to 0.81 in the asphalt pavement; σ is the Stefan–Boltzmann constant, which is equal to 56,697 $\times$ 10^{-8}, $W/(m^2 \cdot K^4)$; $T_1|_{Z=0}$ is the pavement surface temperature, °C; T_a is the air temperature, °C; and T_Z is the absolute zero value, which is equal to -273 (°C).

3.1.3. Solution Equation of Pavement Temperature Field

The temperature field refers to the temperature distribution of each point in an instantaneous object, which is generally a function of space and time coordinates. In the rectangular coordinate system, the temperature field can be expressed as:

$$T = f(x, y, z, t) \quad (10)$$

The heat conduction differential equation of pavement structure can be expressed as:

$$\frac{\partial T_i(z,t)}{\partial t} = \alpha_i \frac{\partial^2 T_i(z,t)}{\partial z^2} \quad (11)$$

When the interlayer structure is good without thermal resistance, the interlayer interface satisfies:

$$T_i(z,t)|_{z=z_i} = T_{i+1}(z,t)|_{z=z_i} \quad (12)$$

$$\lambda_i \frac{\partial T_i(z,t)}{\partial t}|_{z=z_i} = \lambda_{i+1} \frac{\partial^2 T_{i+1}(z,t)}{\partial z}|_{z=z_i} \quad (13)$$

where α_i, λ_i, and h_i represent the thermal conductivity, thermal conductivity, and structural layer thickness of the first layer pavement structure layer, respectively.

3.2. Basic Assumption

In the establishment of the temperature field model, the following assumptions were made to facilitate the analysis and calculation of the problem:

(1) The pavement surface, subgrade, bedding layer, and soil base are considered homogeneous, isotropic, and elastic materials.
(2) The layers of the pavement structure are tightly connected to each other and the stresses, displacement, and heat fluxes vary continuously among the layers.
(3) The inhomogeneity of the lateral distribution of the pavement temperature field is neglected.

(4) The transfer of heat flow is assumed to be one-dimensional conduction perpendicular to the pavement.
(5) The parameters are considered not to vary with temperature, except for the relevant parameters mentioned in the paper.
(6) Due to the asphalt mixture's inhomogeneity and discontinuities between layers could have an effect on the localized calculation results of the temperature field of asphalt pavement, but it has little effect on the overall calculation results. If further accurate local temperature field calculations are desired, a 3D temperature field model that takes into account the inhomogeneities and discontinuities can be built. The calculation accuracy of the model used in this paper has met the needs of the study.

3.3. *Structure and Parameters of the Pavement*

This section lists the various parameters used for modeling. The thermophysical parameters of the pavement structure and pavement materials according to the typical sections of the QTP are listed in Table 2. The elastic parameters and temperature shrinkage coefficients of the materials used in this study are given in Tables 3 and 4.

Table 2. Thermodynamic parameters of materials.

Structural Layer	Above Layer AC-13	Lower Layer AC-20	Base Cement Stabilized Macadam	Cushion Graded Gravel	Soil
Thickness (cm)	4	5	20	20	
Thermal conductivity ($J/(m \cdot h \cdot °C)$)	4200	4000	5400	5200	5400
Density (kg/m^3)	2400	2500	2100	2000	1800
Specific heat ($J/(kg \cdot °C)$)	950	900	900	950	1000
Solar radiation absorptivity			0.9		
Road surface reflectance			0.81		
Absolute zero (°C)			−273		
Stefan–Boltzmann constant ($J/(h \cdot m^2 \cdot K^4)$)			2.041×10^{-4}		

Table 3. Elastic parameters of materials.

Structural Layer	Material Type	Temperature (°C)	Elastic Modulus (MPa)	Poisson's Ratio
Above layer	AC-13	10	900	0.3
		0	1400	0.3
		−10	4500	0.25
		−20	9000	0.25
		−30	14,000	0.25
Lower layer	AC-20	10	1000	0.3
		0	1200	0.3
		−10	4200	0.25

Table 4. Temperature shrinkage coefficient.

Material Type		Temperature Shrinkage Coefficient at Different Temperatures ($\times 10^{-5}$)			
		0 °C	−10 °C	−20 °C	−30 °C
Asphalt mixture	AC-13	2.5	2.2	1.9	1.6
	AC-20	2.5	2.2	1.9	1.6
Cement stabilized macadam		0.98			
Graded gravel		0.5			
Soil		50			

According to the meteorological data, this study was conducted in January after excluding the effect of cold waves. The meteorological parameters of the pavement structure of QTP under the periodic temperature of the cold season are shown in Table 5. The total daily solar radiation is 15.1 MJ/m^2; the daily mean wind speed is 2.0 m/s; and the effective duration of sunshine is 7.5 h.

Table 5. The temperature for 24 h on behalf of the day in winter (January).

Time (h)	0	1	2	3	4	5	6	7	8	9	10	11
Temperature (°C)	−9.1	−10.1	−11.1	−12.1	−12.9	−13.5	−13.7	−13.4	−12.7	−11.5	−9.9	−8.2
Time (h)	12	13	14	15	16	17	18	19	20	21	22	23
Temperature (°C)	−6.4	−4.8	−3.6	−2.9	−2.6	−2.8	−3.4	−4.2	−5.2	−6.2	−7.2	−8.2

3.4. Initial Temperature Field of Asphalt Pavement

The initial temperature field of the pavement structure is ascertained to enhance the simulation reliability of the temperature field [30].

The calculation formula of pavement temperature can be calculated by:

$$T_0 = 64.9e^{0.014T_{a1}} + e^{1.16Q_4} - 66.4 \tag{14}$$

where T_0 is the pavement surface temperature, °C; T_{a1} is the atmospheric temperature an hour ago, °C; and Q_4 is the cumulative solar radiation intensity in the first 4 h, kW/m^2.

The pavement temperature at different depths can be calculated by:

$$T_h = T_0 + (7.01e^{-0.07h} - 7.01) \times \sin(0.011h - 0.316t - 0.659) - 0.003h \times T_0 \tag{15}$$

where T_h is the temperature at different depths of the pavement, °C; T_0 is the pavement surface temperature, °C; and h is the depth from road surface, cm.

3.5. Determination of Cooling Range

The cold wave in QTP often occurs in January, and the cooling amplitude of the cold wave ranges from a few degrees to a dozen degrees. Therefore, in order to make the simulated cooling situation closer to the actual situation, the temperature stresses in the pavement structure under different cooling amplitudes are calculated by considering the cooling amplitude values of 5 °C, 10 °C, and 15 °C, respectively.

After studying and analyzing the pavement structure under low-temperature conditions [31], it was obtained that in excluding the cold wave cooling process, the following equation can be fitted:

$$\Delta t_a = \Delta t_{Ma}\left(t'/t'_{Ma}\right)^{n_a} \tag{16}$$

where Δt_{Ma} is the maximum amount of cooling; t' is the time that from the beginning of cooling is zero point, $0 \le t' \le t'_{Ma}$; t'_{Ma} is the total time of cooling process; and n_a is the fitting coefficient, set to 0.5.

The cooling process within 24 h without considering the daily temperature fluctuation is shown in Figure 7. It can be seen from the figure that the temperature drop rate gradually decreases during the day. The actual ambient temperature of the road surface is the result of the superposition of the daily temperature change and the cooling process. Therefore, the 24 h actual temperature of the pavement is the sum of the daily temperature in Table 5 and the temperature in Figure 7. The actual temperature of the road surface after the superposition of the cooling process is shown in Figure 8.

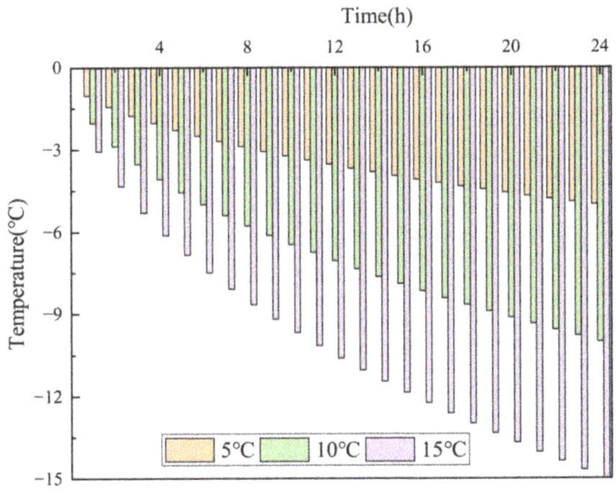

Figure 7. Cooling process diagram.

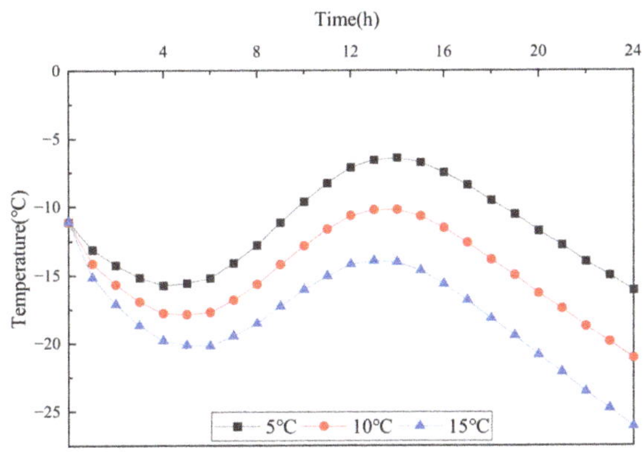

Figure 8. Temperature change curve.

3.6. Temperature Field Model

The pavement of the normal national and provincial roads in QTP is typically 6 m wide. Accordingly, the length of the module in this study was set to 6 m and the thickness was set to 3 m. The model of the temperature field is shown in Figure 9.

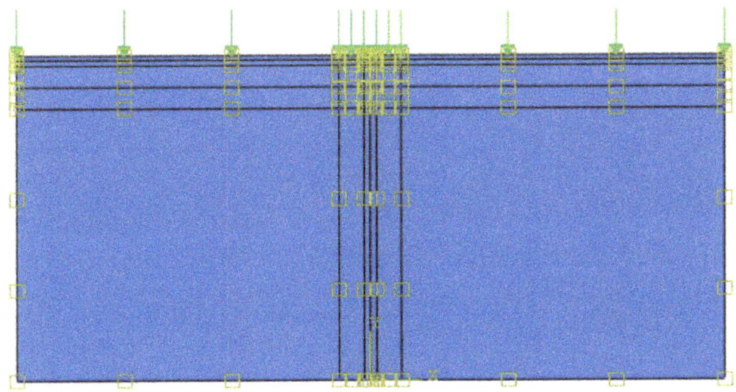

Figure 9. Solar radiation and initial temperature field loading. (The yellow squares represent interaction and the green arrow represents thermal radiation).

3.7. Thermal Stress Model

In ABAQUS, the elastic modulus, Poisson's ratio, and temperature shrinkage coefficient in the material properties were set as elastic parameters that vary with temperature. The boundary conditions were defined as XSYMM for the left and right and ZASYMM for the bottom. The results of the temperature field were imported into the thermal stress calculation model and the grid division was consistent with the temperature field model. The element type was set to a four-node temperature surface strain quadrilateral reduced integral element (CPE4RT). The thermal stress model is displayed in Figure 10.

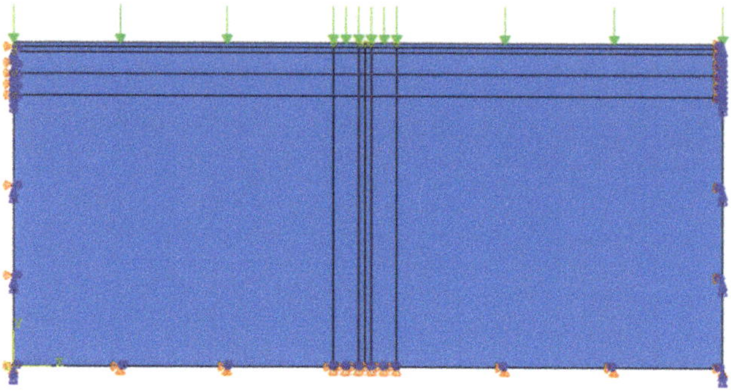

Figure 10. Thermal stress model (the green arrow represents thermal radiation).

4. Result and Discussion

4.1. Verification of Numerical Modeling

For typical sections of the QTP, the calculated temperature data and observed road surface data are shown in Figure 11.

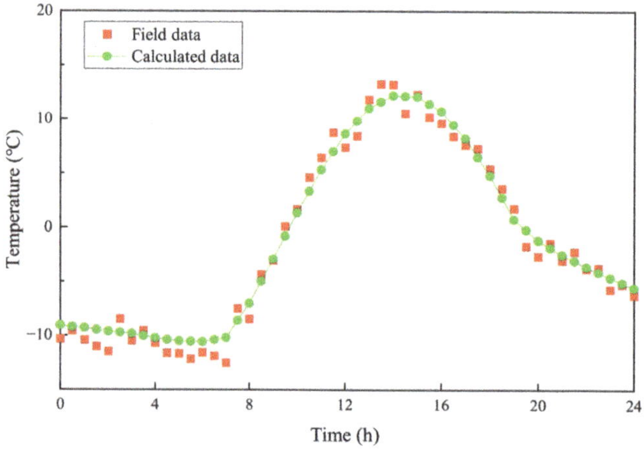

Figure 11. Calculated and measured road surface temperature data.

Due to the continuous and stable temperature of numerical simulation, the measured temperature is severely affected by the environment, resulting in discontinuous and large fluctuations in the data. The simulated temperature data are not identical to the measured data. However, they are similar to each other, and more importantly, they have similar variation trends. The mean value of the residuals between the finite element simulated and measured values of the road surface temperature is −0.27, and the standard deviation of the residuals is 0.49, which shows that the simulation accuracy is good. The validity of the finite element model for the response of pavement temperature field and thermal stress is proved.

4.2. Variation of the Heat Flow Vector

Figure 12 shows the variation of the heat flux vector of the pavement structure during the day.

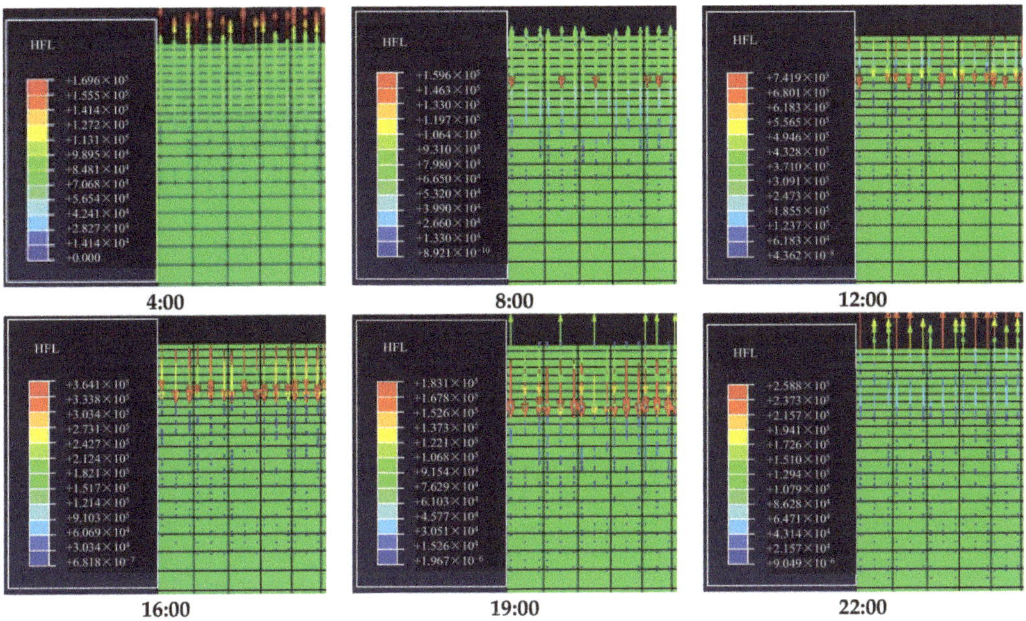

Figure 12. Heat flow vector HFL cloud at different times.

As can be seen in Figure 12, towards the evening, the pavement can release heat to the surroundings due to the weakening of solar radiation. Consequently, the temperature of the pavement structure can gradually decrease. However, due to the solar radiation during the day, the pavement can attract the heat transferred from the atmosphere and the temperature of the pavement structure can grow gradually. Repeated heating and cooling can lead to the accumulation of fatigue damage in the pavement structure. Eventually, temperature cracks are generated.

4.3. Variation of Pavement Temperature

The temperature changes in the pavement structure at different times of the day are displayed in Figure 13, from which it can be revealed that the maximum temperature of the asphalt pavement in the QTP is about 12 °C at 15:00, while the maximum temperature of the representative day in January is −2.6 °C, which is due to the strong solar radiation in the QTP. The maximum pavement temperature usually appears 1–2 h after the maximum solar radiation and the maximum solar radiation in the QTP is at 14:00 in winter, so the simulation results are consistent with the pavement temperature. The lowest pavement temperature occurs around 6:00, which is about −11 °C. The temperature difference between day and night in winter causes a large change in pavement temperature. In the long term, the pavement is vulnerable to fatigue cracks.

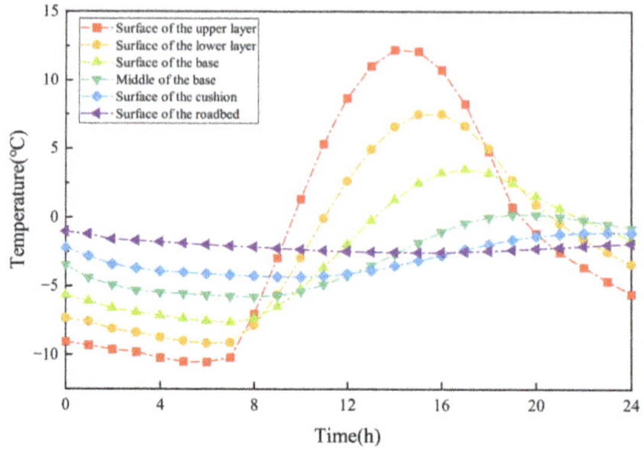

Figure 13. Variation of road surface temperature with time.

4.4. Variation of Temperature Gradient

The variation of the pavement temperature field with depth in the QTP is shown in Figure 14.

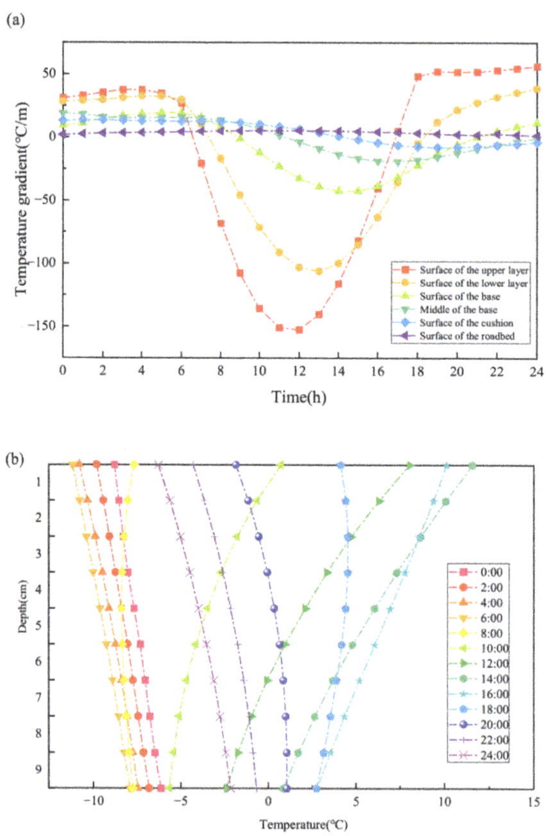

Figure 14. Variation of pavement temperature with depth: (**a**) variation of road surface temperature gradient, and (**b**) variation of surface temperature with depth.

From Figure 14a, it can be concluded that the surface layer is significantly influenced by temperature and radiation. The extremes of positive and negative temperature gradients on the pavement reach 52.2 °C/m and −152.6 °C/m, respectively. The negative temperature gradient changes faster due to the influence of radiation during the daytime, and the large changes in the temperature gradient can lead to vertical fatigue damage to the pavement structure. It can be noticed from Figure 14b that the temperature field of the surface layer is complicated. At 12:00 and 14:00, the temperature difference between the surface layer and the bottom of the lower layer reaches more than 10 °C. Solar radiation is the main factor contributing to this phenomenon, which can be ameliorated by solar heat reflective coating.

4.5. Analysis of Thermal Stress in One Day

The pavement temperature field is channeled into the thermal stress model and the thermal stresses of the corresponding structural layers are exported through the field variables. The variation of daily thermal stresses is shown in Figure 15.

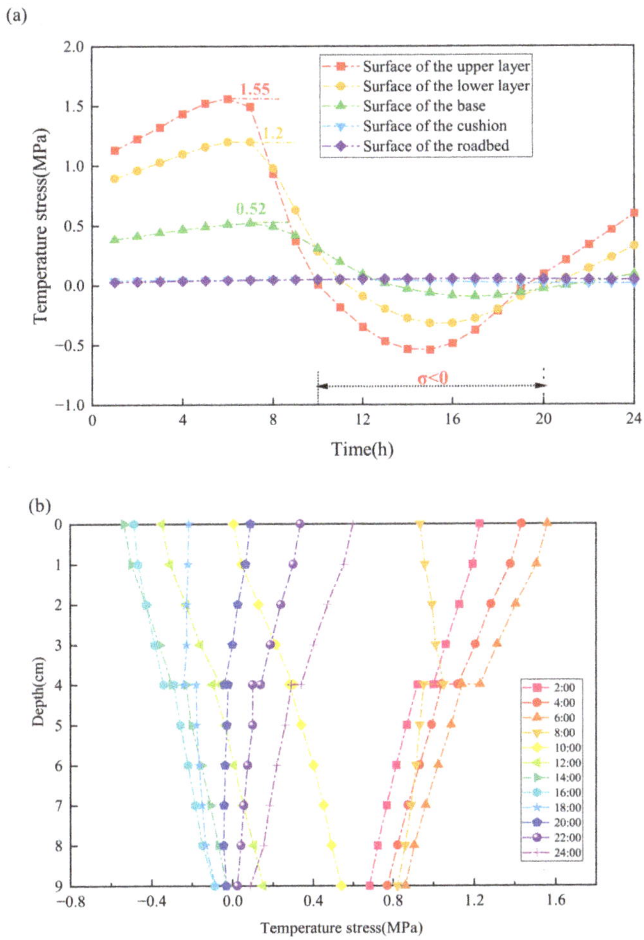

Figure 15. Variation of pavement thermal stress on the day: (**a**) thermal stress curve of the pavement structure, and (**b**) variation of surface thermal stress.

From Figure 15a, it can be noted that the maximum peak of thermal stress in the pavement occurs at 6:00 a.m., which is 1.55 MPa. The greatest thermal stress on the lower surface is 1.2 Mpa, which occurs at 7:00 a.m. Also, from the figure, it can be revealed that the tensile stress period in the pavement under the role of low temperature is from 20:00 to 10:00, and fatigue cracks can occur in the long-term tensile state.

As can be seen in Figure 15b, the stresses in the pavement gradually decrease along the depth. There is a certain hysteresis in the change in thermal stress between the upper and lower layers. The reason for this is that the modulus and temperature shrinkage coefficient of the upper and lower layers are temperature dependent, and the thermal load has a certain response time.

4.6. Analysis of Thermal Stress in Cold Wave

In the study, the temperature conditions of 5 °C, 10 °C, and 15 °C of cooling were selected to simulate the thermal stress changes in asphalt pavement under different cooling amplitudes. And it was selected 0 cm, 4 cm, 9 cm, and 14 cm below the pavement for the analysis. The thermal stress distribution of the pavement structure layer at different depths is illustrated in Figure 16.

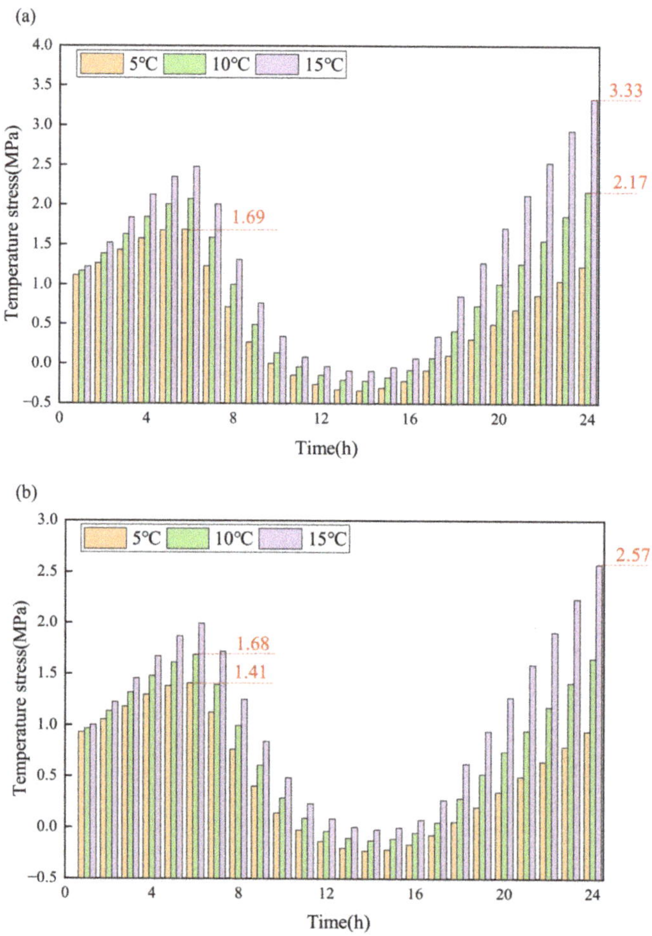

Figure 16. *Cont.*

(c)

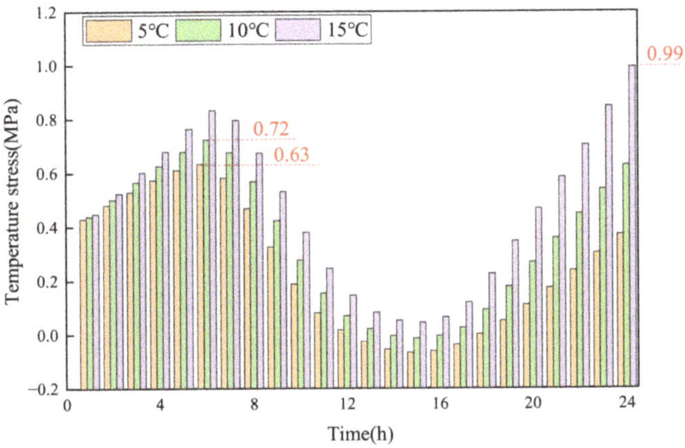

(d)

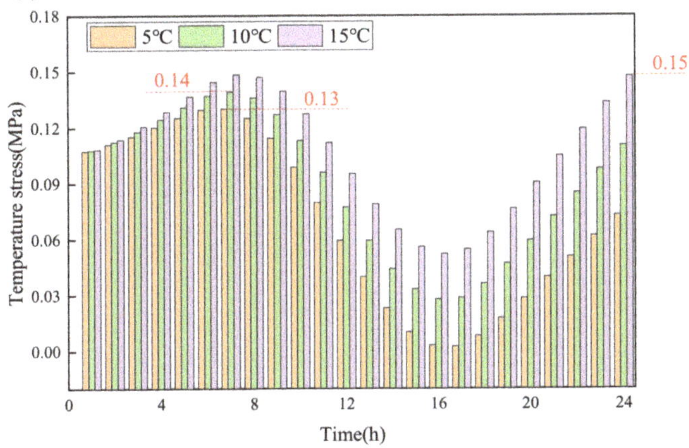

Figure 16. Thermal stress curve of pavement under different cooling amplitude: (**a**) 0 cm; (**b**) 4 cm; (**c**) 9 cm; and (**d**) 14 cm.

It can be revealed from Figure 16 that the greater the cooling amplitude, the greater the thermal stress generated in the pavement and the longer the pavement is in a tensile state. When the cooling amplitude is 15 °C, the maximum thermal stress of the pavement reaches 3.33 MPa, which easily exceeds the splitting tensile strength of the AC-13 asphalt mixture. It leads to the low-temperature shrinkage cracking of the pavement.

The asphalt pavement surface layer is most obviously subjected to cooling as seen in Figure 17. When the cooling range of the pavement reaches 15 °C, the maximum thermal stress reaches 3.33 MPa and the thermal stress increases linearly with a maximum increase of 53%. The maximum thermal stress also reaches 2.57 MPa when the cooling range of the lower pavement reaches 15 °C, and the greatest percentage increase in the thermal stress also reaches more than 50%. When the temperature is quite low, the modulus of asphalt concrete increases rapidly, showing brittle and stiff characteristics. Due to the changeable weather of the QTP, under the combined influence of various unfavorable weather factors, the thermal stress is very likely to exceed the tensile strength of the asphalt mixture, resulting in pavement cracking.

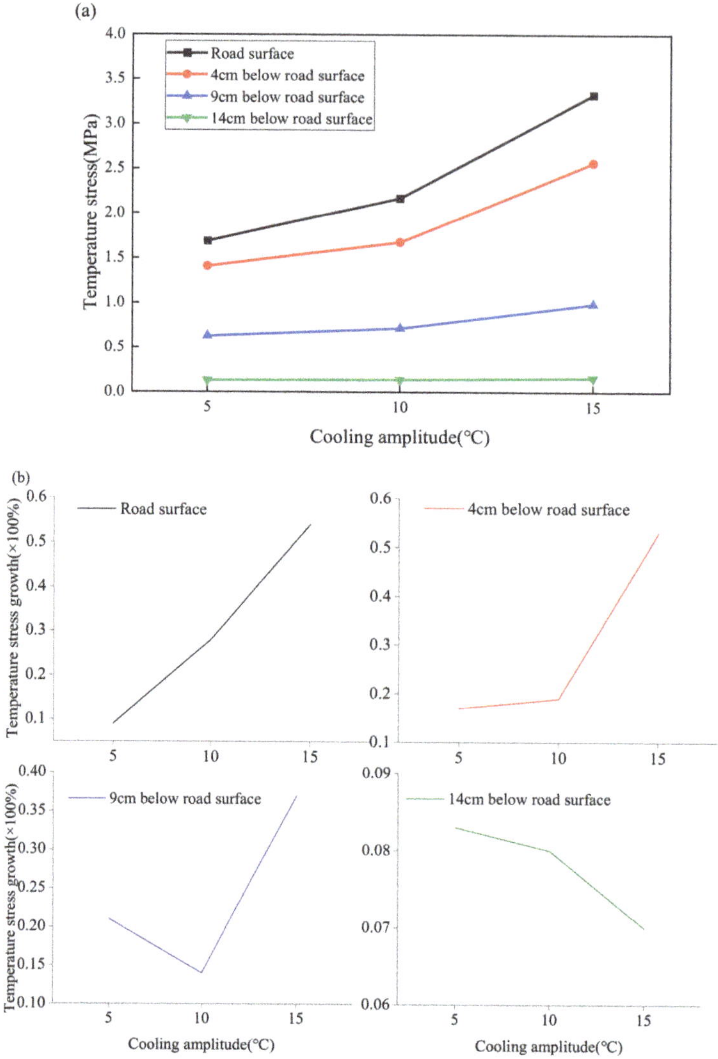

Figure 17. Maximum thermal stress under different cooling amplitudes: (**a**) maximum thermal stress at different depths, and (**b**) thermal stress amplification at different depths.

4.7. Parameter Sensitivity Analysis

A sensitivity analysis of the factors influencing surface stress was performed in this research by designing an orthogonal test. Then, the main factors affecting the surface thermal stresses were found. It can offer a basis for the design of pavement structures against cracking.

4.7.1. Orthogonal Experimental Design

To analyze the influencing factors of the shear strength, orthogonal tests were utilized. Orthogonal tests are an experimental design method that evaluates the effects of multiple factors and levels on a single target variable [32]. Four influencing factors of surface modulus, surface thickness, surface temperature shrinkage coefficient, and base modulus were selected, respectively. They are highly effective in evaluating the influence of multiple factors on the target variable and identifying the optimal level for each factor. The experimental factor levels and design for orthogonal tests are presented in Tables 6 and 7 respectively. Each factor takes four levels, and the values of each factor and level are shown in Table 6.

Table 6. Orthogonal analysis factor level table.

Factor/Level	Surface Modulus (MPa)	Surface Thickness (cm)	Surface Temperature Shrinkage Coefficient (10^{-5}/°C)	Base Modulus (MPa)
1	3500	9	1.5	1200
2	4500	12	2	1500
3	5500	15	2.5	1800
4	6500	18	3	2100

Table 7. Thermal stress calculation results.

Test Serial Number	Surface Modulus (MPa)	Surface Thickness (cm)	Surface Temperature Shrinkage Coefficient (10^{-5}/°C)	Base Modulus (MPa)	Blank Column	Thermal Stress (MPa)
1	1 (3500)	1 (9)	1 (1.5)	1 (1200)	1	0.99
2	1	2 (12)	2 (2.0)	2 (1500)	2	1.32
3	1	3 (15)	3 (2.5)	3 (1800)	3	1.65
4	1	4 (18)	4 (3.0)	4 (2100)	4	1.98
5	2 (4500)	1	2	3	4	1.70
6	2	2	1	4	3	1.27
7	2	3	4	1	2	2.54
8	2	4	3	2	1	2.12
9	3 (5500)	1	3	4	2	2.59
10	3	2	4	3	1	3.11
11	3	3	1	2	4	1.55
12	3	4	2	1	3	2.07
13	4 (6500)	1	4	2	3	3.68
14	4	2	3	1	4	3.06
15	4	3	2	4	1	2.45
16	4	4	1	3	2	1.84

The orthogonal test was performed with five factors and four levels. L16(45) was chosen, and the fifth column was blank. The orthogonal test table was generated by Minitab software. Altogether, 16 tests were calculated. The results of the maximum thermal stress calculation are listed in Table 7.

4.7.2. Extremum Difference Analysis

In this study, Minitab was used to analyze the maximum thermal stress of the pavement at each factor level. The results are illustrated in Table 8.

Table 8. Range analysis results of the orthogonal test.

Level	Surface Modulus	Surface Thickness	Surface Temperature Shrinkage Coefficient	Base Modulus
1	1.485	2.240	1.412	2.165
2	1.907	2.190	1.885	2.167
3	2.330	2.047	2.355	2.075
4	2.758	2.002	2.827	2.072
Extreme difference	1.273	0.237	1.415	0.095

From the results, it can be concluded that the surface modulus and surface temperature shrinkage coefficient have the greatest effect on the thermal stress of the pavement. The order of influence of various factors on thermal stresses is surface temperature shrinkage coefficient > surface modulus > surface thickness > base modulus. Consequently, the material parameters of the pavement structure should be mainly considered in the design of the pavement structure in the QTP to ensure the crack resistance of the pavement structure. The mean response of each factor at different levels is shown in Figure 18.

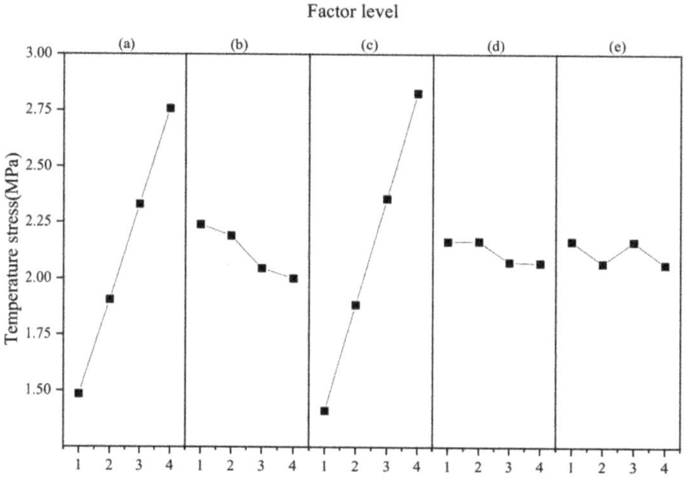

Figure 18. Mean response of thermal stress: (**a**) surface modulus; (**b**) surface thickness; (**c**) surface temperature shrinkage coefficient; (**d**) base modulus; and (**e**) blank column.

Figure 18 indicates that reducing the modulus and temperature shrinkage coefficient of the surface layer can effectively diminish the thermal stress of the pavement. Hence, materials with small modulus and temperature shrinkage coefficient should be selected as much as possible in the design of pavement structure, and the thickness of the surface layer can be enlarged appropriately. The optimal pavement structure combination obtained by polarization is listed in Table 9.

Table 9. Optimum combination of the pavement structure.

Surface Modulus (MPa)	Surface Thickness (cm)	Surface Temperature Shrinkage Coefficient (10^{-5}/°C)	Base Modulus (MPa)	Base Thickness (cm)
3500	18	1.5	2100	20

4.7.3. Analysis of Variance

Variance analysis results of the orthogonal test are shown in Table 10.

Table 10. Thermal stress variance analysis.

Project	Surface Modulus	Surface Thickness	Surface Temperature Shrinkage Coefficient	Base Modulus
Degree of freedom (DF)	3	3	3	3
The sum of squares deviation from the mean (MS)	3.596	0.153	4.446	0.034
Mean square (SS)	1.198	0.051	1.482	0.011
F-valued	3.08	0.08	4.66	0.02
Critical value	F0.05(3,12) = 3.49; F0.1(3,12) = 2.56			
Significance	Obvious	Insignificant	Obvious	Insignificant

The variance analysis reveals that the pavement thermal stress is considerably subject to the pavement structure material parameters. For this purpose, the modulus and temperature shrinkage coefficient of pavement materials at low temperatures should be fully considered in the design of pavement structures.

5. Summary and Conclusions

This investigation concentrates on the temperature field and the distribution of thermal stresses in QTP. Several conclusions can be obtained as follows:

(1) There is a significant contribution of temperature and solar radiation to the pavement temperature field, and this effect decreases with depth. The portion below the subgrade is almost unaffected. This is because it takes time for heat to conduct, and the deeper the depth, the longer it takes for its heat to conduct, making it insensitive to changes in the external environment. The temperature gradient under the effects of solar radiation reaches a maximum of 152.6 °C/m.

(2) The pavement temperature field has a substantial bearing on the pavement thermal stress, and the trend of the pavement thermal stress is heightened to resemble the temperature field. When the pavement temperature reaches its minimum, the tensile stress in the pavement reaches its maximum.

(3) The cooling amplitude has a dramatic effect on the thermal stress of the pavement when cold snaps hit. For every 5 °C increase in cooling amplitude, the thermal stress rises by more than 50%. This is because the greater the cooling, the greater the temperature gradient of the pavement, making the temperature-induced strains at different depths of the pavement vary widely, ultimately resulting in higher temperature stresses.

(4) The surface temperature shrinkage coefficient and surface modulus have the strongest impact on the thermal stress of the pavement. Where the change in the temperature shrinkage coefficient directly affects the temperature-induced strain, and the surface modulus affects the thermal stress due to strain. The thermal stress can be reduced by more than 0.4 MPa when the surface temperature shrinkage coefficient is reduced by $5 \times 10^{-6}/°C$ or the surface modulus is reduced by 1000 Mpa. In the design and construction of pavement in large temperature difference areas, the surface temperature shrinkage coefficient and surface modulus should be suitably lowered to diminish the thermal stress of the pavement.

(5) It is recommended to employ a structure with a surface modulus of 3500 Mpa, a surface thickness of 18 cm, a surface temperature shrinkage coefficient of $1.5 \times 10^{-5}/°C$, a base modulus of 2100 Mpa, and a base thickness of 20 cm as the anti-cracking structure of asphalt pavement.

Author Contributions: Conceptualization, X.Z.; Methodology, B.L. and Y.X.; Software, X.Z.; Validation, X.Z.; Formal Analysis, F.T.; Investigation, Y.X., Y.B., F.T., and Z.C.; Resources, Y.B.; Writing—Original Draft Preparation, Y.X.; Writing—Review and Editing, B.L.; Visualization, B.L., F.T., and Z.C.; Supervision, Z.C.; Project Administration, X.Z.; Funding Acquisition, B.L. and Y.B. All authors have read and agreed to the published version of the manuscript.

Funding: The work described in this paper is supported by the National Natural Science Foundation of China (5220082882), Natural Science Foundation of Chongqing (2022NSCQ-MSX1939), Chongqing Municipal Education Commission Foundation (KJQN202200745, KJQN202300728), Chongqing Transport Bureau (Z33220003), and Chongqing Jiaotong University (F1220032).

Data Availability Statement: The raw data supporting the conclusions of this article will be made available by the authors on request.

Acknowledgments: Thanks to all the individuals and organizations that have contributed to this article and experiments.

Conflicts of Interest: Author Zhimin Cong was employed by the company Shanghai Municipal Engineering Design Institute (Group) Co., Ltd. The remaining authors declare that the research was conducted in the absence of any commercial or financial relationships that could be construed as a potential conflict of interest.

References

1. Lou, P.; Wu, T.; Yang, S.; Wu, X.; Chen, J.; Zhu, X.; Chen, J.; Lin, X.; Li, R.; Shang, C.; et al. Deep learning reveals rapid vegetation greening in changing climate from 1988 to 2018 on the Qinghai-Tibet Plateau. *Ecol. Indic.* **2023**, *148*, 110020. [CrossRef]
2. Li, Z.; Wang, S. Transferring heat accumulated by asphalt pavement from inside to outside through carbon fibers. *Constr. Build. Mater.* **2023**, *383*, 131314. [CrossRef]
3. Jiang, G.; Zhao, H.; Liu, Y.; Wu, Q.; Gao, S. Discrepancies of permafrost variations under thermal impacts from highway and railway on the Qinghai-Tibet Plateau. *Cold Reg. Sci. Technol.* **2023**, *208*, 103784. [CrossRef]
4. Sha, A.; Ma, B.; Wang, H.; Hu, L.; Mao, X.; Zhi, X.; Chen, H.; Liu, Y.; Ma, F.; Liu, Z.; et al. Highway constructions on the Qinghai-Tibet Plateau: Challenge, research and practice. *J. Road Eng.* **2022**, *2*, 1–60. [CrossRef]
5. Șerban, R.-D.; Bertoldi, G.; Jin, H.; Șerban, M.; Luo, D.; Li, X. Spatial variations in ground surface temperature at various scales on the northeastern Qinghai-Tibet Plateau, China. *Catena* **2023**, *222*, 106811. [CrossRef]
6. Pszczoła, M.; Szydłowski, C. Influence of Bitumen Type and Asphalt Mixture Composition on Low-Temperature Strength Properties According to Various Test Methods. *Materials* **2018**, *11*, 2118. [CrossRef] [PubMed]
7. Judycki, J. A new viscoelastic method of calculation of low-temperature thermal stresses in asphalt layers of pavements. *Int. J. Pavement Eng.* **2018**, *19*, 24–36. [CrossRef]
8. Cui, T.; Zhang, T.; Wang, J.; Chen, J.; Wang, W.; Wang, H.; Wang, H.; Yang, D. Optimization and regulation of pavement concrete properties in plateau environment with large daily temperature difference: From mixture design to field application. *Constr. Build. Mater.* **2023**, *378*, 131167. [CrossRef]
9. Tang, J.; Fu, Y.; Ma, T.; Zheng, B.; Zhang, Y.; Huang, X. Investigation on low-temperature cracking characteristics of asphalt mixtures: A virtual thermal stress restrained specimen test approach. *Constr. Build. Mater.* **2022**, *347*, 128541. [CrossRef]
10. Adam, Q.F.; Levenberg, E.; Ingeman-Nielsen, T.; Skar, A. Modeling the use of an electrical heating system to actively protect asphalt pavements against low-temperature cracking. *Cold Reg. Sci. Technol.* **2023**, *205*, 103681. [CrossRef]
11. Zhang, J.; Cao, D.; Zhang, J. Three-dimensional modelling and analysis of fracture characteristics of overlaid asphalt pavement with initial crack under temperature and traffic loading. *Constr. Build. Mater.* **2023**, *367*, 130306. [CrossRef]
12. Zeiada, W.; Hamad, K.; Omar, M.; Underwood, B.; Khalil, M.; Karzad, A. Investigation and modelling of asphalt pavement performance in cold regions. *Int. J. Pavement Eng.* **2019**, *20*, 986–997. [CrossRef]
13. Rys, D.; Judycki, J.; Pszczola, M.; Jaczewski, M.; Mejlun, L. Comparison of low-temperature cracks intensity on pavements with high modulus asphalt concrete and conventional asphalt concrete bases. *Constr. Build. Mater.* **2017**, *147*, 478–487. [CrossRef]
14. Guo, Q.; Wang, H.; Gao, Y.; Jiao, Y.; Liu, F.; Dong, Z. Investigation of the low-temperature properties and cracking resistance of fiber-reinforced asphalt concrete using the DIC technique. *Eng. Fract. Mech.* **2020**, *229*, 106951. [CrossRef]
15. Fu, Q.; Chen, X.; Qiu, X. Spatial distribution characterization of the Temperature-induced gradient viscoelasticity inside asphalt pavement. *Constr. Build. Mater.* **2022**, *346*, 128454. [CrossRef]
16. Ma, B.; Zhou, X.; Liu, J.; You, Z.; Wei, K.; Huang, X. Determination of Specific Heat Capacity on Composite Shape-Stabilized Phase Change Materials and Asphalt Mixtures by Heat Exchange System. *Materials* **2016**, *9*, 389. [CrossRef] [PubMed]
17. Han, D.; Liu, G.; Zhao, Y.; Pan, Y.; Yang, T. Research on thermal properties and heat transfer of asphalt mixture based on 3D random reconstruction technique. *Constr. Build. Mater.* **2021**, *270*, 121393. [CrossRef]
18. Shamsaei, M.; Carter, A.; Vaillancourt, M. A review on the heat transfer in asphalt pavements and urban heat island mitigation methods. *Constr. Build. Mater.* **2022**, *359*, 129350. [CrossRef]
19. Barber, E.S. Calculation of Maximum Pavement Temperatures from Weather Reports. *Highw. Res. Board Bull.* **1957**, *168*, 00237705.
20. Christison, J.T.; Anderson, K.O. The Response of asphalt pavements to low temperature climatic environments. In Proceedings of the Third International Conference on the Structural Design of Asphalt Pavements, London, UK, 11–15 September 1972.

21. Roque, R.; Buth, B.E. Materials characterization and response of flexible pavements at low temperatures (with discussion). In *Association of Asphalt Paving Technologists Proceedings*; AAPT: Lino Lakes, MN, USA, 1987.
22. Lytton, R.L.; Pufahl, D.E.; Michalak, C.H.; Liang, H.S.; Dempsey, B.J. An Integrated Model of the Climatic Effects on Pavements. Final Report. 1993. Available online: https://api.semanticscholar.org/CorpusID:126682412.1993.11.1 (accessed on 17 June 2024).
23. Zhao, X.; Shen, A.; Ma, B. Temperature response of asphalt pavement to low temperatures and large temperature differences. *Int. J. Pavement Eng.* **2020**, *21*, 49–62. [CrossRef]
24. Chen, J.; Wang, H.; Zhu, H. Analytical approach for evaluating temperature field of thermal modified asphalt pavement and urban heat island effect. *Appl. Therm. Eng.* **2017**, *113*, 739–748. [CrossRef]
25. Si, W.; Ma, B.; Zhou, X.Y.; Ren, J.P.; Tian, Y.X.; Li, Y. Temperature responses of asphalt mixture physical and finite element models constructed with phase change material. *Constr. Build. Mater.* **2018**, *178*, 529–541. [CrossRef]
26. Tran, T.T.; Nguyen, H.H.; Pham, P.N.; Nguyen, T.; Nguyen, P.Q.; Huynh, H.N. Temperature-related thermal properties of paving materials: Experimental analysis and effect on thermal distribution in semi-rigid pavement. *Road Mater. Pavement Des.* **2023**, *24*, 2759–2779. [CrossRef]
27. Hall, M.R.; Dehdezi, P.K.; Dawson, A.R.; Grenfell, J.; Isola, R. Influence of the thermophysical properties of pavement materials on the evolution of temperature depth profiles in different climatic regions. *J. Mater. Civ. Eng.* **2012**, *24*, 32–47. [CrossRef]
28. Kang, X.; Zheng, Y.; Zhang, H. Finite element analysis of temperature field and stress field of asphalt pavement based on continuous temperature change. *Highway* **2020**, *65*, 191–195.
29. Middel, A.; Turner, V.; Schneider, F.; Zhang, Y.; Stiller, M. Solar reflective pavements—A policy panacea to heat mitigation? *Environ. Res. Lett.* **2020**, *15*, 064016. [CrossRef]
30. Sun, Y.; Du, C.; Gong, H.; Li, Y.; Chen, J. Effect of temperature field on damage initiation in asphalt pavement: A microstructure-based multiscale finite element method. *Mech. Mater.* **2020**, *144*, 103367. [CrossRef]
31. Llopis-Castelló, D.; García-Segura, T.; Montalbán-Domingo, L.; Sanz-Benlloch, A.; Pellicer, E. Influence of Pavement Structure, Traffic, and Weather on Urban Flexible Pavement Deterioration. *Sustainability* **2020**, *12*, 9717. [CrossRef]
32. Hu, J.; Ma, T.; Zhu, Y.; Huang, X.; Xu, J.; Chen, L. High-viscosity modified asphalt mixtures for double-layer porous asphalt pavement: Design optimization and evaluation metrics. *Constr. Build. Mater.* **2021**, *271*, 121893. [CrossRef]

buildings

MDPI

Article

Influence of Three Different Antistripping Agents on Moisture Susceptibility, Stiffness, and Rutting Resistance of Hot-Mix Asphalt

Mario Orozco [1], Jaime Preciado [1], Gilberto Martinez-Arguelles [1,*], Luis Fuentes [1], Lubinda F. Walubita [2] and Rodrigo Polo-Mendoza [1,3]

1 Department of Civil & Environmental Engineering, Universidad del Norte, Barranquilla 081007, Colombia
2 Texas A&M Transportation Institute (TTI), The Texas A&M University System, College Station, TX 77843, USA
3 Faculty of Science, Charles University, 116 36 Prague, Czech Republic
* Correspondence: garguelles@uninorte.edu.co

Abstract: The construction and maintenance of road infrastructure is required for the sustained economic growth of communities and societies. Nonetheless, these activities imply the tangible risk of boosting the depletion of non-renewable resources (e.g., aggregates and binders). A widely used strategy for preserving as much of these natural resources as possible is the design of high-performance composite materials. For instance, antistripping agents (ASAs) are employed to mitigate the loss of adhesive bonding between asphalt binders and aggregates, enhancing the mechanical behaviour of hot-mix asphalts (HMAs). There is still no consensus on the effectiveness of ASAs. In this regard, the present research aims to contribute to the literature by conducting a case study on the influence of three different ASAs (hydrated lime, an amines-based liquid additive, and a silanes-based liquid additive) on the moisture susceptibility, stiffness, and rutting resistance of HMA. For these purposes, indirect tensile strength, indirect tensile stiffness modulus, and uniaxial cyclic compression tests were carried out. Overall, the involved experimental protocol drew the main conclusion that the incorporation of hydrated lime as a mineral filler (at a content of 1.68% by dry weight of aggregates) is capable of improving the mechanical performance of HMAs through decreases in humidity sensitivity and permanent deformation, together with a slight increase in rigidity.

Keywords: antistripping agents; asphalt mixtures; flexible pavements; hot-mix asphalt; moisture susceptibility; rutting resistance; stiffness

1. Introduction

The literature has well documented that the stripping phenomenon undesirably contributes to the deterioration of asphalt pavements by catalyzing/generating distresses such as raveling, aggregate loss, cracking, and rutting [1–5]. These surface distresses can seriously compromise safety, serviceability, and structural performance, particularly at high vehicle-operational speeds [6–8]. In this regard, several research efforts have proposed the modification of asphalt materials with antistripping agents (ASAs) in order to mitigate the loss of adhesive bonding between the asphalt binder and aggregate interface [9–11]. Table 1 presents a summary of several case studies. ASAs employ different mechanisms to enhance the adhesive bonding between the asphalt binder and aggregates, for example: (i) modifying the surface properties of aggregates and (ii) reducing the surface tension of the asphalt binder [3–5,12–15]. The specific physicochemical properties of aggregates and asphalt binders significantly impact the effectiveness of a particular ASA treatment [16–18]. Further, this effectiveness depends on an adequate dosage and the incorporation method [5,19,20]. Remarkably, there are two main groups of ASAs, namely liquid ASAs and mineral (powder) ASAs [10,21,22].

Liquid ASAs have been used for decades in the road infrastructure industry, especially those based on fatty amines, polyamines, amid amines, and liquid polymers [16,23,24]. Liquid ASAs have an adherence-promoting mechanism that reduces the surface tension between the aggregate surface and asphalt binder, thereby enhancing interfacial bonding [14,25,26]. Notably, liquid ASAs are usually used in low dosages (i.e., around 0.1–1% by weight of the asphalt binder) and can be incorporated as follows: (i) added directly to the aggregates, or (ii) mixed with the asphalt binder before mixing it with the aggregates [22,27,28]. Overall, when the liquid ASAs are added directly to the aggregates, a uniform distribution of the additive is not ensured due to the relatively low dosages typically provided, whilst the liquid ASAs may be better distributed in the asphalt matrix if they are incorporated into the asphalt binder at high temperatures [16,17,29]. Regarding mechanical performance, several case studies have proved that hot-mix asphalts (HMAs) modified with liquid ASAs (e.g., organosilane-based ASA and Iterlene In/400-S) exhibit an enhancement in resistance to moisture damage and fatigue life [3,10,21,23].

On the other hand, mineral ASAs are extra-fine solid powders with adhesion/cohesion-promoting properties over time, e.g., fly ash, hydrated lime (HL), and Portland cement [30–35]. Although HL has been traditionally used as a mineral filler rather than as an adhesion enhancer, this agent is the most widely used mineral ASA to enhance HMAs' behaviour [22,35–37]. HL improves the adhesion bonding between the aggregates and asphalt binder by modifying the chemistry of the aggregate surface through the interaction of calcium and silicates from lime and aggregates [38–40]. Notably, the interaction between the HL and asphalt binder differs from that of typical mineral fillers since, unlike typical mineral fillers, HL is highly chemically reactive and presents a relatively low molecular weight [41–43]. Moreover, HL reacts with the acidic components of the asphalt binder, such as carboxylic acids, to form insoluble products such as organic calcium salts [44–47]. Commonly, HL is used in dosages of around 1–2% by weight of dry aggregates [10,21,37]. It is essential to highlight that numerous investigations have agreed that HL can increase the stiffness and fatigue life of HMA while reducing moisture susceptibility [3,9,36,37,48–50].

Table 1. Summary of case studies focused on HMA modified with ASAs.

Additive		Technology	Additive Dosage (%)	Aggregate	Asphalt Binder [3,4]	Reference
Liquid ASA [1]	Wetfix	Alkylamines	0.2–0.6	Main compound: 50.95% CaO	60/70 PenG	[12]
			0.3	Amphibolite, clinkstone, spilite, granulite, basalt, mixed rock (ash rock, metatuf, spilite)	50/70 PenG	[51]
	Lilamin VP 75P	Fatty amines	0.2–0.6	Main compound: 50.95% CaO	60/70 PenG	[12]
	Zycotherm	Organosilanes	0.1	Amphibolite, clinkstone, spilite, granulite, basalt, mixed rock (ash rock, metatuf, spilite)	50/70 PenG	[51]
			0.1	Limestone	60/70 PenG	[52]
			0.1	Siliceous	PG 64-22	[28]
			0.15	Limestone	PG 64-22	[53]
	AdHere LOF 65-00	Amidoamines	0.3	Amphibolite, clinkstone, spilite, granulite, basalt, mixed rock (ash rock, metatuf, spilite)	50/70 PenG	[51]
			0.25, 0.5, 0.75	Grinded limestone	PG 64-22	[54]
	Evonik	Polyamines	0.3	Limestone	60/70 PenG	[52]

Table 1. *Cont.*

Additive		Technology	Additive Dosage (%)	Aggregate	Asphalt Binder [3,4]	Reference
Liquid ASA [1]	Zycosil	Organosilanes	0.5–4.5	Granite	60/70 PenG	[4]
			0.5	Siliceous, limestone	60/70 PenG	[3]
			0.5	Limestone, slag-limestone, quartzite, granite, andesite	60/70 PenG	[10]
	Iterlene ln/400	Alkylamido-imidazo-polyamine	0.2, 0.3, 0.4	Main compounds: 58.97% SiO_2, 17.37% Al_2O_3	85/100 PenG	[21]
			0.2, 0.4, 0.6	Granite, dolomite	50/70 PenG	[55]
Mineral ASA [2]	HL	Dry HL to wet aggregates procedure	1, 1.5, 2	Limestone, basalt	50/70 PenG	[37]
		HL slurry to aggregates procedure	2	Limestone, quartzite, granite, andesite	60/70 PenG	[10]
		Dry HL to wet aggregates procedure	1	45% limestone type, 55% gravel type	PG 70-28 and PG 64-22	[9]
		As mineral filler	1	50% limestone, 15% No. 10 screenings, 25% natural sand, 10% manufactured sand	PG 64-22	[22]
		As mineral filler	1, 2	Limestone	60/70 PenG	[52]
		Mixed with asphalt binder	1, 1.5, 2	58.97% SiO_2, 17.37% Al_2O_3	85/100 PenG	[21]
	Fly ash	Dry fly ash to wet aggregates procedure	1.13	45% limestone type, 55% gravel type	PG 70-28 and PG 64-22	[9]
		As mineral filler	1	50% Limestone, 15% No. 10 screenings, 25% natural sands, 10% manufactured sand	PG 64-22	[22]
		As mineral filler	0.5–5	Limestone	60/70 PenG	[11]
	Portlandcement	As mineral filler	0.5–5	Limestone	60/70 PenG	[11]

Legend: [1] Dosage by weight of asphalt binder; [2] modification by dry aggregates procedure; [3] PenG—penetration grade; [4] PG—performance grade.

Although there are a considerable number of research efforts on ASAs for HMAs, state-of-the-art research is almost entirely focused on moisture susceptibility without considering the interaction with other aspects of HMAs' mechanical performance. In this way, there is a clear gap in the literature. Consequently, in addition to evaluating moisture susceptibility, this research also evaluates the stiffness and rutting resistance of HMAs. Three ASAs were considered for these purposes: amines-based liquid additive (ALA), silanes-based liquid additive (SLA), and HL.

2. Materials

2.1. Aggregates

The natural aggregates (i.e., coarse and fine aggregates) used in this research were obtained from a quarry located in the department of Atlántico (northern region of Colombia). These aggregates mainly comprise sedimentary rocks of marine origin. Table 2 shows the physical properties of these aggregates and their comparison against the Colombian specification limits [56]. Additionally, Figure 1 exhibits the grain size distribution adopted for the design of asphalt concrete, which corresponds to an HMA with a nominal maximum aggregate size of 19 mm [56]. It is worth mentioning that this type of dense HMA is traditionally used in Colombia for highways and other high-traffic roads [57].

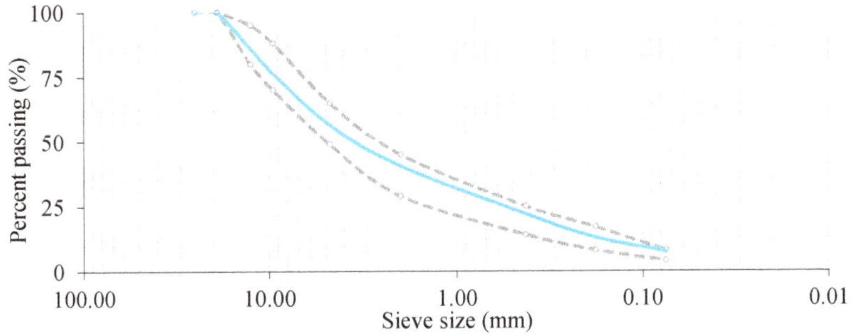

Figure 1. Grain size distribution adopted in this research. Color coding: grey—Colombian requirements for passing percentage; blue—granulometry used.

Table 2. Physical properties of the natural aggregates.

Test	Value (%)	Specification Limits (%)	Standard Method	Reference
LA abrasion loss	18.4	<30	ASTM C131	[58]
Fractured particles in one face	94.7	-	ASTM D5821	[59]
Flakiness index	11.0	<25	BS-EN-933	[60]
Sand equivalent	54.0	>50	AASHTO T176	[61]
Sodium sulfate soundness (fine aggregate)	7.8	<12	AASHTO T104	[62]
Sodium sulfate soundness (coarse aggregate)	5.1	<8	AASHTO T104	[62]
Bulk specific gravity (fine aggregate)	2.65	-	ASTM C128	[63]
Bulk specific gravity (coarse aggregate)	2.53	-	ASTM C127	[64]

2.2. Base Asphalt Binder

In this study, a PG 64-22 asphalt binder was utilized as the base binder. This asphalt binder was refined in the central Colombian refinery, which is located in the department of Santander (northeastern region of Colombia). Table 3 summarizes the fundamental physical properties of this asphalt binder, namely softening point, penetration grade, penetration index, and rotational viscosity. Likewise, Figure 2 shows the rheological properties, expressed in terms of the viscosity and master curve.

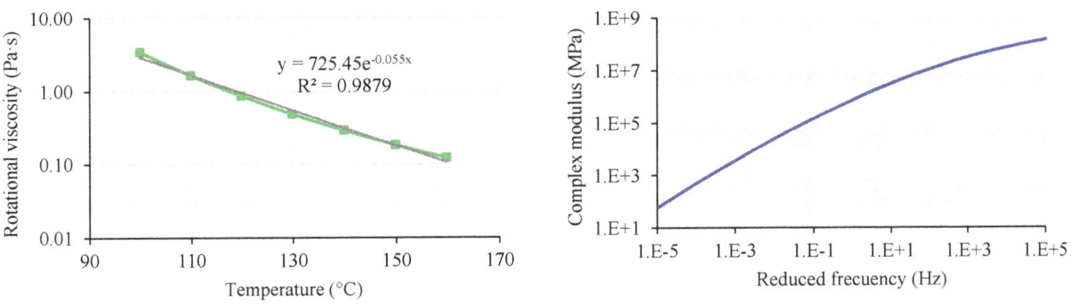

(**a**) Viscosity curve (**b**) Master curve at a reference temperature of 25 °C

Figure 2. Rheological characterization of the base asphalt binder.

Table 3. Physical properties of the base asphalt binder.

Test	Unit	Value (%)	Standard Method	Reference
Softening point	°C	49.7	ASTM D36/D36M	[65]
PenG	dmm	58.3	ASTM D5/D5M	[66]
Penetration index (PI)	-	−0.93	ASTM D5/D5M	[66]
Rotational viscosity @135 °C	Pa·s	0.432	ASTM D4402/D4402M	[67]
Rotational viscosity @160 °C	Pa·s	0.109	ASTM D4402/D4402M	[67]

2.3. ASAs

This research evaluated one mineral ASA (i.e., HL) and two liquid ASAs (i.e., ALA and SLA). In order to avoid commercial advertising, the commercial/trade names of these liquid additives are not reported in this manuscript, but the main generic chemical compositions are informed. The physicochemical properties of the liquid ASAs are presented in Table 4. Also, Figure 3 exhibits the physical appearance of these additives.

(**a**) Amines-based liquid additive

(**b**) Silanes-based liquid additive

Figure 3. Physical appearance of the liquid ASAs.

Table 4. Physical and chemical properties of the liquid ASAs.

	Color	Density (g/cm^3)	Solubility	pH	Freezing Point (°C)	Viscosity (cPs)
ALA	Amber	0.88	Insoluble	10–11	-	-
SLA	Pale yellow	1.01	Miscible with water	10	5	100–500

2.4. HMA Design

For the HMA design, the liquid ASAs were incorporated as asphalt binder modifying agents. Following the manufacturer's recommendations, the ALA and SLA were used at dosages of 0.5 and 0.1% by weight of asphalt binder, respectively. Meanwhile, the HL was included as a mineral filler inside the natural aggregates at three different dosages (0.56, 1.12, and 1.68% by dry weight of aggregates); these HL dosages are equivalent to 10, 20, and 30% by asphalt binder weight, respectively, and were adopted following the manufacturer's suggestions. Only to determine production temperatures, the HL was employed as an additive to the asphalt binder. In this regard, the rotational viscometer test was conducted according to the ASTM D4402/D4402M standard [67]. Table 5 shows the mixing and compaction temperatures based on viscosity limits of 0.17 ± 0.02 and 0.28 ± 0.03 Pa·s, respectively.

Table 5. Production temperature ranges.

Binder	Abbreviation	Mixing Temperatures (°C)	Compaction Temperature (°C)
Base asphalt binder	B-AB	150–155	140–144
HL (0.56%)-modified asphalt binder	HL056-AB	157–162	146–150
HL (1.12%)-modified asphalt binder	HL112-AB	163–168	152–157
HL (1.68%)-modified asphalt binder	HL168-AB	164–169	153–158
ALA-modified asphalt binder	ALA-AB	148–153	138–143
SLA-modified asphalt binder	SLA-AB	148–153	138–143

Table 5 reveals that the asphalt binders modified with liquid ASAs (ALA-AB and SLA-AB) developed almost the same (in fact, slightly lower) production temperatures as those associated with B-AB. Conversely, HL-modified asphalt binders (i.e., HL056-AB, HL112-AB, and HL168-AB) yielded considerably higher mixing and compaction temperatures. Notably, as the HL content augments, the production temperatures also increase incrementally. The preceding is a crucial factor to consider in the environmental and economic performance of HMAs [68–70]. The higher these temperatures are, the greater the potential environmental impacts and the lower the cost-effectiveness of their fabrication [71–73]. Although these aspects are outside the boundaries of this investigation, they must be evaluated in future research efforts (in order to accomplish a comprehensive assessment).

Another potential downside to having higher production temperatures in the HL-modified HMAs is that the asphalt binder (and, subsequently, the fabricated asphalt mixtures) would suffer a more severe short-term thermal aging process [29,74,75]. Unfortunately, increased thermal aging is highly associated with lower cracking resistance and inferior long-term performance [29,76–78]. Therefore, it is recommended that future research lines experimentally investigate the impact of HL on the aging resistance of modified asphalt materials.

In accordance with Colombian regulations [56], the Marshall method was employed (in the laboratory) to produce the HMA samples and determine their optimum asphalt content. It is important to note that the Colombian specification adopts the ASTM D6926-20 standard [79] for these purposes. Thus, the HMA cylindrical specimens (diameter of 101.6 mm and height of 63.5 mm) were manufactured using an automated Marshall hammer. In all cases, the optimum asphalt binder content was 5.6% (by total weight of the mixture). On the other hand, the air voids (also called voids in total mix) were dependent on the type of asphalt binder. Figure 4 shows these results. From this graph, the following findings can be drawn: (i) a low HL content (0.56%) raises the percentage of air voids, while higher dosages (1.12 and 1.68%) reduce it significantly, and (ii) the ALA generates an outstanding increase in air voids, whilst SLA provokes a slight decrease.

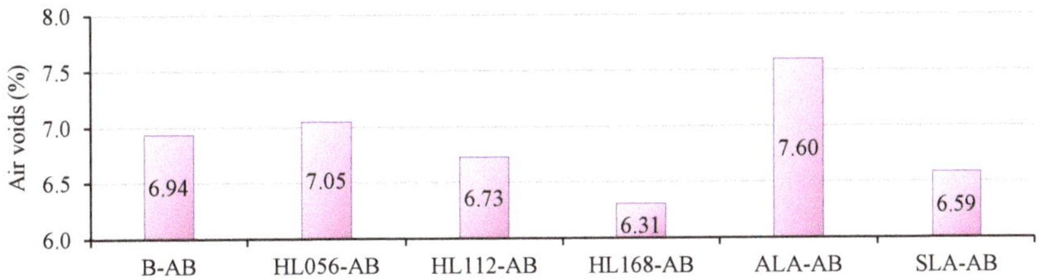

Figure 4. Influence of the type of asphalt binder on the HMA's air voids.

2.5. Modified Asphalt Binders

The rheological characterization of the modified asphalt binders was performed using the dynamic shear rheometer device according to the ASTM D7552 standard [80]. This test was conducted for loading frequencies between 0.016–16 Hz at six temperatures (13,

25, 37, 52, 64, and 76 °C), according to the AASHTO TP-101-14 standard [81]. In this way, a 25 mm diameter plate was employed for temperatures higher than 37 °C, whilst an 8 mm diameter plate was used for the other temperatures. Based on time-superposition principles [82–86], master curves of the complex shear modulus for the asphalt binders were generated. Accordingly, the Christensen–Anderson–Marasteanu model (Equations (1) and (2)) was used to generate these master curves [87–90]. It is essential to highlight that 25 °C was selected as the reference temperature (which represents a close approximation to the ambient temperature). More details on the master curve generation procedure can be found in Walubita et al. [82]. Figure 5 shows the results associated with this test. The reported results represent an average of the three replicate specimens.

$$G^* = G_g\left[1+\left(\frac{f_c}{f}\right)^v\right]^{-\frac{w}{v}} \tag{1}$$

$$R = \frac{\log_{10}(2)}{v} \tag{2}$$

where

- G^* is the complex shear modulus;
- G_g is the glassy modulus;
- f_c is the loading frequency;
- f is the reduced frequency;
- w, v are the experimental curve fitting coefficients;
- R is the rheological index.

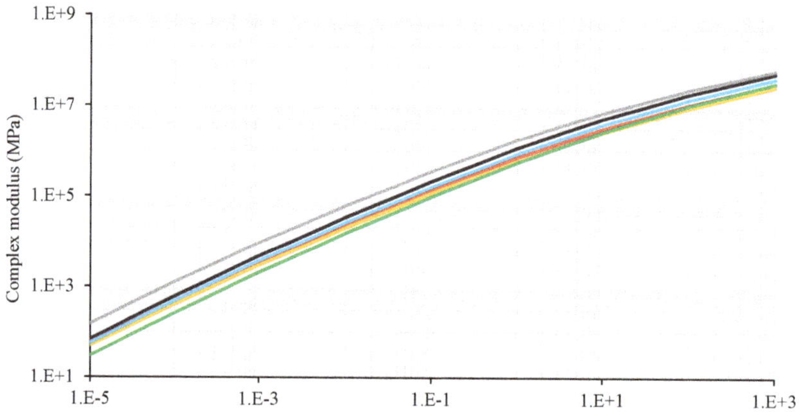

Figure 5. Rheological characterization for the studied asphalt binders. Color coding: red—B-AB; blue—HL056-AB; black—HL112-AB; grey—HL168-AB; green—ALA-AB; orange—SLA-AB.

According to Figure 5, the addition of liquid ASAs does not significantly affect the rheological behaviour of the asphalt binder. However, a slight decrease in the complex shear modulus can be seen when modifying the asphalt binder with ALA. As theoretically expected, the HL168-AB exhibited the highest magnitude of the complex shear modulus, consecutively followed by HL112-AB, HL056-AB, B-AB, SLA-AB, and ALA-AB. The previous trend can be explained by the theory of suspension rheology, which states that the increase in the stiffness of the colloidal systems (e.g., an asphalt binder) upon the addition of solid particles is governed by the volume fraction of the solid particles [91–94]. On the other hand, it is also notable that the augment of the stiffness (in terms of the complex shear modulus) is not directly proportional to the concentration of HL. Notably, higher variations in the complex modulus were observed in the range of 1.12–1.68% of HL than in the range of 0.56–1.12%. This increase in the complex shear modulus is strongly related to particle contacts and interaction at high solid particle concentrations [15,95]. Overall, Figure 5 also indicates that the effects of the ASAs on the rheological behaviour of the base asphalt binder are more pronounced at low loading frequencies, corresponding to the high-temperature domain [96–98].

3. Methods and Results

3.1. Moisture Susceptibility

One of the main aspects that influence the service life of asphalt pavements is the resistance to moisture damage reached by the HMA layers [12,29,99,100]. Notably, moisture susceptibility is associated with loss of adhesion between the asphalt binder and the aggregates, also known as the stripping phenomenon [22,26,52,101]. The ASTM D4867 standard [102] was used to evaluate the moisture susceptibility of the HMAs by determining the tensile strength ratio (TSR). Commonly, a TSR of 80% is used as the minimum acceptance criterion according to the requirements of different agencies [103–106], including the Colombian specifications [56]. Hence, the indirect tensile strength (ITS) test was conducted on HMA samples (cylindrical specimens with 101.6 mm of diameter and 63.5 mm of thickness) in dry and wet conditions. Notably, the ITS test involves loading a cylindrical specimen with a vertical diametric compressive force at a monotonic deformation rate of 50 mm/min until the crack failure occurs [107–110]. The ITS of each specimen is determined using Equation (3) [102]. Meanwhile, the TSR is determined using Equation (4) [102]. These outcomes are presented in Figure 6; the reported results are the average value regarding three samples.

$$ITS = \frac{2 \cdot P}{\pi \cdot d \cdot t} \quad (3)$$

$$TSR = \frac{ITS_{wet}}{ITS_{dry}} \times 100\% \quad (4)$$

where

- P is the peak load;
- d is the diameter of the specimen;
- t is the thickness of the specimen;
- ITS_{wet} is the indirect tensile strength measured in the wet state;
- ITS_{dry} is the indirect tensile strength measured in the dry state.

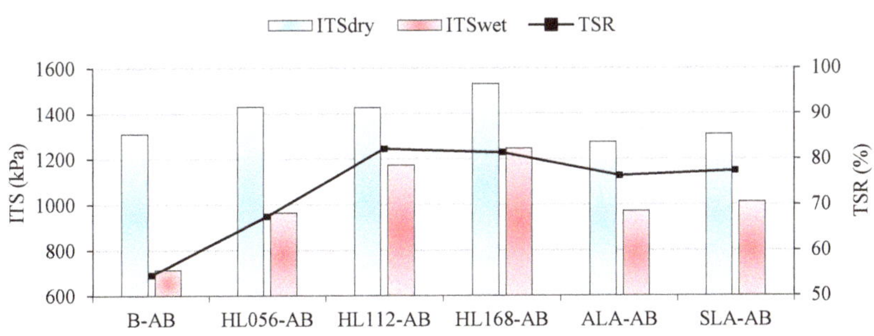

Figure 6. ITS and TSR results.

According to Figure 6, only the HL112-AB and HL168-AB mixtures achieve a TSR value higher than 80%. The preceding means that the mineral ASA technology provokes a better moisture resistance than the liquid ASAs, even when the control HMA performs poorly. For the ITS_{dry}, the HL168-AB and HL112-AB mixtures presented an increase of 17% and 8% regarding the control HMA (i.e., B-AB mixture), respectively. Among the HL mixtures, the HL056-AB exhibited the most deficient performance with lower ITS_{wet} and

TSR values. Overall, Figure 6 shows that adding ASAs causes some positive effects in reducing moisture susceptibility. However, the liquid ASAs yield a similar impact, and both fail to meet an acceptable TSR value.

3.2. Stiffness

For the dynamic characterization of the HMAs' stiffness, the indirect tensile stiffness modulus test was conducted following the BS-EN-12697-26 standard [111]. This non-destructive test can measure the rigidity of HMAs at different temperatures and loading frequencies. This study tested the HMA specimens at three temperatures (5, 25, and 40 °C) and four different load frequencies (1, 2, 4, and 8 Hz). Thus, the stiffness modulus was computed using Equation (5) [111].

$$E^* = \frac{F \times (v + 0.27)}{t \times \Delta H} \tag{5}$$

where

- E^* is the stiffness modulus;
- F is the peak value of the applied vertical load;
- v is the Poisson's ratio;
- t is the mean thickness of the cylindrical HMA specimen;
- ΔH is the amplitude of the horizontal deformation.

For the data interpretation, master curves of the stiffness modulus were constructed. Master curves are traditionally used to describe the time–temperature dependency of the HMA and were fitted using the sigmoidal model and the time–temperature superposition principle [82,112,113]. The principle for generating master curves consists of fitting a sigmoidal curve to the measured stiffness modulus using nonlinear least square regression techniques [114–116]. The amount of shifting required at each temperature to match the master curve describes the temperature dependency of the HMA [117–119]. These shifting factors at each temperature are determined simultaneously with other coefficients of the sigmoidal function, as shown in Equations (6) and (7) [120,121]. In this regard, the Arrhenius shifting model (Equation (8)) was used to determine $\log a_T$ for each test temperature [82,122]. For these purposes, a reference temperature of 25 °C (which is a close representation of the ambient temperature) was adopted in this research.

$$\log_{10}(|E^*|) = \delta + \frac{\alpha}{1 + e^{\beta - \gamma \times \log_{10}(f_r)}} \tag{6}$$

$$\log_{10}(f_r) = \log_{10}(f) + \log_{10}(a_T) \tag{7}$$

$$\log_{10}(a_T) = C_a\left(\frac{1}{T} - \frac{1}{T_r}\right) \tag{8}$$

where

- α, β, δ, and γ are the sigmoidal functional coefficients (fitting parameters);
- f_r is the reduced frequency;
- f is the frequency at the test temperature;
- a_T is the horizontal shift factor at temperature T;
- T is the test temperature (K);
- T_r is the reference temperature (K);
- C_a is a material constant, a function of the activation energy and ideal gas constant.

Figure 7 exhibits the master curves at a reference temperature of 25 °C for all the evaluated HMAs. This graph reveals no significant difference in the stiffness modulus of the HMAs in the high-frequency domain, where all the curves are nearly overlapping each other. Nonetheless, for small magnitudes (mainly at lower frequencies), the HL168-AB mixture yields the higher stiffness modulus, closely followed by the HL112-AB mixture. Notably, the alterations caused by the ASAs are considerably less significant in the high-frequency domain. In order to guarantee proper visualization, Figure 7 only shows a reference temperature (25 °C); however, the other temperatures maintain a behaviour consistent with that previously described. Table 6 highlights the average effect of ASAs on the stiffness modulus of the modified HMA regarding the control mixture for the three considered temperatures (5, 25, and 40 °C). Through this table, it is possible to draw the following findings: (i) the effect of the ALA on the stiffness modulus is negative and more evident at high temperatures; (ii) the SLA induces a complex behaviour without a clear trend; (iii) all the HL-modified HMAs exhibited an increase (i.e., positive percentages) in the stiffness modulus, but the variation in the HL056-AB mixture was below 10%; and (iv) the rigidity augments caused by the HL are more prominent at high temperatures, which is desirable to generate an elevated resistance to the rutting phenomena.

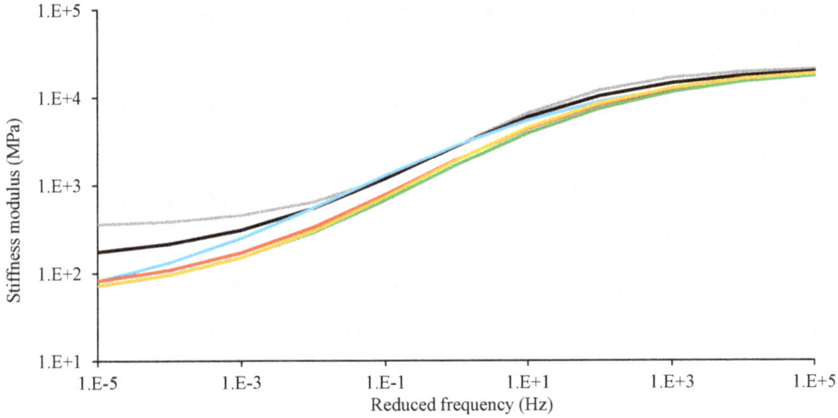

Figure 7. Master curves at 25 °C for the different HMAs. Color coding: red—B-AB; blue—HL056-AB; black—HL112-AB; grey—HL168-AB; green—ALA-AB; orange—SLA-AB.

Table 6. Average variations of stiffness modulus regarding the B-AB mixture.

Temperature (°C)	Frequency (Hz)	ALA-AB	SLA-AB	HL056-AB	HL112-AB	HL168-AB
5	8	+0%	+3%	+4%	+14%	+28%
	4	−3%	+2%	+4%	+16%	+26%
	2	−4%	+1%	+4%	+15%	+27%
	1	−5%	+0%	+4%	+12%	+27%
25	8	−16%	−1%	+29%	+33%	+36%
	4	−14%	−1%	+33%	+36%	+40%
	2	−12%	+0%	+40%	+42%	+46%
	1	−11%	+0%	+45%	+45%	+51%
40	8	−13%	+1%	+43%	+59%	+97%
	4	−17%	+3%	+49%	+66%	+101%
	2	−17%	+0%	+49%	+68%	+98%
	1	−18%	−7%	+52%	+65%	+95%

3.3. Rutting Resistance

The uniaxial cyclic compression test was employed to evaluate the rutting resistance (i.e., accumulation of permanent deformation) of the HMAs following the BS-EN-12697-25 standard [123]. Thus, a preload consisting of 600 s and axial stress of 10 kPa was applied. Subsequently, a square pulse was applied with a 500 ms pulse width and a 500 ms rest period for 3600 cycles. Notably, the uniaxial cyclic compression test was performed at 40 °C (i.e., a relatively high temperature at which the HMA is most susceptible to rutting phenomena) with an axial stress of 100 kPa. The strain was measured using two linear variable displacement transducers. As in previous tests, for each HMA, three replicates were conducted on cylindrical specimens (with a diameter of 101.6 mm and a thickness of 63.5 mm). Figure 8 shows the results from this test, i.e., a plot that related the load cycles versus the permanent deformation accumulated throughout the time. Remarkably, the three HL-modified HMAs exhibited the same behaviour, so this graph only shows an asphalt mixture with HL content (i.e., HL168-AB mixture).

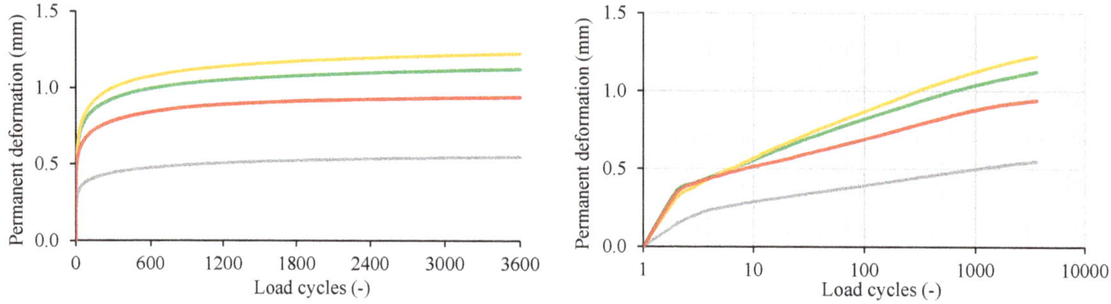

(**a**) Results on a linear scale (**b**) Results on logarithmic scale on the *x*-axis

Figure 8. Results of the uniaxial cyclic compression test. Color coding: red—B-AB; grey—HL168-AB; green—ALA-AB; orange—SLA-AB.

The main results obtained through Figure 8 are described below. Overall, all the HMAs evaluated in this case study exhibited typical characteristics of permanent deformation response under repeated loading, i.e., initial aggregate rearrangement and consolidation with the material hardening stage followed by a steady stage [124–126]. As expected, HL-modified HMA exhibited the lowest permanent deformation magnitude (at all load cycles), indicating more elevated rutting resistance than other designs (namely, the control mixture and HMAs modified with liquid ASAs). In fact, incorporating HL in the HMAs reduced about 41% of the permanent deformation (compared with the control mixture), suggesting a lower rutting potential and improvement in long-term performance. On the other hand, the liquid ASAs appear to have undesirably increased the permanent deformation (and, consequently, the rutting potential) of the HMAs. On average, the addition of the liquid ASAs increased the permanent deformation potential of the HMAs by about 20–57% throughout the load cycles. Notably, the SLA technology yields the worst performance, closely followed by the ALA technology.

4. Discussion

The laboratory tests conducted in this study showed the effects of HL, ALA, and SLA on the properties and performance of asphalt materials (i.e., both the asphalt binder and HMA). The different ASAs did not significantly alter the complex shear modulus of the asphalt binder; however, these small differences were more noticeable in the HL-modified asphalt binders. Regarding the HMA's moisture resistance, the two highest dosages of HL yield the higher TSR value. Meanwhile, in terms of stiffness, only the HL-modified mixtures present a notable enhancement; contrariwise, the liquid ASAs negatively affect the HMAs' rigidity. Similar results were reported for the rutting resistance, in which only the asphalt concrete with HL has less accumulated permanent deformation than the control mixture. The preceding laboratory evaluations indicate that the mineral ASA technology is superior to the liquid ASA technologies, at least under the conditions and boundaries adopted in this case study. Remarkably, these results are compatible with the recent literature, for instance, Abuawad et al. [5] and Sengul et al. [127].

Although the results of this investigation are promising, it is still early to establish definitive conclusions because (due to the study's boundaries) the cracking and fracture characterization of the asphalt materials were not performed. The preceding aspects must be addressed in future research lines in order to ensure holistic and inclusivity concerning all the field-performance indicators. Furthermore, it is also necessary to evaluate the mechanisms involved in the asphalt-HL interaction and the influence (at a molecular level) of the physicochemical composition of the raw materials. Identifying these aspects (those that must be explored in detail for the incoming literature) is the main contribution generated by this case study to advancing the state-of-the-art in HMAs.

Finally, it is essential to note that obtaining HMAs with better performance than conventional alternatives could significantly reduce the thicknesses of the HMA layers in asphalt pavements. Therefore, implementing antistripping agents generates environmental and economic implications that would be appropriate to evaluate. For these purposes, the integration of "life cycle assessment" and "life cycle cost analysis" methodologies would be suitable.

5. Conclusions

There is still no consensus in the literature on the impact of ASAs on the properties and performance of HMAs. In this regard, it is still necessary to carry out investigations that address particular cases in order to identify possible trends and improvement opportunities. Thus, this case study makes an important contribution to the state of the art by evaluating the effects of three different ASAs (an amines-based liquid additive, a silanes-based liquid additive, and hydrated lime) on the behaviour of typical Colombian asphalt mixtures. For the assessed materials and considered test conditions, the following key findings and conclusions were drawn:

- As measured and quantified through the complex shear modulus master curve, the HL additions improve the asphalt binder's temperature susceptibility and resistance to plastic shear flow, increasing the magnitude of the complex shear modulus. By contrast, the liquid ASAs did not significantly impact the asphalt binder's rheological response behaviour and temperature sensitivity;
- All the ASAs evaluated exhibited enhanced resistance to moisture susceptibility of the HMA, with the following order of superiority: HL168-AB, HL112-AB, SLA-AB, ALA-AB, and HL056-AB;
- Overall, the liquid ASAs tended to soften the asphalt materials, whilst using HL increased its stiffness;
- It was observed in the uniaxial cyclic compression test that the HL168-AB mixture had the highest resistance to permanent deformation, with minor rutting potential. On the other hand, adding ALA and SLA yields an undesirable increase in the HMA's permanent deformation by 20% and 57%, respectively;
- All the laboratory tests identified the HL168-AB mixture as the best performer among all the HMAs evaluated in the study;
- In addition to moisture susceptibility, it was evident that the inclusion of ASAs can affect other mechanical properties of the HMAs, such as stiffness and rutting resistance. The preceding prompts a balanced strategic approach (comprising multiple laboratory tests) for evaluating HMA performance. This means that the selection of ASAs should not be solely based on humidity sensitivity evaluation but a balanced consideration of other interactive factors and properties that would optimize the overall performance of the HMA.

Author Contributions: Conceptualization, G.M.-A. and L.F.; methodology, L.F.W.; software, M.O.; validation, G.M.-A., L.F. and L.F.W.; formal analysis, J.P.; investigation, M.O.; resources, G.M.-A.; data curation, M.O.; writing—original draft preparation, M.O.; writing—review and editing, J.P. and R.P.-M.; visualization, R.P.-M.; supervision, L.F.; project administration, G.M.-A.; funding acquisition, G.M.-A. All authors have read and agreed to the published version of the manuscript.

Funding: This research was funded by Administrative Department of Science, Technology, and Innovation (COLCIENCIAS), grant number 745/2016, and the APC was funded by Universidad del Norte.

Data Availability Statement: Data are contained within the article.

Conflicts of Interest: The authors declare no conflicts of interest.

References

1. Haghshenas, H.F.; Khodaii, A.; Saleh, M. Long Term Effectiveness of Anti-Stripping Agents. *Constr. Build. Mater.* **2015**, *76*, 307–312. [CrossRef]
2. Caro, S.; Masad, E.; Bhasin, A.; Little, D.N. Moisture Susceptibility of Asphalt Mixtures, Part 1: Mechanisms. *Int. J. Pavement Eng.* **2008**, *9*, 81–98. [CrossRef]
3. Ameri, M.; Kouchaki, S.; Roshani, H. Laboratory Evaluation of the Effect of Nano-Organosilane Anti-Stripping Additive on the Moisture Susceptibility of HMA Mixtures under Freeze-Thaw Cycles. *Constr. Build. Mater.* **2013**, *48*, 1009–1016. [CrossRef]
4. Behbahani, H.; Ziari, H.; Kamboozia, N.; Mansour Khaki, A.; Mirabdolazimi, S.M. Evaluation of Performance and Moisture Sensitivity of Glasphalt Mixtures Modified with Nanotechnology Zycosoil as an Anti-Stripping Additive. *Constr. Build. Mater.* **2015**, *78*, 60–68. [CrossRef]
5. Abuawad, I.M.A.; Al-Qadi, I.L.; Trepanier, J.S. Mitigation of Moisture Damage in Asphalt Concrete: Testing Techniques and Additives/Modifiers Effectiveness. *Constr. Build. Mater.* **2015**, *84*, 437–443. [CrossRef]
6. Fuentes, L.G.; Gunaratne, M.; de León Izeppi, E.; Flintsch, G.W.; Martinez, G. Determination of Pavement Macrotexture Limit for Use in International Friction Index Model. *Transp. Res. Rec.* **2012**, *16*, 138–143. [CrossRef]
7. Abudinen, D.; Fuentes, L.G.; Muñoz, J.S.C. Travel Quality Assessment of Urban Roads Based on International Roughness Index: Case Study in Colombia. *Transp. Res. Rec.* **2017**, *2612*, 1–10. [CrossRef]
8. Fuentes, L.; Camargo, R.; Arellana, J.; Velosa, C.; Martinez, G. Modelling Pavement Serviceability of Urban Roads Using Deterministic and Probabilistic Approaches. *Int. J. Pavement Eng.* **2021**, *22*, 77–86. [CrossRef]
9. Kim, Y.-R.; Pinto, I.; Park, S.-W. Experimental Evaluation of Anti-Stripping Additives in Bituminous Mixtures through Multiple Scale Laboratory Test Results. *Constr. Build. Mater.* **2012**, *29*, 386–393. [CrossRef]
10. Amelian, S.; Abtahi, S.M.; Hejazi, S.M. Moisture Susceptibility Evaluation of Asphalt Mixes Based on Image Analysis. *Constr. Build. Mater.* **2014**, *63*, 294–302. [CrossRef]
11. Likitlersuang, S.; Chompoorat, T. Laboratory Investigation of the Performances of Cement and Fly Ash Modified Asphalt Concrete Mixtures. *Int. J. Pavement Res. Technol.* **2016**, *9*, 337–344. [CrossRef]
12. Aksoy, A.; Şamlioglu, K.; Tayfur, S.; Özen, H. Effects of Various Additives on the Moisture Damage Sensitivity of Asphalt Mixtures. *Constr. Build. Mater.* **2005**, *19*, 11–18. [CrossRef]
13. Wasiuddin, N.M.; Howell, D.C.; Fogle, C.M.; Zaman, M.M.; O' Rear, E.A. Acid-Base Characteristics of an Asphalt Binder with and without Anti-Strip Additives. In *Airfield and Highway Pavement: Meeting Today's Challenges with Emerging Technologies*; ASCE: Reston, VA, USA, 2006; pp. 412–424. [CrossRef]
14. Xu, L.; Magar, S.; Zhao, Z.; Xiang, Q.; Xiao, F. Rheological and Anti-Moisture Characteristics of Rubberized Reclaimed Asphalt Pavement with Interfacial Bond Behavior. *J. Clean. Prod.* **2023**, *391*, 136172. [CrossRef]
15. Zhou, L.; Wang, S.; Zhang, J.; Zou, B.; Wang, M.; Zhang, W.; Lv, X.; Meng, D.; Hu, X.; Yao, Z.; et al. Developing and Testing of the Principle Prototype for Efficient Micro-Damage Fine Stripping of Asphalt on the Surface of Reclaimed Asphalt Pavement. *Machines* **2023**, *11*, 367. [CrossRef]
16. Putman, B.J.; Amirkhanian, S.N. *FHWA-SC-06-07: Laboratory Evaluation of Anti-Strip Additives in Hot Mix Asphalt*; Department of Civil Engineering, Clemson University: Clemson, SC, USA, 2006; pp. 1–78.
17. Al-Kheetan, M.J. Waste Not, Want Not: Sustainable Use of Anti-Stripping-Treated Waste Ceramic in Superpave Asphalt Mixtures. *Sustainability* **2023**, *15*, 7579. [CrossRef]
18. Gulzar, S.; Fried, A.; Preciado, J.; Castorena, C.; Underwood, S.; Habbouche, J.; Boz, I. Towards Sustainable Roads: A State-of-the-Art Review on the Use of Recycling Agents in Recycled Asphalt Mixtures. *J. Clean. Prod.* **2023**, *406*, 136994. [CrossRef]
19. Curtis, C. *SHRP-A/UIR-90-016: A Literature Review of Liquid Antistripping and Tests for Measuring Stripping*; Strategic Highway Research Program, National Research Council: Washington, DC, USA, 1990; pp. 1–19.
20. Epps, J.; Berger, E.; Anagnos, J.N. *Moisture Sensitivity of Asphalt Pavements—A National Seminar. Topic 4—Treatments*; Transportation Research Board: Washington, DC, USA, 2003; pp. 1–360.
21. Nazirizad, M.; Kavussi, A.; Abdi, A. Evaluation of the Effects of Anti-Stripping Agents on the Performance of Asphalt Mixtures. *Constr. Build. Mater.* **2015**, *84*, 348–353. [CrossRef]
22. Huang, B.; Shu, X.; Dong, Q.; Shen, J. Laboratory Evaluation of Moisture Susceptibility of Hot-Mix Asphalt Containing Cementitious Fillers. *J. Mater. Civ. Eng.* **2010**, *22*, 667–673. [CrossRef]
23. Moghadas Nejad, F.; Azarhoosh, A.R.; Hamedi, G.H.; Azarhoosh, M.J. Influence of Using Nonmaterial to Reduce the Moisture Susceptibility of Hot Mix Asphalt. *Constr. Build. Mater.* **2012**, *31*, 384–388. [CrossRef]
24. Mansour, F.; Vahid, V. Effect of Liquid Nano Material and Hydrated Lime in Improving the Moisture Behaviour of HMA. *Transp. Res. Procedia* **2016**, *17*, 506–512. [CrossRef]

25. Zhang, G.; Qiu, J.; Zhao, J.; Wei, D.; Wang, H. Development of Interfacial Adhesive Property by Novel Anti-Stripping Composite between Acidic Aggregate and Asphalt. *Polymers* **2020**, *12*, 473. [CrossRef] [PubMed]
26. Xu, F.; Nie, X.; Gan, W.; Hongzhi, E.; Xu, P.; Cao, H.; Gong, R.; Zhang, Y. Moisture Sensitivity Evaluation of the Asphalt Mortar-Aggregate Filler Interface Using Pull-Out Testing and 3-D Structural Imaging. *Coatings* **2023**, *13*, 868. [CrossRef]
27. Turbay, E.; Martinez-Arguelles, G.; Navarro-Donado, T.; Sanchez-Cotte, E.; Polo-Mendoza, R.; Covilla-Valera, E. Rheological Behaviour of WMA-Modified Asphalt Binders with Crumb Rubber. *Polymers* **2022**, *14*, 4148. [CrossRef]
28. Mansourkhaki, A.; Aghasi, A. Performance of Rubberized Asphalt Containing Liquid Nanomaterial Anti-Strip Agent. *Constr. Build. Mater.* **2019**, *214*, 468–474. [CrossRef]
29. Polo-Mendoza, R.; Martinez-Arguelles, G.; Walubita, L.F.; Moreno-Navarro, F.; Giustozzi, F.; Fuentes, L.; Navarro-Donado, T. Ultraviolet Ageing of Bituminous Materials: A Comprehensive Literature Review from 2011 to 2022. *Constr. Build. Mater.* **2022**, *350*, 128889. [CrossRef]
30. Behbahani, H.; Hamedi, G.H.; Najafi Moghaddam Gilani, V. Predictive Model of Modified Asphalt Mixtures with Nano Hydrated Lime to Increase Resistance to Moisture and Fatigue Damages by the Use of Deicing Agents. *Constr. Build. Mater.* **2020**, *265*, 120353. [CrossRef]
31. Zhao, X.; Ge, D.; Wang, J.; Wu, D.; Liu, J. The Performance Evaluation of Asphalt Mortar and Asphalt Mixture Containing Municipal Solid Waste Incineration Fly Ash. *Materials* **2022**, *15*, 1387. [CrossRef]
32. Paul, D.; Suresh, M.; Pal, M. Utilization of Fly Ash and Glass Powder as Fillers in Steel Slag Asphalt Mixtures. *Case Stud. Constr. Mater.* **2021**, *15*, e00672. [CrossRef]
33. Dong, Q.; Yuan, J.; Chen, X.; Ma, X. Reduction of Moisture Susceptibility of Cold Asphalt Mixture with Portland Cement and Bentonite Nanoclay Additives. *J. Clean. Prod.* **2018**, *176*, 320–328. [CrossRef]
34. Guha, A.H.; Assaf, G.J. Effect of Portland Cement as a Filler in Hot-Mix Asphalt in Hot Regions. *J. Build. Eng.* **2020**, *28*, 101036. [CrossRef]
35. Wei, Q.; Ashaibi, A.A.; Wang, Y.; Albayati, A.; Haynes, J. Experimental Study of Temperature Effect on the Mechanical Tensile Fatigue of Hydrated Lime Modified Asphalt Concrete and Case Application for the Analysis of Climatic Effect on Constructed Pavement. *Case Stud. Constr. Mater.* **2022**, *17*, e01622. [CrossRef]
36. Little, D.N.; Epps, J.A.; Sebaaly, P.E. *The Benefits of Hydrated Lime in Hot Mix Asphalt*; National Lime Association: Arlington, VA, USA, 2006; pp. 1–79.
37. Gorkem, C.; Sengoz, B. Predicting Stripping and Moisture Induced Damage of Asphalt Concrete Prepared with Polymer Modified Bitumen and Hydrated Lime. *Constr. Build. Mater.* **2009**, *23*, 2227–2236. [CrossRef]
38. Ismael, M.Q.; Ahmed, A.H. Effect of Hydrated Lime on Moisture Susceptibility of Asphalt Mixtures. *J. Eng.* **2019**, *25*, 89–101. [CrossRef]
39. Al Ashaibi, A.; Wang, Y.; Albayati, A.; Byzyka, J.; Scholz, M.; Weekes, L. Thermal Properties of Hydrated Lime-Modified Asphalt Concrete and Modelling Evaluation for Their Effect on the Constructed Pavements in Service. *Sustainability* **2022**, *14*, 7827. [CrossRef]
40. Liang, Y.; Bai, T.; Zhou, X.; Wu, F.; Chenxin, C.; Peng, C.; Fuentes, L.; Walubita, L.F.; Li, W.; Wang, X. Assessing the Effects of Different Fillers and Moisture on Asphalt Mixtures' Mechanical Properties and Performance. *Coatings* **2023**, *13*, 288. [CrossRef]
41. Little, D.N.; Petersen, J.C. Unique Effects of Hydrated Lime Filler on the Performance-Related Properties of Asphalt Cements: Physical and Chemical Interactions Revisited. *J. Mater. Civ. Eng.* **2005**, *17*, 207–218. [CrossRef]
42. Iwański, M.M. Effect of Hydrated Lime on Indirect Tensile Stiffness Modulus of Asphalt Concrete Produced in Half-Warm Mix Technology. *Materials* **2020**, *13*, 4731. [CrossRef] [PubMed]
43. İskender, C.; İskender, E.; Aksoy, A.; Şengül, C.E. Effect of Glass Cullet Size and Hydrated Lime—Nanoclay Additives on the Mechanical Properties of Glassphalt Concrete. *Sustainability* **2021**, *13*, 3284. [CrossRef]
44. Hesami, S.; Roshani, H.; Hamedi, G.H.; Azarhoosh, A. Evaluate the Mechanism of the Effect of Hydrated Lime on Moisture Damage of Warm Mix Asphalt. *Constr. Build. Mater.* **2013**, *47*, 935–941. [CrossRef]
45. Han, S.; Dong, S.; Liu, M.; Han, X.; Liu, Y. Study on Improvement of Asphalt Adhesion by Hydrated Lime Based on Surface Free Energy Method. *Constr. Build. Mater.* **2019**, *227*, 116794. [CrossRef]
46. Diao, H.; Ling, T.; Zhang, Z.; Peng, B.; Huang, Q. Multiscale Fatigue Performance Evaluation of Hydrated Lime and Basalt Fiber Modified Asphalt Mixture. *Materials* **2023**, *16*, 608. [CrossRef] [PubMed]
47. Albayati, A.; Wang, Y.; Haynes, J. Size Effect of Hydrated Lime on the Mechanical Performance of Asphalt Concrete. *Materials* **2022**, *15*, 3751. [CrossRef] [PubMed]
48. Diab, A.; You, Z.; Othman, A.M.; Ahmed, H.Y. Effect of Hydrated Lime Application Method on Mechanical and Fatigue Properties of HMA. In Proceedings of the 12th International Conference of Transportation Professionals (CICTP), Beijing, China, 3–6 August 2012; pp. 3327–3334.
49. Kok, B.V.; Yilmaz, M. The Effects of Using Lime and Styrene-Butadiene-Styrene on Moisture Sensitivity Resistance of Hot Mix Asphalt. *Constr. Build. Mater.* **2009**, *23*, 1999–2006. [CrossRef]
50. Buttlar, W.G.; Bozkurt, D.; Al-Khateeb, G.G.; Waldhoff, A.S. Understanding Asphalt Mastic Behavior through Micromechanics. *Transp. Res. Rec.* **1999**, *1681*, 157–169. [CrossRef]
51. Valentová, T.; Altman, J.; Valentin, J. Impact of Asphalt Ageing on the Activity of Adhesion Promoters and the Moisture Susceptibility. *Transp. Res. Procedia* **2016**, *14*, 768–777. [CrossRef]

52. Ameri, M.; Vamegh, M.; Chavoshian Naeni, S.F.; Molayem, M. Moisture Susceptibility Evaluation of Asphalt Mixtures Containing Evonik, Zycotherm and Hydrated Lime. *Constr. Build. Mater.* **2018**, *165*, 958–965. [CrossRef]
53. Faramarzi, M.; Golestani, B.; Lee, K.W. Improving Moisture Sensitivity and Mechanical Properties of Sulfur Extended Asphalt Mixture by Nano-Antistripping Agent. *Constr. Build. Mater.* **2017**, *133*, 534–542. [CrossRef]
54. Zhu, C.; Xu, G.; Zhang, H.; Xiao, F.; Amirkhanian, S.; Wu, C. Influence of Different Anti-Stripping Agents on the Rheological Properties of Asphalt Binder at High Temperature. *Constr. Build. Mater.* **2018**, *164*, 317–325. [CrossRef]
55. Paliukaite, M.; Vorobjovas, V.; Bulevičius, M.; Andrejevas, V. Evaluation of Different Test Methods for Bitumen Adhesion Properties. *Transp. Res. Procedia* **2016**, *14*, 724–731. [CrossRef]
56. *INVIAS Especificaciones Generales de Construcción de Carreteras*; Instituto Nacional de Vías: Bogotá, Colombia, 2022; pp. 1–1269.
57. Polo-Mendoza, R.; Mora, O.; Duque, J.; Turbay, E.; Martinez-Arguelles, G.; Fuentes, L.; Guerrero, O.; Perez, S. Environmental and Economic Feasibility of Implementing Perpetual Pavements (PPs) against Conventional Pavements: A Case Study of Barranquilla City, Colombia. *Case Stud. Constr. Mater.* **2023**, *18*, e02112. [CrossRef]
58. *ASTM C131/C131M-20*; Standard Test Method for Resistance to Degradation of Small-Size Coarse Aggregate by Abrasion and Impact in the Los Angeles Machine. ASTM International: West Conshohocken, PA, USA, 2020; pp. 1–5. [CrossRef]
59. *ASTM D5821-13*; Standard Test Method for Determining the Percentage of Fractured Particles in Coarse Aggregate. ASTM International: West Conshohocken, PA, USA, 2017; pp. 1–6. [CrossRef]
60. *BSI BS-EN-933*; Tests for Geometrical Properties of Aggregates. British Standards Institution: London, UK, 2012; pp. 1–8.
61. *AASHTO T-176*; Plastic Fines in Graded Aggregates and Soils by the Use of the Sand Equivalent Test. The American Association of State Highway and Transportation Official: Madison, WI, USA, 2017; pp. 1–10.
62. *AASHTO T-104-99*; Standard Method of Test for Soundness of Aggregate by Use of Sodium Sulfate or Magnesium Sulfate. The American Association of State Highway and Transportation Official: Madison, WI, USA, 2007; pp. 1–10.
63. *ASTM C128-22*; Standard Test Method for Relative Density (Specific Gravity) and Absorption of Fine Aggregate. ASTM International: West Conshohocken, PA, USA, 2022; pp. 1–6. [CrossRef]
64. *ASTM C127-15*; Standard Test Method for Density, Relative Density (Specific Gravity), and Absorption of Coarse Aggregate. ASTM International: West Conshohocken, PA, USA, 2015; pp. 1–5. [CrossRef]
65. *ASTM D36/D36M-14*; Standard Test Method for Softening Point of Bitumen (Ring-and-Ball Apparatus). ASTM International: West Conshohocken, PA, USA, 2020; pp. 1–5. [CrossRef]
66. *ASTM D5/D5M-20*; Standard Test Method for Penetration of Bituminous Materials. ASTM International: West Conshohocken, PA, USA, 2020; pp. 1–4. [CrossRef]
67. *ASTM D4402/D4402M-23*; Standard Test Method for Viscosity Determination of Asphalt at Elevated Temperatures Using a Rotational Viscometer. ASTM International: West Conshohocken, PA, USA, 2023; pp. 1–4. [CrossRef]
68. Polo-Mendoza, R.; Martinez-Arguelles, G.; Peñabaena-Niebles, R. Environmental Optimization of Warm Mix Asphalt (WMA) Design with Recycled Concrete Aggregates (RCA) Inclusion through Artificial Intelligence (AI) Techniques. *Results Eng.* **2023**, *17*, 100984. [CrossRef]
69. Polo-Mendoza, R.; Martinez-Arguelles, G.; Peñabaena-Niebles, R. A Multi-Objective Optimization Based on Genetic Algorithms for the Sustainable Design of Warm Mix Asphalt (WMA). *Int. J. Pavement Eng.* **2023**, *24*, 2074417. [CrossRef]
70. Sharma, A.; Rongmei Naga, G.R.; Kumar, P.; Rai, P. Mix Design, Development, Production and Policies of Recycled Hot Mix Asphalt: A Review. *J. Traffic Transp. Eng.* **2022**, *9*, 765–794. [CrossRef]
71. Polo-Mendoza, R.; Peñabaena-Niebles, R.; Giustozzi, F.; Martinez-Arguelles, G. Eco-Friendly Design of Warm Mix Asphalt (WMA) with Recycled Concrete Aggregate (RCA): A Case Study from a Developing Country. *Constr. Build. Mater.* **2022**, *326*, 126890. [CrossRef]
72. Calabi-Floody, A.T.; Valdés-Vidal, G.A.; Sanchez-Alonso, E.; Mardones-Parra, L.A. Evaluation of Gas Emissions, Energy Consumption and Production Costs of Warm Mix Asphalt (WMA) Involving Natural Zeolite and Reclaimed Asphalt Pavement (RAP). *Sustainability* **2020**, *12*, 6410. [CrossRef]
73. Polo-Mendoza, R.; Turbay, E.; Martinez-Arguelles, G.; Peñabaena-Niebles, R. Environmental Efficiency Assessment of RCA Inclusion as Partial Replacement of Virgin Aggregates in HMA and WMA. In *Roads and Airports Pavement Surface Characteristics*; Crispino, M., Toraldo, E., Eds.; CRC Press: Milan, Italy, 2023; pp. 778–787, ISBN 9781003429258.
74. Abouelsaad, A.; White, G. The Combined Effect of Ultraviolet Irradiation and Temperature on Hot Mix Asphalt Mixture Aging. *Sustainability* **2022**, *14*, 5942. [CrossRef]
75. Yang, X.; Shen, A.; Jiang, Y.; Meng, Y.; Wu, H. Properties and Mechanism of Flame Retardance and Smoke Suppression in Asphalt Binder Containing Organic Montmorillonite. *Constr. Build. Mater.* **2021**, *302*, 124148. [CrossRef]
76. Lyu, L.; Fini, E.H.; Pei, J.; Poulikakos, L.D. Aging Evolution and Sustainability Implications of Crumb Rubberized Asphalt Binder: A State-of-the-Art. *J. Clean. Prod.* **2024**, *434*, 140202. [CrossRef]
77. Liu, K.; Xu, Y.; Gao, D.; Yang, S. Rheological Properties and Microstructural Analysis of Asphalt Binder Doped with Anti-Stripping Agent after Aging. *Coatings* **2022**, *12*, 1848. [CrossRef]
78. Ghavibazoo, A.; Abdelrahman, M. Effect of Crumb Rubber Dissolution on Low-Temperature Performance and Aging of Asphalt-Rubber Binder. *Transp. Res. Rec.* **2014**, *2445*, 47–55. [CrossRef]
79. *ASTM D6926-20*; Standard Practice for Preparation of Asphalt Mixture Specimens Using Marshall Apparatus. ASTM International: West Conshohocken, PA, USA, 2020; pp. 1–8. [CrossRef]

80. *ASTM D7552-22*; Standard Test Method for Determining the Complex Shear Modulus (G*) of Asphalt Mixtures Using Dynamic Shear Rheometer. ASTM International: West Conshohocken, PA, USA, 2022; pp. 1–9. [CrossRef]
81. *AASHTO TP-101-14*; Standard Method of Test for Estimating Damage Tolerance of Asphalt Binders Using the Linear Amplitude Sweep. The American Association of State Highway and Transportation Official: Madison, WI, USA, 2012; pp. 1–7.
82. Walubita, L.F.; Alvarez, A.E.; Simate, G.S. Evaluating and Comparing Different Methods and Models for Generating Relaxation Modulus Master-Curves for Asphalt Mixes. *Constr. Build. Mater.* **2011**, *25*, 2619–2626. [CrossRef]
83. Alamdary, Y.A.; Singh, S.; Baaj, H. Laboratory Simulation of the Impact of Solar Radiation and Moisture on Long-Term Age Conditioning of Asphalt Mixes. *Road Mater. Pavement Des.* **2019**, *20*, S521–S532. [CrossRef]
84. Xu, H.; Sun, Y.; Chen, J.; Li, J.; Yu, B.; Qiu, G.; Zhang, Y.; Xu, B. Investigation into Rheological Behavior of Warm-Mix Recycled Asphalt Binders with High Percentages of RAP Binder. *Materials* **2023**, *16*, 1599. [CrossRef]
85. Naderi, K.; Nejad, F.M.; Khodaii, A. Time-Temperature-Age Superposition Validation for Linear Viscoelastic Properties of Bituminous Materials. *J. Mater. Civ. Eng.* **2018**, *30*, 04017292. [CrossRef]
86. Luo, R.; Hou, Q. Development of Time-Temperature-Humidity Superposition Principle for Asphalt Mixtures. *Mech. Mater.* **2021**, *156*, 103792. [CrossRef]
87. Zhu, Y.; Zhang, J.; Si, C.; Yan, T.; Li, Y. Laboratory Evaluation on Performance of Recycled Asphalt Binder and Mixtures under Short-Term Aging Conditions. *Sustainability* **2021**, *13*, 3404. [CrossRef]
88. Xi, L.; Luo, R.; Ma, Q.; Tu, C.; Shah, Y.I. An Improved Method to Establish Continuous Relaxation Spectrum of Asphalt Materials. *Constr. Build. Mater.* **2022**, *354*, 129182. [CrossRef]
89. Yusoff, N.I.M.; Hainin, M.R.; Airey, G.D. A Comparative Study of Phase Angle Predictive Equations Using Bituminous Binder Data. *Arab. J. Sci. Eng.* **2012**, *37*, 1571–1583. [CrossRef]
90. Vestena, P.M.; Schuster, S.L.; de Almeida, P., Jr.; Faccin, C.; Specht, L.P.; da Silva Pereira, D. Dynamic Modulus Master Curve Construction of Asphalt Mixtures: Error Analysis in Different Models and Field Scenarios. *Constr. Build. Mater.* **2021**, *301*, 124343. [CrossRef]
91. Lesueur, D. The Colloidal Structure of Bitumen: Consequences on the Rheology and on the Mechanisms of Bitumen Modification. *Adv. Colloid Interface Sci.* **2009**, *145*, 42–82. [CrossRef]
92. Nazockdast, E.; Morris, J.F. Microstructural Theory and the Rheology of Concentrated Colloidal Suspensions. *J. Fluid Mech.* **2012**, *713*, 420–452. [CrossRef]
93. Grmela, M.; Maîtrejean, G.; Chinesta, F.; Ammar, A. Kinetic Theory of Colloidal Suspensions: Morphology, Rheology, and Migration. *Rheol. Acta* **2013**, *52*, 557–577. [CrossRef]
94. Suzuki, K.; Hayakawa, H. Theory for the Rheology of Dense Non-Brownian Suspensions: Divergence of Viscosities and μ-J Rheology. *J. Fluid Mech.* **2019**, *864*, 1125–1176. [CrossRef]
95. Underwood, B.S.; Kim, Y.R. A Four Phase Micro-Mechanical Model for Asphalt Mastic Modulus. *Mech. Mater.* **2014**, *75*, 13–33. [CrossRef]
96. Cho, S.H.; Safavizadeh, S.A.; Kim, Y.R. Verification of the Applicability of the Time–Temperature Superposition Principle to Interface Shear Stiffness and Strength of GlasGrid-Reinforced Asphalt Mixtures. *Road Mater. Pavement Des.* **2017**, *18*, 766–784. [CrossRef]
97. Zou, G.; Zhang, X.; Xu, J.; Chi, F. Morphology of Asphalt Mixture Rheological Master Curves. *J. Mater. Civ. Eng.* **2010**, *22*, 806–810. [CrossRef]
98. Tan, G.; Wang, W.; Cheng, Y.; Wang, Y.; Zhu, Z. Establishment of Complex Modulus Master Curves Based on Generalized Sigmoidal Model for Freeze–Thaw Resistance Evaluation of Basalt Fiber-Modified Asphalt Mixtures. *Polymers* **2020**, *12*, 1698. [CrossRef] [PubMed]
99. Xiao, R.; Ding, Y.; Polaczyk, P.; Ma, Y.; Jiang, X.; Huang, B. Moisture Damage Mechanism and Material Selection of HMA with Amine Antistripping Agent. *Mater. Des.* **2022**, *220*, 110797. [CrossRef]
100. Xie, J.; Wu, S.; Pang, L.; Lin, J.; Zhu, Z. Influence of Surface Treated Fly Ash with Coupling Agent on Asphalt Mixture Moisture Damage. *Constr. Build. Mater.* **2012**, *30*, 340–346. [CrossRef]
101. Polo-Mendoza, R.; Navarro-Donado, T.; Ortega-Martinez, D.; Turbay, E.; Martinez-Arguelles, G.; Peñabaena-Niebles, R. Properties and Characterization Techniques of Graphene Modified Asphalt Binders. *Nanomaterials* **2023**, *13*, 955. [CrossRef]
102. *ASTM D4867/D4867M-22*; Standard Test Method for Effect of Moisture on Asphalt Mixtures. ASTM International: West Conshohocken, PA, USA, 2022; pp. 1–6. [CrossRef]
103. Xu, S.; Xiao, F.; Amirkhanian, S.; Singh, D. Moisture Characteristics of Mixtures with Warm Mix Asphalt Technologies—A Review. *Constr. Build. Mater.* **2017**, *142*, 148–161. [CrossRef]
104. Sapkota, K.; Yaghoubi, E.; Wasantha, P.L.P.; Van Staden, R.; Fragomeni, S. Mechanical Characteristics and Durability of HMA Made of Recycled Aggregates. *Sustainability* **2023**, *15*, 5594. [CrossRef]
105. Pasandín, A.R.; Pérez, I. Fatigue Performance of Bituminous Mixtures Made with Recycled Concrete Aggregates and Waste Tire Rubber. *Constr. Build. Mater.* **2017**, *157*, 26–33. [CrossRef]
106. Brondani, C.; Menezes Vestena, P.; Faccin, C.; Lisboa Schuster, S.; Pivoto Specht, L.; da Silva Pereira, D. Moisture Susceptibility of Asphalt Mixtures: 2S2P1D Rheological Model Approach and New Index Based on Dynamic Modulus Master Curve Changes. *Constr. Build. Mater.* **2022**, *331*, 127316. [CrossRef]

107. Zhang, J.; Li, P.; Sun, C.; Liang, M.; Li, Y.; Yao, Z.; Zhang, X. Effects of Composite Warm Mix Additive (CAR) on the Physical and Rheological Performance of Bitumen and the Pavement Performance of Its Concrete. *Materials* **2019**, *12*, 3916. [CrossRef] [PubMed]
108. Ali, A.; Abbas, A.; Nazzal, M.; Alhasan, A.; Roy, A.; Powers, D. Effect of Temperature Reduction, Foaming Water Content, and Aggregate Moisture Content on Performance of Foamed Warm Mix Asphalt. *Constr. Build. Mater.* **2013**, *48*, 1058–1066. [CrossRef]
109. Sanchez-Cotte, E.; Fuentes, L.; Martinez-Arguelles, G.; Rondon, H.; Walubita, L.; Cantero, J. Influence of Recycled Concrete Aggregates from Different Sources in Hot Mix Asphalt Design. *Constr. Build. Mater.* **2020**, *259*, 120427. [CrossRef]
110. Cantero-Durango, J.; Polo-Mendoza, R.; Martinez-Arguelles, G.; Fuentes, L. Properties of Hot Mix Asphalt (HMA) with Several Contents of Recycled Concrete Aggregate (RCA). *Infrastructures* **2023**, *8*, 109. [CrossRef]
111. *BSI BS-EN-12697-26*; Bituminous Mixtures—Test Methods for Hot Mix Asphalt. Part 26: Stiffness. British Standards Institution: London, UK, 2012; pp. 1–50.
112. Walubita, L.F.; Fuentes, L.; Lee, S.I.; Dawd, I.; Mahmoud, E. Comparative Evaluation of Five HMA Rutting-Related Laboratory Test Methods Relative to Field Performance Data: DM, FN, RLPD, SPST, and HWTT. *Constr. Build. Mater.* **2019**, *215*, 737–753. [CrossRef]
113. Cheng, H.; Sun, L.; Yang, R.; Zhang, Y.; Liu, L. Relating Field Moduli of Asphalt Mixture Layer Under Vehicular Loading and Its Dynamic Moduli Under Laboratory Loading. *Transp. Res. Rec.* **2022**, *2676*, 567–579. [CrossRef]
114. Ling, M.; Fuentes, L.; Liu, H.; Khalili, M.; Walubita, L.F. Evaluating Fatigue Resistance Based on Viscoelastic Properties of Asphalt Mixtures. *Int. J. Pavement Eng.* **2022**, *24*, 2126976. [CrossRef]
115. Dan, H.; Yang, P.; Cao, W.; Shan, H.; Zhang, Z. The Stiffness Behavior of Asphalt Mixtures with Different Compactness under Variable Confinement. *Materials* **2023**, *16*, 771. [CrossRef]
116. Booshehrian, A.; Mogawer, W.S.; Bonaquist, R. How to Construct an Asphalt Binder Master Curve and Assess the Degree of Blending between RAP and Virgin Binders. *J. Mater. Civ. Eng.* **2013**, *25*, 1813–1821. [CrossRef]
117. Gómez-Meijide, B.; Pérez, I.; Airey, G.; Thom, N. Stiffness of Cold Asphalt Mixtures with Recycled Aggregates from Construction and Demolition Waste. *Constr. Build. Mater.* **2015**, *77*, 168–178. [CrossRef]
118. Hasheminejad, N.; Vuye, C.; Margaritis, A.; Van den Bergh, W.; Dirckx, J.; Vanlanduit, S. Characterizing the Complex Modulus of Asphalt Concrete Using a Scanning Laser Doppler Vibrometer. *Materials* **2019**, *12*, 3542. [CrossRef]
119. Forough, S.A.; Nejad, F.M.; Khodaii, A. An Investigation of Different Fitting Functions to Accurately Model the Compressive Relaxation Modulus Master Curve of Asphalt Mixes. *Road Mater. Pavement Des.* **2015**, *16*, 767–783. [CrossRef]
120. Witczak, M.W.; Bari, J. *Development of a Master Curve (E*) Database for Lime Modidied Asphaltic Mixtures*; Arizona State University Research Report: Tempe, AZ, USA, 2004; pp. 1–29.
121. Pellinen, T.K.; Witczak, M.W.; Bonaquist, R.F. Asphalt Mix Master Curve Construction Using Sigmoidal Fitting Function with Non-Linear Least Squares. In *Recent Advances in Materials Characterization and Modeling of Pavement Systems*; ASCE: Reston, VA, USA, 2004; pp. 83–101. [CrossRef]
122. Nobakht, M.; Sakhaeifar, M.S. Dynamic Modulus and Phase Angle Prediction of Laboratory Aged Asphalt Mixtures. *Constr. Build. Mater.* **2018**, *190*, 740–751. [CrossRef]
123. *BSI BS-EN-12697-25*; Bituminous Mixtures—Test Methods for Hot Mix Asphalt. Part 25: Cyclic Compression Test. British Standards Institution: London, UK, 2005; pp. 1–29.
124. Mohammad, L.N.; Herath, A.; Rasoulian, M.; Zhang, Z. Laboratory Evaluation of Untreated and Treated Pavement Base Materials: Repeated Load Permanent Deformation Test. *Transp. Res. Rec.* **2006**, *1967*, 78–88. [CrossRef]
125. Zhang, J.; Alvarez, A.E.; Lee, S.I.; Torres, A.; Walubita, L.F. Comparison of Flow Number, Dynamic Modulus, and Repeated Load Tests for Evaluation of HMA Permanent Deformation. *Constr. Build. Mater.* **2013**, *44*, 391–398. [CrossRef]
126. Walubita, L.F.; Zhang, J.; Das, G.; Hu, X.; Mushota, C.; Alvarez, A.; Scullion, T. Hot-Mix Asphalt Permanent Deformation Evaluated by Hamburg Wheel Tracking, Dynamic Modulus, and Repeated Load Tests. *Transp. Res. Rec.* **2012**, *2296*, 46–56. [CrossRef]
127. Sengul, C.E.; Aksoy, A.; Iskender, E.; Ozen, H. Hydrated Lime Treatment of Asphalt Concrete to Increase Permanent Deformation Resistance. *Constr. Build. Mater.* **2012**, *30*, 139–148. [CrossRef]

buildings

MDPI

Article

Understanding Toughening Mechanisms and Damage Behavior in Hybrid-Fiber-Modified Mixtures Using Digital Imaging

Yaohui Yang [1,2], Yinzhang He [3,*], Rui Fu [4], Xiaokang Zhao [3], Hongfa Shang [1,2] and Chuanyi Ma [2]

1 Shandong Key Laboratory of Highway Technology and Safety Assessment, Jinan 250013, China; yangyaohui1905@163.com (Y.Y.); sdgsshf@163.com (H.S.)
2 Innovation Research Institute, Shandong Hi-Speed Group Co., Ltd., Jinan 250013, China; machuanyi@163.com
3 School of Highway, Chang'an University, Xi'an 710064, China; zhaoxk@chd.edu.cn
4 Shandong Road and Bridge Group Co., Ltd., Jinan 250013, China; 13272008498@163.com
* Correspondence: heyinzhang@chd.edu.cn

Abstract: Pavement cracking is a primary cause of early damage in asphalt pavements, and fiber-reinforcement technology is an effective method for enhancing the anti-cracking performance of pavement mixtures. However, due to the multi-scale dispersed structure of pavement mixtures, it is challenging to address cracking and damage with a single fiber type or fibers of the same scale. To investigate the toughening mechanisms and damage behavior of hybrid-fiber-modified mixtures, we analyzed the fracture process and damage behavior of these mixtures using a combination of basalt fiber and calcium sulfate whisker hybrid fiber modification, along with semicircular bending tests. Additionally, digital imaging was employed to examine the fracture interface characteristics, revealing the toughening mechanisms at play. The results demonstrated that basalt fibers effectively broaden the toughness range of the modified mixture at the same temperature, reduce mixture stiffness, increase residual load at the same displacement, and improve crack resistance in the mixture matrix. While calcium sulfate whiskers enhanced the peak load of the mixture, their high stiffness modulus was found to be detrimental to the mixture's crack toughness. The fracture interface analysis indicated that the three-dimensionally distributed fibers form a spatial network within the mixture, restricting the relative movement of cement and aggregate, delaying crack propagation, and significantly improving the overall crack resistance of the mixture.

Keywords: hybrid fiber; digital image; characteristics of fracture interface; crack resistance; damage behavior

1. Introduction

Asphalt pavement is currently one of the most widely used forms of pavement globally. However, asphalt pavement is susceptible to environmental influences, which can lead to poor high-temperature stability, cracking, and water damage during service. These issues result in a decline in pavement service quality and durability, negatively impacting economic development and traffic operations [1–4]. Fiber additives have been shown to significantly enhance the high- and low-temperature performance as well as the fatigue resistance of asphalt pavement [5–7]. For instance, Liu et al. [8] found that anhydrous calcium sulfate whisker (CSW) can improve the high-temperature performance of asphalt mixtures, although it may have some negative effects on low-temperature performance. Similarly, Guan et al. [9] investigated the effects of CSW on the water stability and fatigue properties of asphalt mixtures, concluding that it could significantly enhance both properties, with a recommended dosage of 0.4%.

While fiber products can improve the performance of asphalt mixtures, the use of single fibers often fails to address the needs across different scales of research. Additionally, composite fibers of the same scale lack complementarity, resulting in suboptimal modification effects in fiber-reinforced asphalt concrete and affecting engineering quality [10,11]. This underscores the urgent need to develop multi-scale hybrid fiber asphalt concrete pavement materials that offer high strength and toughness to achieve sustainable asphalt pavements.

Fiber-reinforced materials are classified into three types: nano-scale, micro-scale, and millimeter-scale, corresponding to common nanomaterials, whisker materials, and fiber materials, respectively. For example, Javad Tanzadeh et al. [12] used basalt fiber (BF) and glass fiber to address drainage issues in nano-SiO_2-SBS modified open-graded friction courses, finding that BF was more effective than glass fiber in reducing drainage. Gong et al. [13] optimized the mix ratio of nano-$TiO_2/CaCO_3$ (NTC)-BF-composite-modified asphalt mixtures using a response surface method and Box-Behnken Design (BBD) model. The analysis showed that the optimal content of NTC was 5.1%, BF was 3.9%, and the asphalt–aggregate ratio was 5.67%. Ren et al. [14] found that the interfacial adhesion between fiber and asphalt is stronger after fiber treatment with a coupling agent, which enhances the crack resistance of fiber asphalt mixtures. Despite the widespread interest in hybrid-fiber-modified asphalt mixtures, most research focuses on different types and scales of fiber combinations, and there is limited documentation on the modification mechanisms and crack resistance enhancement of multi-scale hybrid fiber asphalt concrete [11].

There are various test methods used to evaluate the crack resistance of asphalt mixtures both domestically and internationally, including indirect tensile tests (IDT), disc compact tension tests (DCT), thermal stress restrained stress tests (TSRST), low-temperature bending beam tests (BBT), semi-circular bending tests (SCB), and asphalt concrete cracking devices (ACCD) [10,14]. Among these, SCB is gaining increasing attention due to its simplicity, repeatability, and flexibility in testing and evaluation. Researchers primarily use SCB to evaluate the crack resistance of asphalt mixtures, applying fracture mechanics theory to characterize low-temperature and fatigue crack resistance. The specific parameters used include fracture toughness, fracture energy, and crack resistance index [15]. Compared to IDT and DCT tests, SCB testing concentrates stress at the notch, reducing the peak load and mitigating the local damage effect of cylindrical steel bars on the specimen. Unlike the TSRST test, SCB can use semi-circular specimens made from Marshall specimens, pavement drilling specimens, or rotary compaction specimens, making it widely applicable and reducing the difficulty of indoor test operations. Compared to BBT tests, SCB tests have a slower loading rate and a larger specimen cross-sectional area, making them more representative of actual pavement cracking conditions [10].

The use of digital image technology to study hybrid-fiber-modified mixtures is driven by its high precision, non-contact, and full-field measurement capabilities, which are crucial in studying the deformation and failure behavior of asphalt mixtures. Digital image technology provides detailed images of the internal structure, aiding in the observation of interfacial adhesion effects between fibers, asphalt, and aggregates, thereby revealing the influence of mixed fibers on asphalt mixture performance. For instance, Li et al. [16] applied digital image technology to SCB tests to observe the cracking behavior at the asphalt-coarse aggregate interface under load, offering valuable data for understanding the impact of fiber modification on crack development. Similarly, Kutuk-Sert et al. [17] used digital image technology to study the adhesion effect of nano-olivine in warm-mix modified asphalt, demonstrating the technology's effectiveness in this area. This shows that digital image technology has clear advantages in revealing fiber distribution, adhesion, and structural changes in modified asphalt materials. Consequently, it can more accurately analyze microstructural changes, crack propagation, and fiber reinforcement effects in hybrid-fiber-modified mixtures, providing more comprehensive research data and helping to optimize asphalt mixture performance.

Moreover, current research on the toughening mechanism of hybrid-fiber-modified mixtures predominantly focuses on rheological perspectives. There is a notable lack of studies on the internal reinforcement and toughening fracture mechanisms of BF-CSW modified asphalt mixtures based on the digital image technology used in SCB tests.

In response to these research gaps, this study selects micron-scale calcium sulfate whiskers and millimeter-scale basalt fibers to prepare hybrid fiber asphalt mixtures. The crack resistance and toughening mechanisms of different types of asphalt mixtures are compared using pre-opened SCB tests. During the tests, an electronic extensometer and a camera are used to record the bending and tensile fracture processes of SCB specimens, allowing for the analysis of load–displacement–time trends and crack propagation speeds at different temperatures. Based on fracture mechanics theory, the crack resistance of various asphalt mixtures is comprehensively evaluated using fracture toughness, fracture energy, and crack resistance index. Additionally, the fracture interface characteristics of specimens are compared and analyzed to characterize the toughening mechanisms of hybrid-fiber-modified mixtures.

2. Materials and Methods

2.1. Raw Material

2.1.1. Calcium Sulfate Whisker

Calcium sulfate whisker (CSW) is a micron-sized fibrous powder prepared from chemical gypsum or industrial waste gypsum, which can be divided into three types according to the content of water bound by CSW: dihydrate CSW, hemihydrate CSW and anhydrous CSW. Because the dihydrate CSW and hemihydrate CSW would lose water in high-temperature asphalt, the interfacial adhesion between CSW and asphalt is weakened, so KH-550 (3-aminopropyltriethoxysilane) is used to modify the surface of anhydrous CSW to improve the compatibility of CSW with asphalt [9,18–20]. Anhydrous CSW is produced in Jiangxi Fengzhu New Materials Technology Co., Ltd., located in Yichun, China, and its basic physical indexes are shown in Table 1.

Table 1. Basic physical indexes of anhydrous CSW.

Project	Result	Unit
Appearance	White powder	--
Loose density	0.15–0.3	$g\cdot cm^{-3}$
Density	2.69	$g\cdot cm^{-3}$
Diameter	1–15	μm
Average length	20–35	μm
Aspect ratio	20–35	--
Mohs hardness	3	--

The macro and micro morphology of CSW are shown in Figure 1. It can be found that CSW has obvious edges and corners under SEM, showing a rod-like structure; its surface is rough and uneven, which is mainly caused by the rapid precipitation of calcium sulfate whiskers in the growth process [19,21]; and there is partial agglomeration without surface treatment.

(a) (b)

Figure 1. CSW (**a**) macro morphology; (**b**) microscopic morphology under SEM.

2.1.2. Basalt Fiber

At present, basalt fiber (BF) produced by various manufacturers can be divided into bundle BF and flocculent BF according to the dispersion. Among them, chopped strand BF, as a millimeter-grade fiber reinforced material, is made of natural basalt as raw material and cut into chopped fibers by continuous BF at a fixed length, which has the advantages of high tensile strength, excellent thermal stability, excellent corrosion resistance and good surface wettability [5]. In this study, the chopped BF produced by Changzhou Bochao Engineering Materials Co., Ltd., located in Changzhou, China, after surface treatment with silane coupling agent was used. The technical indexes of BF are shown in Table 2.

Table 2. Basic technical indexes of basalt fiber.

Project	Result	Unit
Length	6	Mm
Diameter	15	Mm
Density	2.64	$g{\cdot}cm^{-3}$
Tensile strength	3000	MPa
Elasticity modulus	65.2	GPa
Elongation at break	2.58	%
Melting point	79.1	°C
Mass loss (thermal stability)	0.47	%
Moisture absorption rate (water stability)	0.8	%
Oil absorption multiple (oil absorption)	6.048	--

The macro and micro morphology of BF are shown in Figure 2. From this, it can be found that BF has a slender cylindrical shape under SEM, and its surface is relatively smooth and uniform (Figure 1) compared to CSW, which can improve the stability of the matrix [22]. Additionally, Xing et al. [23] thought that BF can play a role in strengthening and toughening asphalt.

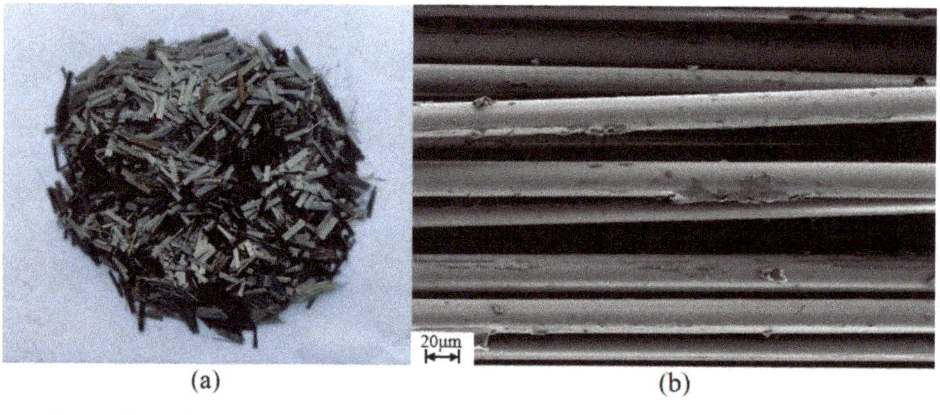

Figure 2. Basalt fiber (**a**) macro morphology; (**b**) microscopic morphology under SEM.

2.2. Design and Preparation of Mixture

2.2.1. Grading Design of Mixture

Asphalt mixtures with different mineral aggregate gradation have different road performance and different pavement layers. Considering that the maximum fiber length is 9 mm, AC-13C gradation was used for asphalt mixture in the test, and the synthetic gradation curve was the curve close to the median gradation in the Technical Specifications for Construction of Highway Asphalt Pavements (JTG F40-2004) [24], so the pass rate of key sieve holes (2.36 mm) was less than 40%.

Marshall compaction test was carried out in accordance with the Test Method of Aggregates for Highway Engineering (JTG E42-2005-T0702) [25]. With the median asphalt-aggregate ratio of 4.0%, it was divided into five groups according to 0.5%—that is, the molding asphalt–aggregate ratios were 3.0%, 3.5%, 4.0%, 4.5%, and 5%, respectively. The relative density of the gross volume of the specimen was measured by surface drying method, and the theoretical maximum relative density, stability, flow value, porosity, saturation, mineral aggregate porosity and other indicators were calculated, and the Marshall test was carried out in groups. The aggregate drying temperature and asphalt heating temperature were selected at 180 °C and 160 °C respectively. The mixing temperature and compaction temperature were 170 °C and 148 °C respectively. Marshall test results are shown in Table 3.

Table 3. Marshall test results of AC-13C asphalt mixture.

Oil-Stone Ratio (%)	Relative Density of Gross Volume (g/cm^3)	Void Ratio (%)	Gap Rate of Mineral Aggregate (%)	Saturation (%)	Stability (kN)	Flow Value (0.1 mm)
3.0	2.386	7.1	13.90	48.3	6.54	3.59
3.5	2.401	5.8	13.72	57.6	8.02	2.79
4.0	2.420	4.3	13.43	67.5	10.23	2.38
4.5	2.445	3.0	13.17	78.1	9.57	2.70
5.0	2.468	1.9	13.06	86.2	7.06	3.48

According to the Technical Specifications for Construction of Highway Asphalt Pavements (JTG F40-2004) [24], the optimal asphalt–aggregate ratio was determined. When there was no peak in the density or stability of the asphalt–aggregate ratio selected in the experiment, the asphalt–aggregate ratio a_3 corresponding to the target void ratio could be directly taken as OAC_1, and the median value of OAC_{min}~OAC_{max} was taken as OAC_2, so the optimal asphalt–aggregate ratio of AC-13C matrix asphalt mixture was 4.3%.

2.2.2. Preparation of SCB Test Sample

In this study, four types of asphalt mixture samples were prepared: matrix asphalt mixture (70#), BF modified asphalt mixture (BF), CSW modified asphalt mixture (CSW), and hybrid fiber modified mixture (CSW + BF). Six SCB specimens were prepared for each group to conduct parallel experiments. Based on the preliminary research by the team, the basalt fiber content was set at 0.32% (by aggregate weight) and the calcium sulfate whisker content at 5% (by asphalt weight). The CSW-modified asphalt was first prepared using the wet process, followed by the preparation of the CSW + BF asphalt mixture using the dry process. The detailed preparation process is outlined in references [5,9].

In this study, a superpave rotary compaction tester was used to form cylindrical specimens with radius: height = 75 mm × 130 mm, as shown in Figure 3a. Considering the difference in density between the top and bottom materials of rotary compaction specimens and the middle materials, before cutting semi-circular bending specimens, 15 mm was cut off from the upper and lower surfaces of cylindrical specimens; the remaining cylindrical specimens were cut according to the sample size of radius: height = 75 mm × 50 mm; and the two cut specimens were divided radially to make four semicircular specimens. Finally, the middle seam was cut at the bottom of each semicircular specimen, and the size of the middle seam was 15 mm × 1.5 mm. The semicircular bending specimen and its preparation process are shown in Figure 3b,c.

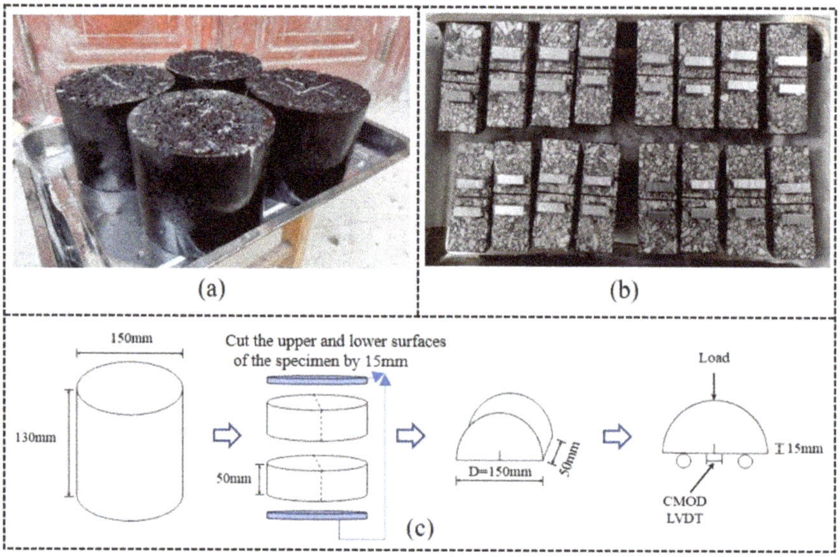

Figure 3. (**a**) Rotary compaction test piece; (**b**) SCB specimen; (**c**) specimen preparation process.

2.3. Experimental Design of SCB Based on Digital Image Method

In this study, an electronic extensometer produced by Steel Research and Testing Technology Co., Ltd., based in Shanghai, China, was used to record the central mouth opening displacement (CMOD) of SCB specimens. The extensometer model was YYU-10/25, with a range of 10 mm. To ensure that the crack path of the specimen was clearly recorded during the test, the surface of the prepared SCB specimen was manually dyed with red hand paint, as shown in Figure 4a,b. The extensometer was designed to move in real time with the opening of the middle seam, which was achieved by fixing an extended staple on both sides of the seam with AB glue and securing the extensometer clip to the staple with a rubber band.

Test temperature is a key factor affecting the low-temperature stiffness of asphalt materials. To prevent brittle fracture, rapid crack propagation, increased difficulty in test operation, and an incomplete load–displacement curve during the SCB test—factors that are not conducive to comparing different types of asphalt mixtures—the insulation temperatures of the specimens were set at −5 °C, 10 °C, and 25 °C. Each specimen was insulated using a material testing machine (MTS) temperature control box, and each group of specimens was insulated for 6 h to ensure that the internal and external temperatures were consistent, as shown in Figure 4c.

During the test, a semi-circular specimen with a middle seam was placed on a three-point bending bracket with adjustable spacing, and the distance between the two supporting points at the bottom of the specimen was adjusted to 0.8D (where D is the diameter of the specimen). After heat preservation, the specimen was quickly removed, the extensometer was fixed, and the specimen was placed on the loading bracket. The upper pressure head of the universal MTS was aligned with the middle seam of the specimen, ensuring the load direction and middle seam were in the same straight line, with a constant loading rate of 2 mm/min, as shown in Figure 4d. The real-time load, vertical displacement, and CMOD of the specimen were recorded by MTS software. Additionally, a high-resolution camera was positioned 1.5 m away from the MTS instrument, with its visual axis perpendicular to the specimen's surface, to record the entire fracture process during loading, capturing all the damage and changes throughout, as shown in Figure 4e.

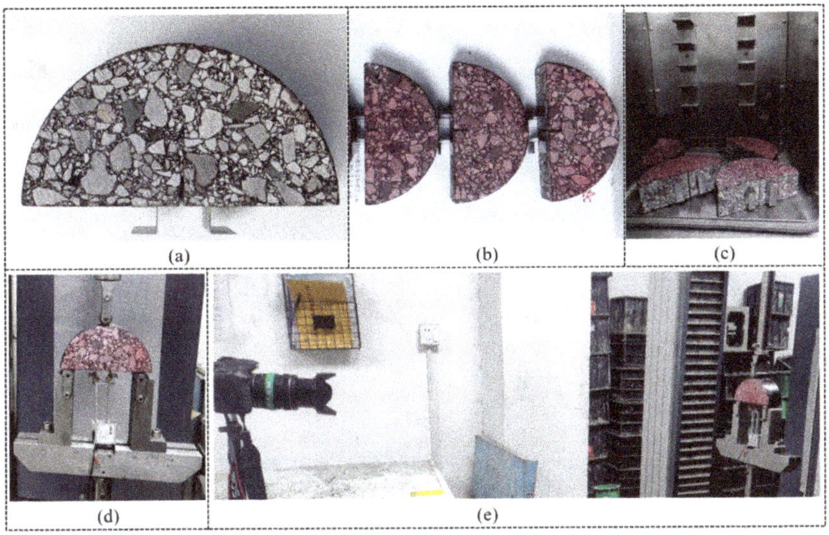

Figure 4. SCB test process (**a**) before specimen dyeing; (**b**) after specimen dyeing; (**c**) heat preservation of the test piece; (**d**) installation of extensometer; (**e**) camera video recording to record the cracking process.

3. Results and Discussion

3.1. Analysis of Fracture Process of Mixture Based on SCB Test

3.1.1. Load–Vertical Displacement Curve

The results of the SCB test, specifically the load–vertical displacement for different types of asphalt mixtures, are plotted on separate curves. The representative data from parallel tests are used as examples, as shown in Figure 5. The lower the temperature of SCB test, the smaller the vertical displacement and the greater the peak load when the specimen breaks under load. Among them, 70# asphalt and CSW show brittle fracture at −5 °C, and the low-temperature crack resistance is the worst. With the increase in test temperature, the elastic modulus of asphalt mixture decreases gradually, and the curve transitions to gentle type, and the vertical displacement corresponding to peak load increases gradually, and the toughness increases, indicating that temperature is the key factor to determine the viscoelasticity of asphalt mixture. There are four kinds of asphalt mixtures at different temperatures: the peak load of matrix asphalt mixture is the smallest and the curve is smoother; the peak load of BF is larger, and the vertical displacement under peak load is also larger; the peak load of CSW is basically the same as that of BF, but the vertical displacement at peak load is the smallest; the peak load of CSW + BF changes little, but the vertical displacement is larger, and the curve develops most gently after the peak load.

At low temperature, with the increase of the load at the top of the semi-circular specimen, the tensile stress at the bottom of the specimen gradually increases, and micro-cracks are generated at the tip of the middle seam, which quickly and gradually extend upward and reach the peak load in a very short time. At this time, the mixture is mainly elastic and shows brittle fracture. At room temperature, the cracking speed of the specimen is obviously slow, the crack breakthrough time is delayed, and the specimen can still bear a large residual load after reaching the peak load. At this time, the mixture is mainly in a viscoelastic state, showing ductile fracture.

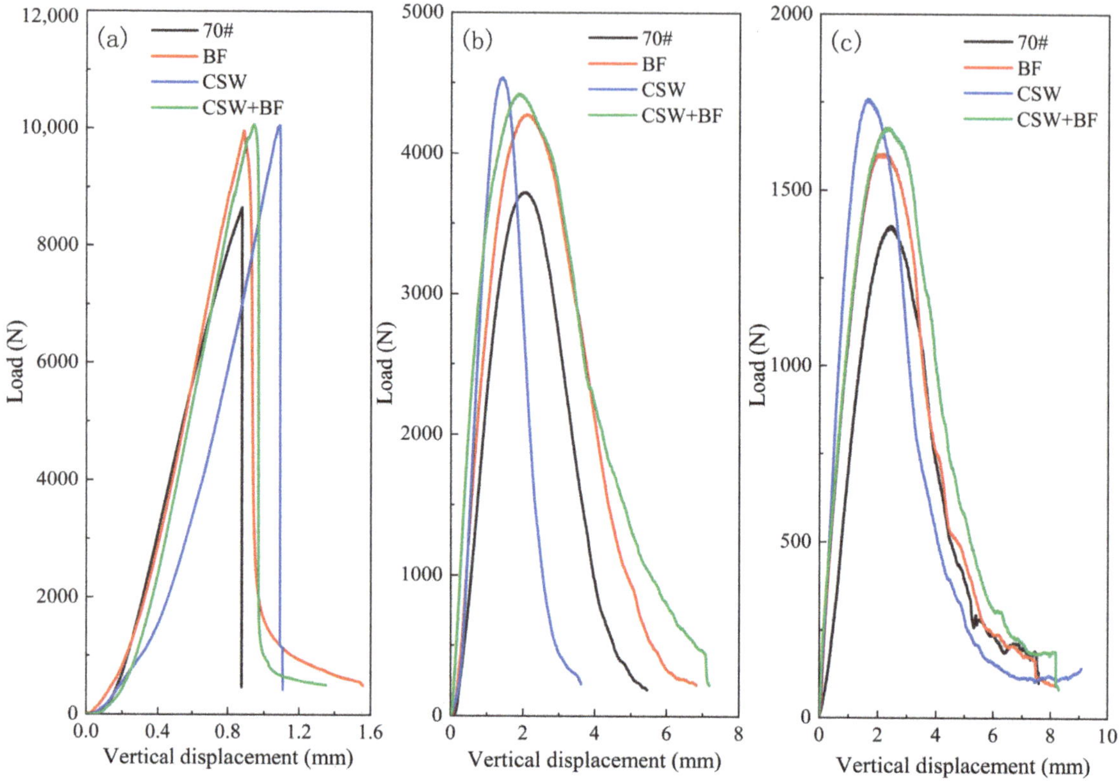

Figure 5. Load–vertical displacement curve (**a**) −5 °C; (**b**) 10 °C; (**c**) 25 °C.

At the same temperature, the peak load of fiber asphalt mixture is larger and the vertical displacement at the peak is larger, which shows that fiber materials can enhance the tensile strength of the mixture and reduce the crack propagation speed after the peak load of the mixture. When the fiber is tested in a specimen with a large cross-sectional area, it shows excellent reinforcement and crack resistance. Before the peak load, the fibers in the micro-cracks adhere to the asphalt material, fixing the aggregate in the mixture, and the peak load is enhanced. After the peak load, the crack width increases and expands into the main crack. Because BF has excellent tensile strength and is distributed perpendicularly to the crack interface, even if the interface adhesion fails, the pull-out effect still exists, and the macroscopic performance is that the residual load increases [19,23]. Compared to the matrix asphalt mixture, the curve slope of CSW modified asphalt mixture before peak load is larger and the vertical displacement under peak load is smaller, which shows that CSW material can strengthen the interface adhesion between asphalt and aggregate and increase the stiffness modulus of asphalt. However, the modulus is too large, which also leads to faster crack propagation and poor toughness performance of the mixture.

3.1.2. Load–CMOD Displacement–Time Curve

Time is the key factor of fracture toughness in a semi-circular bending test, and CMOD displacement is also an important evaluation index of fracture opening rate of specimens. The longer the fracture process, the smaller the crack opening rate of specimens and the stronger the low-temperature fracture toughness of asphalt mixture. Therefore, in this part, the load–time curve and CMOD displacement–time curve of different types of asphalt mixtures at different temperatures in SCB test were combined and analyzed, taking the representative data in parallel test as an example, and the results are shown in Figure 6.

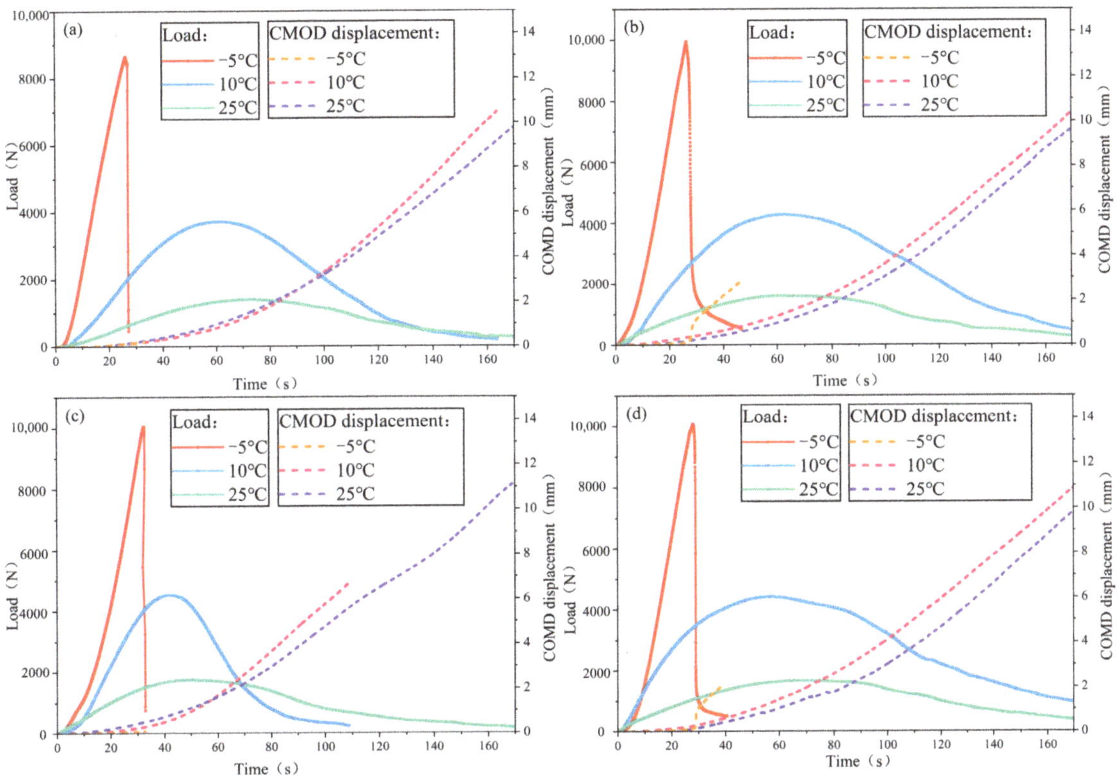

Figure 6. Load and CMOD displacement–time curve (**a**) 70#; (**b**) BF; (**c**) CSW; (**d**) CSW + BF.

Under a constant rate load, the load-time curves of different asphalt mixtures closely resemble the load-vertical displacement curves, displaying a pattern where the load initially increases with time and then decreases. The CMOD displacement also increases over time, and a steeper curve indicates a faster crack width expansion rate at the middle crack. The CMOD displacement curves show that as the temperature increases, the crack opening rate slows down. Notably, the specimen at −5 °C fractures too quickly, resulting in fewer recorded CMOD data points, which obscures long-term trends. By comparing the abrupt changes in CMOD displacement between the 10 °C and 25 °C curves, we can gauge the ability of different asphalt mixtures to delay crack growth as the temperature rises. The latest mutation point occurs in the 70# asphalt, followed by CSW, with BF and CSW + BF showing the earliest mutation points. This suggests that fibers effectively reduce the fracture rate through their adhesion with aggregate and asphalt [26], making them a valuable material for enhancing the toughness of asphalt mixtures.

By analyzing the characteristic points of load–time curve and CMOD displacement-time curve, it is found that the addition of BF delays the time corresponding to the peak load of load–time curve and reduces the fracture rate of asphalt mixture. The absolute value of the post-peak slope of the load–time curve is reduced, the stiffness modulus after cracking is reduced, the displacement–time curve of CMOD is shifted downward, and the residual load in the fracture process is increased. Combining Figures 6 and 7 reveals that the addition of CSW increases the peak load, but the time to reach this peak is shortened, and the CMOD displacement at the same time increases. This suggests that CSW negatively impacts the low-temperature toughness of the asphalt mixture.

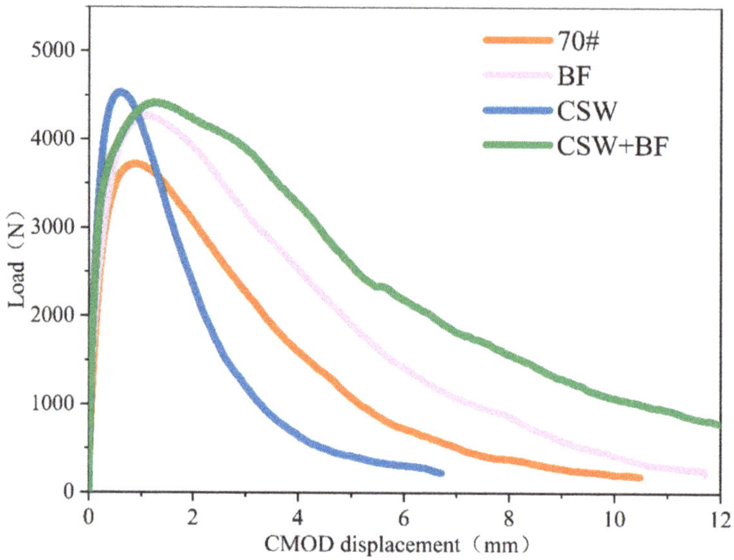

Figure 7. Load–CMOD displacement curves of different types of asphalt mixtures.

As can be seen from Figure 7, with the increase of CMOD displacement, the load on SCB specimens first rises and then falls, and the slope before peak load is greater than that after peak load. The order of peak load of different mixtures is CSW > CSW + BF > BF > 70#, which shows that CSW and BF can improve the tensile strength of asphalt mixture matrix, and the material is not easy to produce through cracks under the action of temperature stress and load. However, the CMOD displacement corresponding to the peak load of CSW decreases, which shows that CSW will reduce the flexural-tensile strain of asphalt mixture and increase the stiffness of asphalt, which is unfavorable to the crack toughness of materials at low temperature. The CMOD displacement of BF corresponding to the peak load of CSW + BF is large, which shows that fiber materials can toughen asphalt concrete and reduce its crack propagation speed.

The absolute value of the slope of the load-CMOD displacement curve after the peak load is ranked as CSW > 70# > BF > CSW + BF. The order shows that CSW increases the stiffness at the crack tip, CSW modified asphalt has greater stiffness, and low temperature conditions reduce the interfacial adhesion between asphalt and aggregate, which makes CSW's crack resistance poor. The elastic modulus of BF and CSW + BF modified asphalt is the smallest at the crack tip. The reason may be that the temperature sensitivity of fiber is weak. Under the condition of 10 °C, the tensile strength of fiber is greater than the viscosity of asphalt itself. After the interface of asphalt cracks, the bridging and reinforcement of fiber improve the tensile strength at the crack, so the elastic modulus decreases under the condition of constant deformation. However, under the condition of −5 °C, the tensile strength of fiber is less than the interfacial adhesion between fiber and asphalt, and the brittle fracture of fiber and asphalt occurs directly, so the elastic modulus has little difference.

3.1.3. Crack Propagation Speed

In this study, the Potplayer video player v1.6.63891 was used to capture the images of SCB specimens at different cracking moments in the video of specimen cracking process. Taking the cracking process of CSW + BF modified asphalt mixture at 10 °C as an example, six representative moments were selected as the crack state images at different cracking stages, as shown in Figure 8. It can be found that macro cracks did not appear in the specimen for a long time after loading, and cracks appeared for the first time after loading for 52 s, but they were not clear, indicating that micro cracks had been produced. After loading for 84 s, the specimen has a clear crack propagation path, which indicates that it is in the stress attenuation stage after peak load, and at this time, the micro-cracks have expanded upward, and micro-cracks interweave at the upper part of the specimen. After loading for 103 s, the surface crack of the specimen has spread more than half, and the middle crack at the bottom has obvious deformation, and the crack width increases rapidly at this stage. After loading for 132 s, the cracks have basically penetrated, but the residual load of the material is still large, which does not meet the test minimum load condition and continues to be loaded. After loading for 164 s, the crack runs up to the top, which means that the specimen is completely broken.

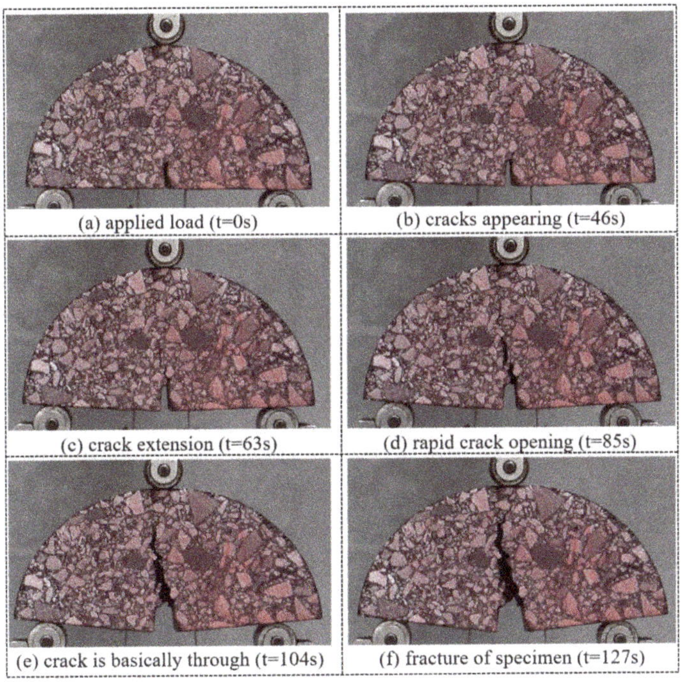

Figure 8. Cracking process of CSW + BF modified mixture specimen (10 °C).

During the SCB test, a Canon D80 high-resolution digital camera, manufactured by Canon Inc. in Japan, was utilized to capture the surface cracking process of the specimen. The width of open cracks at different cracking moments can be measured by ImageJ 1.54j software, and the crack resistance of the asphalt mixture can be evaluated by the crack propagation speed, which is the crack propagation path length per unit time [27,28], and the calculation formula is as follows:

$$v = \frac{\Delta s}{\Delta t} \tag{1}$$

where Δs is the crack propagation path length, mm; Δt is the load application time, s.

In this study, ImageJ 1.54j software was used to calibrate and measure the paths of the pictures of the specimens during fracture, as shown in Figure 9. The yellow box in the figure marks the ductile zone of the crack, and the width of the ductile zone is small, indicating that the specimen is an obvious type I open crack at 10 °C. The fulcrum on both sides of the bottom of SCB specimen was selected as the calibration size, and its length was calibrated to be 120 mm. The crack path was marked using the software's segmented line tool, and the length of the continuous broken line mark was read, which is the crack propagation path length. The crack propagation speed can be calculated from the crack propagation path length and load application time.

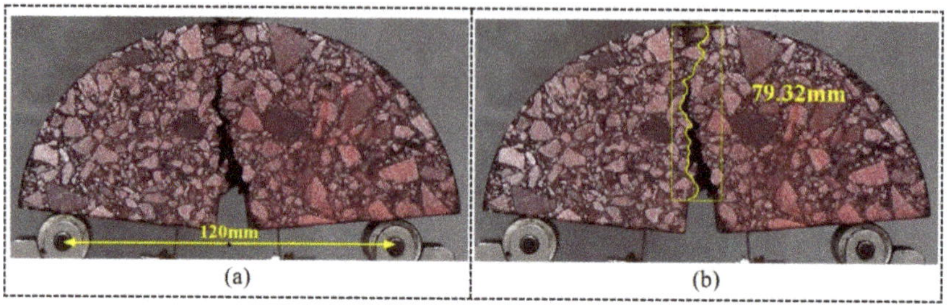

Figure 9. Calibration and measurement of crack propagation path of SCB specimen (**a**) calibration size; (**b**) route marking measurement.

Figure 10 shows the crack propagation speed of different specimens. It can be observed that as the test temperature decreases, the crack propagation speed of SCB specimens with different asphalt mixtures gradually increases. The crack propagation speed of fiber asphalt mixtures remains relatively constant. In contrast, the CSW-modified asphalt mixture exhibits the fastest crack propagation speed, suggesting that while CSW alone may negatively impact the crack resistance of the mixture, the addition of BF can significantly enhance crack resistance during the crack development stage and improve the overall toughness of the mixture. Temperature is a key factor affecting the crack shape of SCB specimens; the lower the temperature, the more the crack tends to form a straight line connecting the middle crack and the apex of the specimen. Consequently, the characteristics of the crack propagation path in different asphalt mixtures are also consistent with the observed changes in crack propagation speed.

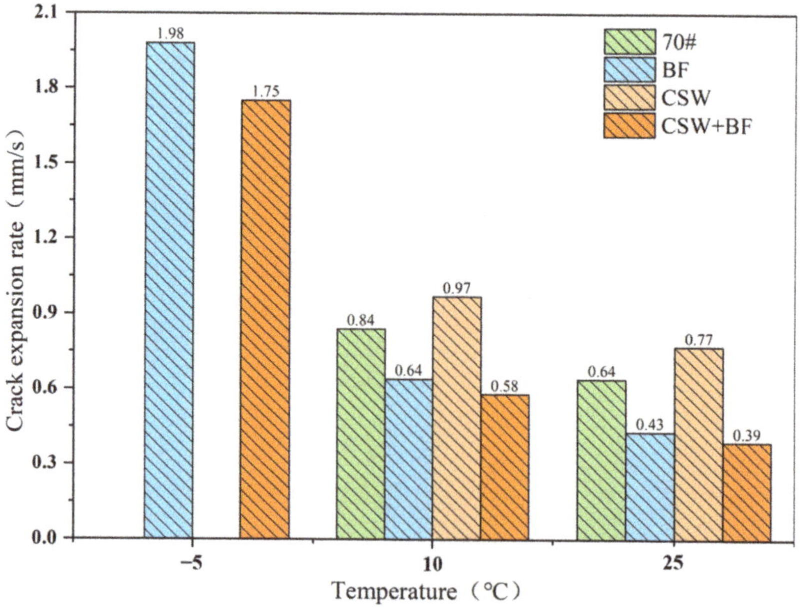

Figure 10. Crack growth rate of SCB test at different temperatures.

Drawing from the analyses in Sections 3.1.1 and 3.1.2, the lowest crack propagation speed observed in the CSW + BF mixture may be attributed to the synergistic interaction between the different scales of CSW and BF. This synergy enhances the stability and toughness of the mixture, resulting in a slower crack propagation speed. It is also worth noting that at −5 °C, both 70# asphalt and CSW exhibit brittle fracture, characterized by a short fracture process and distorted crack propagation speed data, which have not been included in the figure.

3.2. *Analysis of Crack Resistance of Mixture Based on Fracture Mechanics*

3.2.1. Fracture Toughness

The stress intensity factor K_I is often used to describe the stress field at the crack tip during the fracture of the specimen, which is related to the load stress value, the width of the middle crack, the size of the specimen, and other factors. As the crack width increases, the stress intensity factor K_I increases. When K_I reaches the critical strength value K_{IC}, cracks appear on the surface of the specimen, which means that the crack tip has become unstable [29]. The greater the K_{IC}, the greater the energy absorbed by the material during the fracture process, and the stronger the crack resistance and toughness. The calculation formula of K_{IC} is:

$$K_{IC} = Y_{I(0.8)}\sigma_0\sqrt{\pi a} \tag{2}$$

$$\sigma_0 = \frac{P_{\max}}{2rh} \tag{3}$$

$$Y_{I(0.8)} = 4.782 + 1.219(\frac{a}{r}) + 0.063e^{7.045\frac{a}{r}} \tag{4}$$

where σ_0 is the stress corresponding to the peak load, kN; $P_{\max}$ is the peak load, mN; $Y_{I(0.8)}$ is the standard stress intensity factor, dimensionless; r is the radius of the specimen, m; h is the thickness of the specimen, m; and a is the length of the middle seam, m.

Figure 11 depicts the change curves of fracture toughness of different mixture specimens at different temperatures. It can be found that the fracture toughness decreases significantly with the decrease of temperature, and the stiffness modulus of the mixture is improved at low temperature, which makes the mixture absorb more energy during the fracture process, indicating that the lower the temperature, the stronger the bearing capacity of the mixture. Among different types of asphalt mixtures, the fracture toughness of 70# asphalt is the worst, and that of CSW is the largest, which shows that CSW improves the low-temperature stiffness of asphalt mixtures and makes the physical properties of materials mainly elastic. The fracture toughness of BF and CSW + BF is similar to that of CSW, and the crack strength of the mixture is obviously improved. In addition, the fracture toughness is determined by the peak load P_{max} without considering the displacement factor of specimen fracture, which cannot fully reflect the advantages and disadvantages of different types of asphalt mixtures.

3.2.2. Fracture Energy

Fracture energy, G_f, also known as capacity release rate, is a main index to evaluate the crack resistance of asphalt mixture SCB specimens. The ratio of fracture work to crack propagation area is fracture energy, and the calculation formula of G_f is as follows:

$$G_f = \frac{W_f}{A_{lig}} \tag{5}$$

where W_f is the fracture work, J; $W_f = \int pd(u)$, P is the load, N; u is the average load line displacement, m; A_{lig} is the crack propagation area, m^2; $A_{lig} = (r - a)h$, r is the radius of the specimen, m; a is the length of the middle seam, m; and h is the thickness of the specimen, m.

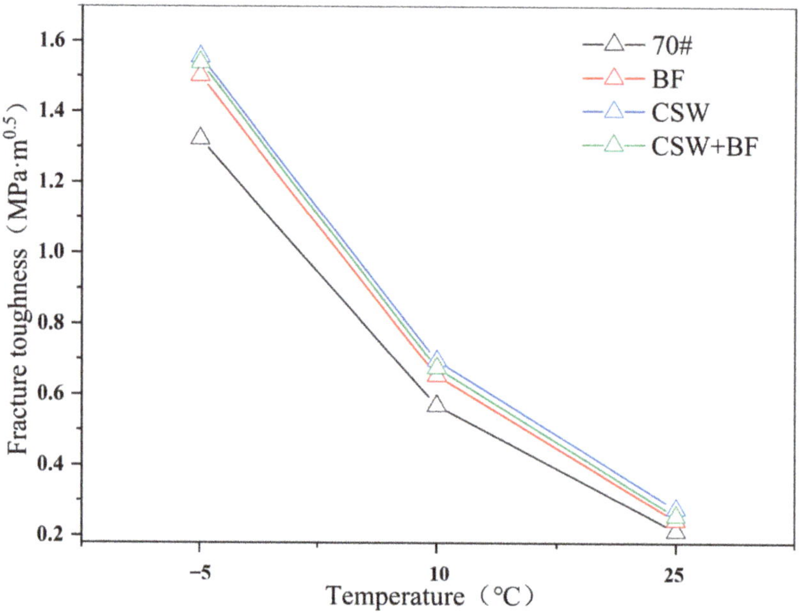

Figure 11. Effect of temperature on fracture toughness of specimen.

The fracture work is the total work done by the external load on the SCB specimen, that is, the *X*-axis integral of the load–displacement curve from the beginning of deformation to the moment when the residual load is 0.1 kN. In this study, Origin software 2021 was used to fit the curve before the peak with cubic polynomial, and the curve after the peak with exponential function. Finally, the integral of the whole curve on the X axis was calculated, and the area of the graph surrounded by the curve and the X axis was obtained. The calculation formula is as follows:

$$W_f = \int_0^{u_0} P_1(u)du + \int_{u_0}^{u_{final}} P_2(u)du \tag{6}$$

where $P_1(u)$ is the fitting curve before the peak; $P_2(u)$ is the fitting curve after peak; u_0 is the displacement corresponding to the peak load, mm; and u_{final} is the displacement corresponding to 0.1 kN residual load, mm.

Figure 12 shows the change in fracture energy of different mixture specimens at different temperatures. It can be seen that with the increase in temperature, the fracture energy of SCB specimen first rises and then falls. It is considered that with the increase in temperature, the specimen has better viscoelasticity and elastic–plastic deformation increases, but the peak load decreases and the vertical displacement of the load increases, so the fracture energy of the specimen at different temperatures cannot be compared. Under the condition of 5 °C, the specimen was mostly brittle fracture, the vertical displacement was too short, the work done by the external force during the whole process of fracture was the least, and the fracture energy was the smallest. Under the condition of 25 °C, the specimens were predominantly tough fracture, and the specimen cracks were not necessarily typical type I fracture but might be I–II composite fracture, which led to the peak load being too small. As a result, the external force did less work, and the fracture energy decreased. Among different types of asphalt mixtures, BF and CSW + BF have significantly larger fracture energies, and 70# is not much different from CSW, indicating that the fiber mesh structure in the interstices of the mineral material plays a role in reinforcing and resisting cracking, slowing down the crack development rate and improving the toughness of the asphalt mixtures.

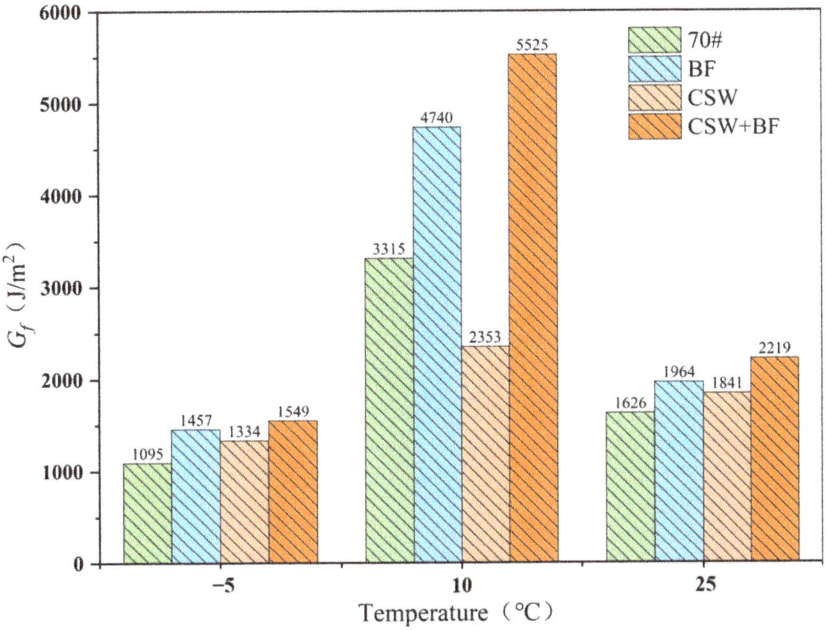

Figure 12. Effect of temperature on fracture energy of specimens.

3.2.3. Crack Resistance Index

Once the brittle asphalt mixture reaches the peak load, it will break, resulting in no post-peak displacement data to calculate the flexibility index *FI* and m. The fracture work W_f and fracture energy G_f are determined by the peak load (P_{max}) and final displacement (u_{final}) of the specimen during fracture. Two kinds of mixtures with obviously different toughness characteristics can have similar G_f values, but because of the greater stiffness of brittle mixtures, they usually have higher peak load and smaller displacement than flexible mixtures. Therefore, in addition to G_f, peak load or final displacement can also be used to distinguish different types of asphalt mixtures. Zhu et al. [30] studied the calculation of fracture strength index (S_f) by using P_{max}, and then normalized G_f with respect to S_f to obtain fracture strain tolerance value (*FST*). *FST* can distinguish mixtures with similar fracture energy but different peak load and post-peak displacement in the load–displacement curve. Subsequently, Fawaz Kaseer et al. [31] used Pmax to simplify the calculation method of cracking resistance index (*CRI*) and put forward *CRI* as an alternative SCB cracking parameter, which is defined as the ratio of total fracture energy to peak load. The calculation formulas of *FST* and *CRI* are as follows [32]:

$$S_f = \frac{2P_{\max}(2r+b)}{h(r-b)^2} \tag{7}$$

$$FST = \frac{G_f}{S_f} \tag{8}$$

$$CRI = \frac{G_f}{P_{\max}} \tag{9}$$

where P_{max} is the peak load, kN; r is the radius of the specimen, m; b is the length of ductile zone, and $b = r - a$, m; h is the thickness of the specimen, m; G_f is the fracture energy, J/m^2.

Figure 13 describes the change of *CRI* of different mixture specimens at different temperatures. It can be seen that with the increase in temperature, the *CRI* of the mixture shows an upward trend, and the *CRI* of CSW + BF is the largest at 10 °C and 25 °C, indicating that it has strong toughness and good crack resistance. The overall order of CRI value is: CSW + BF > BF > 70# > CSW, which shows that mixing CSW alone will reduce the toughness of asphalt mixture, but hybrid fiber will improve the overall crack resistance of asphalt mixture. This is different from the addition of single material characteristics. The mixing law in composite material theory has many application conditions, and many factors need to be considered in the performance prediction of fiber reinforced concrete materials.

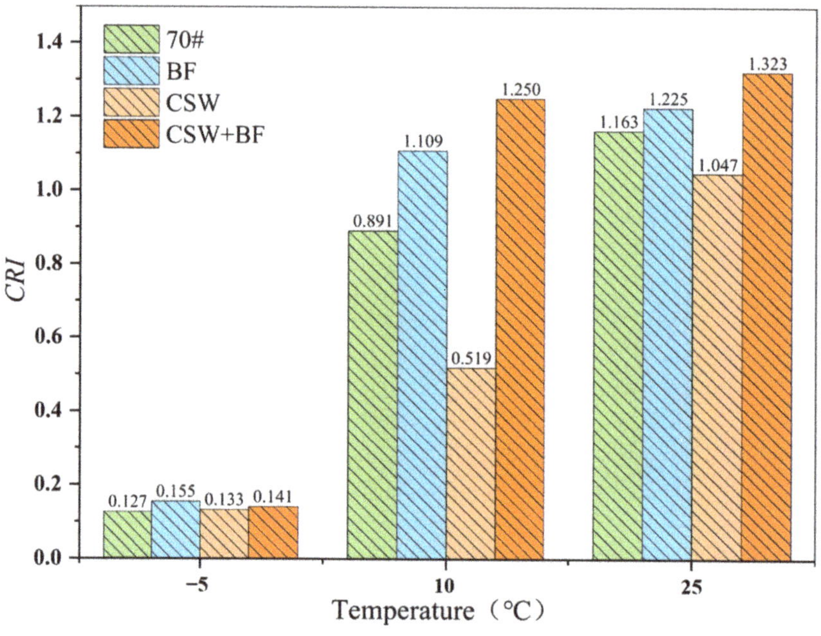

Figure 13. Effect of temperature on *CRI* of specimen.

3.3. Characterization of Fracture Interfaces on the Toughening Mechanism of Hybrid Fiber Modified Mixture

The toughness of fiber asphalt mixture is outstanding in the SCB test, which shows that this test is an effective method for evaluating the toughening mechanism of the fiber asphalt mixture. Correspondingly, the material characteristics of the failure interface of the specimen should also contain the important information of fiber toughening, and the fracture area of SCB specimen is larger, so the material distribution characteristics can be observed more clearly. Therefore, in order to explore the macro-toughening mechanism of hybrid fiber modified mixture, a Canon camera was used to record the fracture interface of SCB fracture specimen, as shown in Figure 14. Generally speaking, the crack propagation path of SCB specimens has three forms: (1) fracture along the internal micro-cracks of asphalt; (2) adhesion failure occurs along the interface between asphalt mastic and aggregate; (3) when there is no obvious stress at the interface tip, the aggregate is directly broken and the cracks pass through the stone. From Figure 14a, it can be found that the first two forms are more common in matrix asphalt mixture. For the third form, cracks in fiber asphalt mixture pass through the stone at low temperature, indicating that fibers transfer dispersed stress in asphalt mastic, which increases the adhesion of asphalt mastic–aggregate interface and makes the stress distribution in asphalt mixture more uniform.

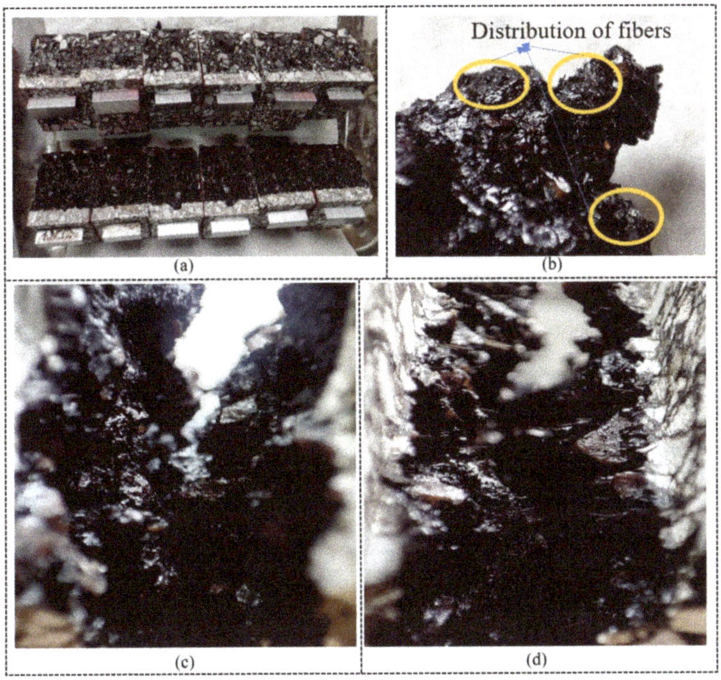

Figure 14. (**a**) Fracture interface of SCB specimen; (**b**) fiber distribution of CSW + BF modified mixture specimen; (**c**) interface between two sides of crack of 70# modified mixture specimen; (**d**) interface between two sides of crack of CSW + BF modified mixture specimen.

Figure 14b reveals that the fibers on the SCB fracture surface are uniformly and randomly dispersed, interlacing within the asphalt mastic. As a millimeter-scale BF material, these fibers not only enhance the adhesion at the asphalt–aggregate interface but also disperse concentrated stress within the micro-cracks of the asphalt mastic. This limits the relative deformation and displacement of the aggregate. It is known that a calcium-containing powder material of micron size improves the adhesion of the asphalt mixture and provides good thermal crack resistance [26]. CSW-modified asphalt further enhances the adhesion between the fiber–asphalt interface and the asphalt–aggregate interface through its excellent viscoelastic properties. These two materials complement each other by reducing the gaps between minerals and aggregates, increasing the overall tensile strength of the specimen, and improving the toughness of the hybrid-fiber-modified mixture.

From Figure 14c, it can be seen that the asphalt film thickness on the aggregate surface of matrix asphalt mixture is insufficient, and there are more irregular aggregates. On the aggregate surface of fiber asphalt mixture, CSW adheres to fibers together, the asphalt film thickness is large, and a large number of fibers are inserted between aggregate particles, dispersing the aggregates formed by asphalt mastic and fine aggregate. BF has great toughness and tensile strength. After the specimen is broken, the fiber is still in the distribution state of tensile stress direction, which shows that mineral fiber with reasonable length and reasonable content can reach the ideal distribution state and stress state in asphalt mixture. It can be seen from Figure 14d that a large number of fibers are randomly distributed in the asphalt mixture in three dimensions, and they are overlapped together to form a spatial network structure, which restricts the movement of asphalt mastic and aggregate. When the fiber asphalt mixture is stressed, the asphalt mixture matrix first deforms and transfers the force to the fiber network structure. The higher the strain of the asphalt mixture matrix, the greater the stress shared by the fiber network, and the more obvious the toughness strengthening effect of the fiber network on the matrix. Therefore, the fiber network can improve the toughness of asphalt mixture by delaying the development speed of cracks, increase the overall tensile strength of asphalt mixture, make asphalt mixture difficult to crack, and effectively improve the crack resistance of asphalt mixture.

4. Conclusions

In this study, the SCB test was used to test the crack resistance of different types of asphalt mixtures. Based on the fracture mechanics theory, three fracture performance parameters were introduced to study the crack resistance of hybrid fiber modified mixtures, and the toughening mechanism of the mixtures was characterized by analyzing the fracture development process and fracture interface morphology of the specimens.

(1) At the same temperature, BF can widen the toughness area of asphalt mixture, reduce the stiffness of mixture, increase the residual load at the same displacement, and improve the crack resistance of asphalt mixture matrix. CSW increases the peak load of the mixture, but the stiffness modulus is large, which is unfavorable to the crack toughness of the mixture.
(2) The hybrid-fiber-modified mixture has the slowest crack propagation speed, the smallest absolute value of post-peak curve slope, the largest toughness band width, the largest fracture energy and crack resistance index, the strongest ability to prevent crack propagation, and the best toughness at low temperature.
(3) CSW-modified asphalt mixture absorbs the most energy in the process of fracture and has the highest fracture toughness, but the fracture energy and crack resistance index are small, the crack propagation speed is the fastest, and the toughness is poor, which indicates that the fracture toughness parameters have strong limitations.
(4) The interface characteristics show that BF can increase the asphalt film thickness on the surface of aggregate, disperse the concentrated stress in asphalt mastic, and limit the relative displacement of fine aggregate. CSW can improve the viscoelastic properties of asphalt, improve the interfacial adhesion between materials, and strengthen the fiber reinforcement.
(5) BF is randomly distributed in asphalt mixture in three dimensions and overlapped with each other to form a spatial network structure, which improves the overall tensile strength of asphalt mixture, delays the development speed of cracks, and effectively improves the crack resistance of asphalt mixture.

The findings of this study demonstrate that incorporating fibers of different scales can significantly enhance the strength and toughness of asphalt mixtures. These results are valuable for road engineering applications and can contribute to extending the service life of asphalt pavements while also supporting the development of a low-carbon, environmentally friendly society. However, the optimal fiber content and types may vary depending on the specific fiber materials and asphalt used. Additionally, the challenge of quantitatively assessing the dispersion of hybrid fibers in asphalt concrete persists. It is important to note that the methods employed in this study are limited to macro-scale approaches. Future research may integrate SEM techniques to investigate microstructural changes, which could provide a more comprehensive understanding of the toughening mechanisms at play. This future work will also focus on improving the uniformity of fiber dispersion through advancements in production processes.

Author Contributions: Methodology, Y.H. and C.M.; Validation, Y.H.; Formal analysis, Y.Y. and X.Z.; Resources, Y.Y. and X.Z.; Data curation, H.S.; Writing—original draft, Y.H.; Writing—review & editing, R.F.; Visualization, X.Z.; Supervision, Y.H.; Project administration, X.Z.; Funding acquisition, X.Z. All authors have read and agreed to the published version of the manuscript.

Funding: This study is supported by the Open Project of Shandong Key Laboratory of Highway Technology and Safety Assessment under Grant No. SH202202, the Shaanxi Natural Science Foundation under Grant No. 2024JC-YBMS-373, to which the authors are very grateful.

Data Availability Statement: Data are contained within the article.

Conflicts of Interest: Authors Yaohui Yang and Chuanyi Ma were employed by the company Shandong Hi-speed Group Co., Ltd. Author Rui Fu was employed by the company Shandong Road and Bridge Group Co., Ltd. The remaining authors declare that the research was conducted in the absence of any commercial or financial relationships that could be construed as a potential conflict of interest.

References

1. He, Y.; Li, Y.; Zhang, J.; Xiong, K.; Huang, G.; Hu, Q.; Zhao, X. Performance evolution mechanism and affecting factors of emulsified asphalt cold recycled mixture performance: A state-of art review. *Constr. Build. Mater.* **2024**, *411*, 134545. [CrossRef]
2. Joumblat, R.; Kassem, H.A.; Basiouni, A.; Masri, A.Z.; Khateeb, G.A.; Absi, J. Performance evaluation of hot-mix asphalt with municipal solid waste incineration fly ash using the stress sweep rutting test. *Innov. Infrastruct. Solutons* **2023**, *8*, 261. [CrossRef]
3. He, Y.; Wang, Q.; Zhang, J.; Li, Y.; He, H.; Huang, G. Foamed crumb rubber asphalt binder: Preparation, rheological properties and adhesion characteristics. *J. Clean. Prod.* **2023**, *396*, 136516. [CrossRef]
4. He, Y.; Zhang, J.; Gao, B.; Wang, L.; Li, Y.; Guo, F.; Huang, G. Experimental investigation of the interface interaction between asphalt binder and mineral filler from the aspects of materials properties. *Constr. Build. Mater.* **2023**, *393*, 132094. [CrossRef]
5. Dai, J.; Jia, X. Fatigue Performance of High-Modulus Basalt Fiber-Reinforced Asphalt Mixture. *J. Appl. Sci. Eng.* **2023**, *26*, 1444. [CrossRef]
6. He, Y.; Rao, R.; Zhang, J.; Li, Y.; Xiong, K.; Li, Y.; Hu, Q.; Huang, G. Investigation on the rheological properties and modification mechanism of waste badminton shuttlecock modified asphalt binder. *Environ. Technol. Innov.* **2024**, *34*, 103561. [CrossRef]
7. Ai, X.; Yi, J.; Pei, Z.; Zhong, J.; Zhou, W.; Feng, D. Application of waste rice straw as fibre or filler in asphalt mixtures: Materials characterisation and performance evaluation. *Int. J. Pavement Eng.* **2024**, *25*, 2334307. [CrossRef]
8. Liu, Y.; Yang, Z.; Luo, H. Experimental Investigation of Eco-Friendly Anhydrous Calcium Sulfate Whisker and Waste Cooking Oil Compound Modified Asphalt Mixture. *Materials* **2023**, *16*, 2409. [CrossRef]
9. Guan, B.; Liu, J.; Wu, J.; Liu, J.; Tian, H.; Huang, T.; Liu, C.; Ren, T.; Grilli, A. Investigation of the Performance of the Ecofriendly Fiber-Reinforced Asphalt Mixture as a Sustainable Pavement Material. *Adv. Mater. Sci. Eng.* **2019**, *2019*, 6361032. [CrossRef]
10. Bai, X.; Wang, L.; Chen, X. Multiscale Study on Low Temperature Crack Resistance Mechanism of Steel Slag Asphalt Mixture. *J. Wuhan Univ. Technol.-Mat Sci Ed.* **2024**, *39*, 705–715. [CrossRef]
11. Wu, J.; Zhao, Z.; Jiang, C.; Yang, Y.; Zhishuai, S.; Jie, Y.; Feipeng, X. Recent development and application of natural fiber in asphalt pavement. *J. Clean. Prod.* **2024**, *449*, 141832. [CrossRef]
12. Tanzadeh, J.; Shahrezagamasaei, R. Laboratory Assessment of Hybrid Fiber and Nano-silica on Reinforced Porous Asphalt Mixtures. *Constr. Build. Mater.* **2017**, *144*, 260–270. [CrossRef]
13. Gong, Y.; Song, J.; Bi, H.; Tian, Z. Optimization Design of the Mix Ratio of a Nano-TiO_2/$CaCO_3$-Basalt Fiber Composite Modified Asphalt Mixture Based on Response Surface Methodology. *Appl. Sci.* **2020**, *10*, 4596. [CrossRef]
14. Ren, D.; Luo, W.; Su, S.; Wang, Z.; Kong, L.; Ai, C. Study on crack resistance of basalt fiber reinforced asphalt mixture modified by titanate coupling agent based on digital image correlation. *Constr. Build. Mater.* **2024**, *437*, 136934. [CrossRef]
15. Saha, G.; Biligiri, K.P. Fracture properties of asphalt mixtures using semi-circular bending test A state-of-the-art review and future research. *Constr. Build. Mater.* **2016**, *105*, 103–112. [CrossRef]
16. Li, C.; Wang, L.; Wang, X. Crack and crack growth behavior analysis of asphalt mixtures based on the digital speckle correlation method. *Constr. Build. Mater.* **2017**, *147*, 227–238. [CrossRef]
17. Kutuk-Sert, T.; Ozturk, M.; Kutuk, S. Digital image processing of warm mix asphalt enriched with nanocolemanite and nanoulexite minerals. *Constr. Build. Mater.* **2023**, *399*, 132542. [CrossRef]
18. Rusli, R.; Shanmuganathan, K.; Rowan, S.J.; Wedar, C.; Eichorn, J.S. Stress Transfer in Cellulose Nanowhisker Composites—Influence of Whisker Aspect Ratio and Surface Charge. *Biomacromolecules* **2011**, *12*, 1363. [CrossRef]
19. Saulat, H.; Cao, M.; Khan, M.M.; Khan, M.; Khan, M.M.; Rehman, A. Preparation and applications of calcium carbonate whisker with a special focus on construction materials. *Constr. Build. Mater.* **2020**, *236*, 117613. [CrossRef]
20. Hu, K.; Zhou, J.; Han, S.; Chen, Y.; Zhang, W.; Fan, C.; Fan, C. Silane coupling agent enhances recycle aggregate/asphalt interfacial properties an experimental and molecular dynamics study. *Mater. Today Commun.* **2024**, *39*, 108681. [CrossRef]
21. Fan, T.; Wang, X. Influence of calcium sulfate whisker on the high temperature performance of asphalt binder. *Pet. Sci. Technol.* **2020**, *38*, 303–308. [CrossRef]
22. Ren, D.; Luo, W.; Wang, Z.; Su, S.; Kong, L.; Ai, C. Enhancing the Adhesion Interface Between Asphalt and Basalt Fibers Through Surface Modification with Titanate Coupling Agents. *Arab. J. Sci. Eng.* **2024**, *49*, 5547–5563. [CrossRef]
23. Xing, X.; Chen, S.; Li, Y.; Pei, J.; Zhang, J.; Wen, Y.; Li, R.; Cui, S. Effect of different fibers on the properties of asphalt mastics. *Constr. Build. Mater.* **2020**, *262*, 120005. [CrossRef]

24. *JTG F40-2004*; Technical Specifications for Construction of Highway Asphalt Pavements. Research Institute of Highway Ministry of Transport: Beijing, China, 2004.
25. *JTG E42-2005*; Test Methods of Aggregate for Highway Engineering. Research Institute of Highway Ministry of Transport: Beijing, China, 2005.
26. Kutuk, S.; Kutuk-Sert, T. An examination of nanoparticle colemanite mineral added warm mix asphalt. *Constr. Build. Mater.* **2020**, *243*, 118252. [CrossRef]
27. Wu, H.; Ji, X.; Song, W.; Deng, Z. Multi-scale analysis on fracture behaviors of asphalt mixture considering moisture damage. *Constr. Build. Mater.* **2024**, *416*, 135234. [CrossRef]
28. Tarefder, R.A.; Kias, E.M.; Ng, T. Factors affecting cracking of notched asphalt concrete. *Int. J. Pavement Res. Technol.* **2009**, *2*, 137–147.
29. He, Y.; Xiong, K.; Zhang, J.; Guo, F.; Li, Y.; Hu, Q. A state-of-the-art review and prospectives on the self-healing repair technology for asphalt materials. *Constr. Build. Mater.* **2024**, *421*, 135660. [CrossRef]
30. Zhu, Y.; Dave, E.V.; Rahbar-Rastegar, R.; Sias, D.J.; Zofka, A. Comprehensive evaluation of low-temperature fracture indices for asphalt mixtures. *Road Mater. Pavement Des.* **2017**, *18*, 467–490. [CrossRef]
31. Kaseer, F.; Yin, F.; Arámbula-Mercado, E.; Amy Epps, M.; Jo Sias, D.; Saman, S. Development of an index to evaluate the cracking potential of asphalt mixtures using the semi-circular bending test. *Constr. Build. Mater.* **2018**, *167*, 286–298. [CrossRef]
32. Nian, T.; Wang, M.; Ge, J.; Guo, R.; Li, P.; Song, J. Enhancing low-temperature crack resistance, A method for establishing meso-models and evaluating steel fiber-reinforced hot recycled asphalt mixtures. *Constr. Build. Mater.* **2024**, *438*, 137026. [CrossRef]

buildings

MDPI

Article

Influence of the Bailey Gradation Method on the Mechanical Behavior of Asphalt Mixture Containing Steel Slag as an Alternative Aggregate

Lisley Madeira Coelho [1,*], William Wilson dos Santos [1], Antônio Carlos Rodrigues Guimarães [1] and Sergio Neves Monteiro [2]

[1] Department of Fortification and Construction, Military Institute of Engineering-IME, Praça General Tibúrcio, 80, Urca, Rio de Janeiro 22290-270, Brazil; william_wilson94@hotmail.com (W.W.d.S.); cap-guimaraes@hotmail.com (A.C.R.G.)

[2] Department of Materials Science, Military Institute of Engineering-IME, Praça General Tibúrcio, 80, Urca, Rio de Janeiro 22290-270, Brazil; snevesmonteiro@gmail.com

* Correspondence: madeiralisley@gmail.com

Abstract: This study evaluates the feasibility of reusing steel slag aggregates in asphalt concrete, analyzing the impact of different gradation methods (Bailey method and conventional Brazilian method) on the mechanical properties of the mix. Using the Marshall methodology and Petroleum Asphalt Concrete (PAC) 30/45, parameters such as Marshall stability, indirect tensile strength, resilient modulus, fatigue life through diametral compression, and permanent deformation (Flow Number) were investigated. Additionally, a simulation for a hypothetical section in the city of Rio de Janeiro, Brazil, was performed using the mechanistic-empirical pavement design software, Medina. The results showed that the mixture produced by the Bailey method outperformed the others in all analyses. This method led to a more compact mix, providing significant advantages, including up to a 35% reduction in final pavement thickness and a 110.6% increase in Flow Number (FN), enabling the mix to withstand extremely heavy traffic, as reported in the literature. Regarding fatigue life, the Bailey mixture achieved a fatigue class of 4, compared to the conventional mixture class 1. These findings indicate that using the Bailey gradation method for producing asphalt mixtures with steel slag can optimize binder content and improve resistance to permanent deformation and fatigue, making it a viable and sustainable alternative for asphalt pavements.

Keywords: alternative aggregate; steel slag; Bailey method; asphalt mixture; mechanical behavior

1. Introduction

There is currently a significant concern about environmental degradation, given that available natural resources are finite and the generated waste must be properly managed to minimize negative impacts on nature. The dependence on natural resources in the road construction sector is undeniable. Crushed materials are used in almost all the layers that make up a pavement.

The construction of pavements involves high costs, particularly when conventional materials are used for the structural layers [1,2]. In this context, steel slag has increasingly been considered an effective alternative to replace conventional materials, contributing to the reduction in environmental impacts caused by the extraction of natural resources. Nascimento [3] and Wang [4] report the use of steel slag by various countries (England, Germany, Poland, France, Japan, etc.) since the early 20th century for various purposes, and in Brazil, data on the use of this material in paving have been available since the 1970s [5]. Moreover, the European Union (EU) recommends prioritizing the reuse of these wastes as byproducts, given their contribution to sustainability and the reduction in environmental impacts, provided they possess the necessary technical characteristics for their intended applications [6,7]. The global production of steel, which has been increasing in countries such as China, the USA, Brazil, Australia, and several EU nations, generates large quantities of steel slag [7], motivating research into its utilization in various applications [8–16].

Studies show that steel slag possesses favorable mechanical characteristics, offering adequate strength and structural performance for different applications, such as railway ballast and road pavements. The mechanical characterization and study of permanent deformation behavior of alternative materials for paving are essential for evaluating their long-term effectiveness [17]. In this context, steel slag stands out by exhibiting robust performance when subjected to cyclic loading, demonstrating significant energy dissipation capacity and resistance to permanent deformation. These attributes are crucial for ensuring the durability and stability of infrastructures over time, without compromising the integrity of the structural layers [8,9,16,18–21]. Furthermore, recent studies have reinforced the potential of steel slag as a viable alternative material for hot mix asphalt. Kumar and Varma [22] emphasize that when properly integrated into asphalt mixtures, steel slag not only meets the required volumetric and mechanical performance standards but also contributes to enhanced durability and performance in road construction. Steel slag, a byproduct of the steel industry, possesses properties that make it a viable alternative to conventional aggregates in the production of asphalt mixtures. Applications in asphalt mixtures have been successfully tested by several studies, such as those presented by Guimarães [23], Loures et al [24], Guimarães et al. [9], Pasetto et al [25], Nascimento [3], and Plati et al. [26]. However, it became evident that special precautions need to be taken when using this type of material in asphalt mixtures, but the mechanical results are encouraging.

Various asphalt mixture designs aim to achieve properties such as durability, permeability, strength, stability, fatigue, and stiffness. A crucial factor influencing these properties is the gradation of the aggregates, including shape, size, angularity, and the flat-elongated characteristic, all of which directly affect the resistance and stability of the mixture [27]. Since aggregates form the skeleton of the asphalt mixture, a well-interlocked structure is essential to withstand most of the traffic load on the pavement. The gradation of the aggregates also plays an important role in the performance of the mixture. For example, an increase in the void content raises the porosity and permeability, negatively impacting tensile strength and pavement durability [28]. Furthermore, the characteristics of the aggregates, such as resistance and shape, directly influence the internal structure of the mixture, impacting its service life [29]. Different gradation types alter the average contact number and internal structure index, affecting both short- and long-term mechanical properties [30]. Gradation is also a determining factor in the fracture performance of the asphalt mixture, with the microstructure of coarse aggregates significantly affecting fracture performance, especially when they have the same gradation [31].

Contrary to conventional volumetric mix design methods, such as Marshall and Superpave, which consider aggregate characteristics indirectly and through trial and error, the Bailey Method proposes particle interlock through a continuous and balanced gradation. Sulaiman [32] highlights that this method optimizes asphalt mixtures to achieve the desired performance characteristics, ensuring consistency during asphalt production. Rahi [33] states that the Bailey method improves stability and resistance to permanent deformation, resulting in better performance and greater resistance to wear in pavements. Additionally, Rafeez [34] emphasizes that the method enhances aggregate compaction, promoting superior performance compared to conventional gradations. Developed by the Illinois Department of Transportation (IDOT), the Bailey method has been used and refined since the 1980s. Several studies confirm its effectiveness in designing ideal mixtures (Komba et al. [35]; Blaauw et al. [36]; Horak et al. [37]; Ghuzlan and Al-mestarehi [38]; Imaninasab [39]); Alsheyab et al. [40]; Joumblabt et al. [41]).

Alongside the recycling of aggregates, the reuse of asphalt binder has gained increasing importance in the pursuit of more sustainable practices in pavement construction. Recent studies have emphasized the role of recycled asphalt binders in reducing the demand for virgin materials, enhancing the sustainability of pavement systems while maintaining performance standards [42–44]. Integrating these practices with alternative aggregates, such as steel slag, further promotes the circular economy in road construction, aligning with global sustainability goals.

The objective of this study was to assess the technical feasibility of using 100% steel slag in asphalt paving by comparing asphalt mixtures designed with the Brazilian conventional method (trial and error method) and the Bailey method for determining the mineral skeleton. To evaluate the performance of these mixtures, tests for indirect tensile strength (ITS), resilient modulus (RM), fatigue life (FL), and permanent deformation (PD) were conducted. Additionally, a simulation for a hypothetical section in the city of Rio de Janeiro, Brazil, was performed using the mechanistic-empirical pavement design software, Medina (v 1.5.0).

2. Gradation Selection Methods

2.1. Bailey Method

The Bailey method was conceived in the 1980s to meet the demands of American agencies regarding the need to provide asphalt coatings with resistance to permanent deformations, fatigue cracks, and wear. This method aims to achieve these objectives through the systematic selection of the gradation of asphalt mixtures. Its primary purpose is to combine aggregates in such a way that, when necessary, interlocking occurs between them, thus forming the structural skeleton of the layer, in addition to employing a continuous and balanced gradation to complete the mixture, as highlighted by Vavrik et al. [45].

The method presents a different way of classifying coarse and fine aggregates based on the Nominal Maximum Size (NMS) of the mixture, defining coarse aggregates as those that form voids when placed in a defined volume and fine aggregates as those capable of filling the voids created by the coarse aggregates present in the mixture.

2.2. Aggregate Size Definition for the Bailey Method

$$PCS = NMS \times 0.22 \tag{1}$$

PCS: Primary Control Sieve.

NMS: Nominal Maximum Size.

Through two-dimensional (2D) and three-dimensional (3D) analyses of the arrangement of aggregates of different shapes, factors ranging from 0.18 to 0.28 were identified for use in Equation (1), without affecting the gradation analysis. However, a standardized value of 0.22 was adopted (Vavrik et al. [45]). Coarse aggregates are those retained on the PCS and are referred to as the Coarse Aggregate Proportion (CA Proportion). The material passing through the PCS represents the fine aggregates, which are further divided into two proportions: the Coarse Portion of the Fine Aggregate (CPFA Proportion) and the Fine Portion of the Fine Aggregate (FPFA Proportion). To determine the two fractions of fine aggregates, the Secondary Control Sieve (SCS) is used, as defined by Equation (2).

$$SCS = PCP \times 0.22 \tag{2}$$

where

SCS: Secondary Control Sieve;

PCS: Primary Control Sieve.

In addition to the two sieves determined previously, we can also mention the following sieves: the median sieve (MS), which is the one closest to half of the nominal maximum size (NMS); and the tertiary control sieve (TCS), which is obtained by multiplying the secondary control sieve (SCS) by a factor of 0.22. With the determination of aggregate fractions and control sieves, the aggregate proportions can be determined using Equations (3)–(5)

$$\text{CA Proportion} = \frac{\text{\% passing through MS} - \text{\% passing through PCS}}{100\% - \text{\% passing through MS}} \tag{3}$$

$$\text{CPFA Proportion} = \frac{\%\ \text{passing through SCS}}{\%\ \text{passing through PCS}} \quad (4)$$

$$\text{FPFA Proportion} = \frac{\%\ \text{passing through TCS}}{\%\ \text{passing through SCS}} \quad (5)$$

According to Vavrik et al. [45], the values for these aggregate proportions should fall within the ranges indicated in Table 1. Mixtures whose values for these proportions do not align with the recommendations of this author may become unstable and exhibit compaction difficulties in the field, for instance.

Table 1. Aggregate proportion relative to nominal maximum size (NMS) in mm.

Aggregate Proportion	Nominal Maximum Size (NMS), in mm					
	37.5	**25.0**	**19.0**	**12.5**	**9.5**	**4.75**
CA	0.80–0.95	0.70–0.85	0.60–0.75	0.50–0.65	0.40–0.55	0.30–0.45
CPFA	0.35–0.50	0.35–0.50	0.35–0.50	0.65–0.50	0.35–0.50	0.35–0.50
FPFA	0.35–0.50	0.35–0.50	0.35–0.50	0.35–0.50	0.35–0.50	0.35–0.50

Densities of Coarse Aggregate

After determining the loose bulk density (LBD) and compacted bulk density (CBD) of the coarse aggregate according to AASHTO T-19 [46], the next step is to determine the chosen bulk density of the coarse aggregate, which is essentially the selection of the void content of the mixture. This choice leads to determining the volume of coarse aggregate and also the degree of interlocking of the mixture. With the determination of the CBD, it is also determined whether the mixture is coarse, where the skeleton is formed by the coarse aggregate, or if it is fine, when the amount of coarse aggregate is not sufficient to form the skeleton and it is formed by fine aggregate [45].

Considerations for determining the LBD, which must align with the desired gradation of the mixture, are illustrated in Figure 1. For dense-graded mixtures, values should be between 95% and 100% of the LBD. Values between 90% and 95% of the LBD should be avoided as they may exhibit significant variation in the degree of interlocking of coarse aggregates in the field. Values above 105% should also be avoided due to the increased risk of aggregate degradation and the difficulty of field compaction.

Figure 1. Steel slag used in this research. (**a**) In wet form; (**b**) in dry form.

2.3. Conventional Trial-and-Error Method

The term "conventional method" or "trial-and-error method" refers to a widely used approach in asphalt mix design, where the designer selects the aggregate gradation based on practical experience and the characteristics of available materials. Although the specific application of this method in Brazil is regulated by the National Department of Transport Infrastructure (DNIT), this approach is globally recognized and applied, regardless of national standards.

In the conventional method, the process of determining the aggregate gradation involves choosing particle size ranges for the aggregates based on project requirements and material specifications. The designer defines the gradation to be used after analyzing characteristics such as the specific gravity and absorption of the aggregates. Then, the binder content is adjusted based on the aggregate characteristics and the selected gradation.

According to Bernucci et al. [47], the specific gravities of the constituent materials of the mixture should first be determined, as well as the particle size distribution of the aggregates to be used. After that, a gradation range is selected according to predetermined intervals or based on the designer's practical experience, ensuring it meets the project's requirements. The binder content is then adjusted to ensure the desired performance of the mixture.

Bernucci et al. [47] further explain that once the appropriate gradation range is selected, the composition of the aggregates must be determined, meaning the mass percentage of each type of aggregate in the mixture. The designer ensures that the percentages of each material fall within the selected range, with permissible variations in the field, according to the specifications.

3. Materials and Methods

To assess the feasibility of using the Bailey Method in asphalt mixtures containing steel slag, two asphalt concrete mixtures were prepared: one in which the mineral skeleton was dosed using the conventional trial-and-error method [48], and another in which the Bailey Method was used to determine the mineral skeleton.

3.1. Material

The slag used in this study is Linz–Donawitz steel slag, supplied by the Ternium steel company located in the Santa Cruz neighborhood in the city of Rio de Janeiro. The particle size distribution analysis of the steel aggregate was conducted according to the DNER-ME 083 standard [49]. Figure 1 presents the slag in wet and dry forms, and Figure 2 represents the particle size distribution curve of the slag used.

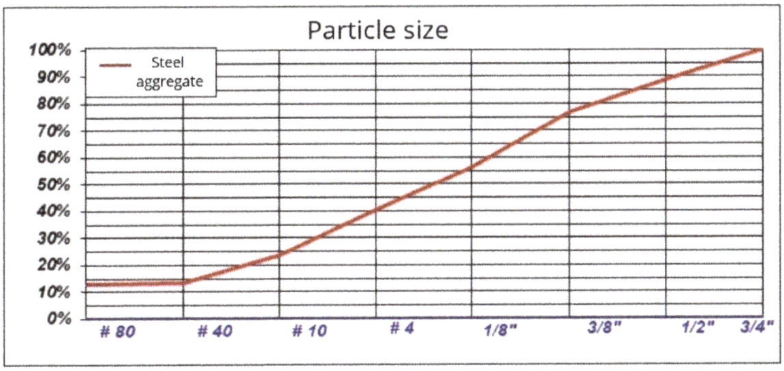

Figure 2. Particle size distribution curve of the steel slag used in this research.

For the characterization of the slag, tests were conducted for density [50], absorption [51], sand equivalent [52], and adhesivity [53], following the technical specifications [54].

The environmental suitability of the steel slag used in this study was assessed by the supplier through standardized tests, which confirmed compliance with all relevant technical and environmental standards, including evaluations of potential leachates. Additionally, expansion tests were conducted, and the results met the limits established by current standards [55], ensuring the material's stability and safety for application in asphalt mixtures.

In addition, tests related to the slag's resistance were conducted, such as the evaluation of shock loss using the Treton equipment [56], and the Los Angeles abrasion test [57].

The asphalt binder used in this research is the Petroleum asphalt concrete (PAC) 30/45, supplied by the Asphalt Plant of Caju (Rio de Janeiro City Hall), as it is the most commonly used type in asphalt pavement works in the region. The characteristics of the binder are presented in Table 2.

Table 2. Characterization tests as per the National Agency of Petroleum, Natural Gas, and Biofuels.

Test	Unit	PAC 30/45 Limit	Standard
Flash Point, min.	°C	min 235	DNER-ME 148 [58]
Penetration (100 g, 5 s, 25 °C)	0.1 mm	30 to 45	DNIT ME 155 [59]
Softening Point	°C	min 52	DNIT ME 131/10 [60]
Thermal Susceptibility Index	-	(−1.5) to (+0.7)	–
Rotational Viscosity	cP	min 374	NBR 15184 [61]
		min 203	
		76–285	
Residue After RTFOT Tests			ASTM 2872 [62]
Mass Variation	% mass	≤ 0.5	–
Softening Point Variation	°C	max 8	DNIT ME 131 [60]
Penetration Retained	%	min 60%	DNIT ME 155 [59]

3.2. Method

To prepare the mixture containing 100% steel slag aggregates, the initial dosing aimed to fit the aggregate gradation into the C range, as specified in the Brazilian standard [48]. The C range refers to a specific gradation interval that ensures the aggregates meet the requirements for optimal performance in asphalt mixtures. This classification defines the particle size distribution, with the objective of achieving proper compaction, adequate void distribution, and good binder adhesion. The ranges listed specify the percent passing through particular sieve sizes, ensuring that the mix is balanced for the intended layer thickness, typically requiring the compacted layer to be at least 2.5 times the nominal maximum particle size of the selected gradation.

Thus, the decision was made to fractionate the raw slag, resulting in the slag being divided into coarse slag I (passing through #3/4 and retained on #1/2), coarse slag II (passing through #1/2 and retained on #10), and fine slag (passing through #10), with their respective particle size distributions presented in Figure 3.

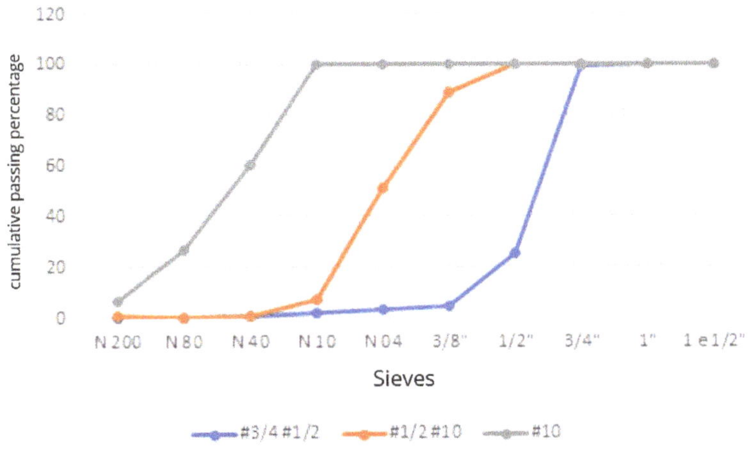

Figure 3. Particle size distributions of the materials used in this study.

The proportioning of aggregates was determined through the trial and error method and was denoted as the conventional mixture. From this conventional mixture, modifications were made following the suggestions of the Bailey method, resulting in a new mixture designated as the Bailey mixture. In the Bailey mixture, it was necessary to add 1.7% limestone filler. The resulting particle size distributions are presented in Table 3.

Table 3. Percentage composition of components for conventional and Bailey mixtures.

Mixture	Coarse Slag I (%)	Coarse Slag II (%)	Fine Slag (%)	Filler (%)
Conventional Mixture	12	60	28	-
Bailey Mixture	24.8	49	24.5	1.7

Table 4 presents the granulometric parameters indicated by the Bailey method for the studied Bailey mixture. It is noted that the mixture exhibited a proportion of coarse aggregate in fine aggregate (GAF) higher than desirable. However, even with the non-compliance of this proportion, it was decided to proceed with the experimental procedures.

Table 4. Aggregate distribution for Bailey mixture with acceptance limits.

Parameter/Mixture	CA	FCA	FFA
Bailey Mixture	0.57	0.56	0.45
Limit	0.50–0.65	0.35–050	0.35–0.50

To establish the optimal design contents for both mixtures (conventional mixture and Bailey mixture), we followed the Marshall method [63], with the application of 75 blows per face. The final results indicated design contents of 8% and 6.5% for the conventional mixture and Bailey mixture, respectively. These values were determined based on a voids condition of 4%, as analyzed in volumetric studies conducted by Coelho [64], which employed the Marshall methodology as a reference.

The results in Tables 3 and 4 highlight the distinct aggregate distribution achieved using the Bailey method compared to the conventional method. Despite the GAF proportion exceeding the desirable range for the Bailey mixture, the decision to proceed with the experimental procedures underscores the adaptability of the Bailey method. The optimized gradation allowed for a lower binder content (6.5%) compared to the conventional mixture (8%), demonstrating its efficiency in achieving target volumetric properties. These findings reinforce the method's potential to accommodate variations in granulometric parameters while maintaining its design principles.

It is worth noting that the conventional mixture failed to meet the required value for voids in bituminous mix (VBM). However, it exceeded the minimum stability specified by the methodology, and the other volumetric parameters were within the recommended limits.

With the determined optimum contents, tests were carried out for indirect tensile strength (ITS), resilient modulus (RM), permanent deformation (PD), and fatigue life (FL) as per the methods outlined in DNIT 136 [65], DNIT 135 [66], DNIT 184 [67], and DNIT 183 [68], respectively. The moisture damage resistance of compacted asphalt mixture was investigated by the ITS test. They were both conducted according to AASHTO T 238 standard methods. In the ITS test, Marshall samples were averagely divided into two groups. One was the control group, and the other was conditioned. The conditioned group was subjected to moisture freeze-thaw damage, as shown in Figure 4. All samples were sent to equipment to measure the tensile strength under room temperature about 25 °C. The tensile strength ratio (TSR) was computed according to Equation (6).

$$TSR = \frac{TSh}{TSo} \times 100\% \tag{6}$$

where TSo is the average tensile strength of the control group, MPa; e TSh is the average tensile strength of the conditioned group after freeze-thaw damage, MPa.

Figure 4. Specimens used in the freeze-thaw cycles.

Mechanistic Analysis

For the analysis and comparison of the mechanical behavior of a flexible pavement structure composed of the studied mixtures (conventional mixture and Bailey mixture), the Brazilian National Pavement Design Method, Medina, was utilized. Medina is the result of research developed between 2015 and 2018 by the Road Research Institute—IPR and the Alberto Luiz Coimbra Institute of Postgraduate and Research in Engineering, Federal University of Rio de Janeiro—COPPE, with the collaboration of the Petrobras Research Center—CENPES and several universities in Brazil. Additionally, it is an update of the SISPAV program developed by Franco [69] in his doctoral thesis.

The software uses as input data the dimensions of the layers, the mechanical properties of the materials (RM, FL model, PD model, and Poisson's ratio) composing the layers and the loads generated by the passage of vehicles. Once the stresses, strains, and displacements are calculated, the software verifies whether the number of load applications will lead to excessive cracking of the asphalt pavement or cemented layers, or to rutting in the wheel track exceeding the established limit.

Moreover, based on the regression coefficients obtained from the fatigue test, the software automatically calculates a fatigue factor of the mixture (FFM) for each model inserted. Comparing this value with the resilient modulus of the mixture, the fatigue class (FC) of the material is calculated. According to the Medina manual [70], the higher the FC (ranging from 1 to 4), the better the mechanical behavior of the asphalt mixture.

The hypothetical structure of the flexible pavement will be composed according to the scheme presented in Table 5, taking into account the characteristics of the base, sub-base, and subgrade materials of the pavement. These characteristics were selected for a standard Brazilian single-wheel axle load and considering a design life of 10 years, based on representative traffic data obtained in Loures' research [24]. The asphalt pavement will be a mixture commonly used in Rio de Janeiro, while the studied slag mixtures will also be incorporated into the design.

Table 5 presents the hypothetical structure of the flexible pavement adopted for analysis, considering the characteristics of the base, sub-base, and subgrade materials, as well as the standard Brazilian single-wheel axle load. A design life of 10 years was assumed, with hypothetical traffic levels representing low, medium, and high traffic volumes. These scenarios were defined to evaluate the adaptability of the mixtures under varying demands, inspired by representative data from Loures' research [24]. For comparison purposes, the RJ PAC 30/45 #12.5mm Sepetiba (M0) mixture, available in the MeDiNa program database, was selected as a reference. The physical and mechanical characteristics of the pavements are presented later in Table 5.

The hypothetical nature of this analysis allows for its application in other countries or regions, providing a basis for future research or as a framework for evaluating similar pavement designs and mixtures in diverse traffic and environmental conditions.

Table 5. Characteristics of materials and traffic for road pavement design.

Layers	Type	RM (MPa)	Thickness (cm)
Base	Granular Material (Gneiss C1)	259	15
Sub-base	Fine, silty, or clayey soil	189	15
Subgrade	Fine, silty, or clayey soil	189	-
	Standard Brazilian Road Axle Data		
Axle load (tonf)		8.2	
Tire pressure (MPa)		0.56	
	Traffic Characteristics		
Light Traffic		(9.96×10^5)	
Medium Traffic		(5×10^6)	
Heavy Traffic		(5×10^7)	

4. Results and Discussion

4.1. Physical Characterization

The results presented in Table 6 demonstrate that the studied steel slag follows the characteristics determined by the standard, except for absorption, which shows higher values than the standardized ones.

Table 6. Results of physical and mechanical properties tests.

Test	Result
Real Density (g/cm^3)	3.30
Apparent Density (g/cm^3)	2.62
Absorption (%)	3.60
Sand Equivalent (%)	68.52
Adhesivity	Satisfactory
Expansion (%)	1.24

In addition, tests for impact loss and Los Angeles abrasion were conducted, and the results are presented in Table 7, demonstrating compliance with the standards used.

Table 7. Results of impact loss and Los Angeles abrasion tests.

Test	Result
Impact Loss (%)	9.20
Los Angeles Abrasion (%)	21.57

4.2. Tensile Strength and Resilient Modulus

Table 8 presents the results of the tests, based on the average (three replicates) values found for each mixture. The ITS values showed no significant differences but met the minimum criterion of the standard (0.65 MPa).

Table 8. RT and RM values for conventional and Bailey mixtures.

Mixture	ITS (MPa)	RM (MPa)
Conventional Mixture	1.16	8700
Bailey Mixture	1.09	10,693

The high RM values observed in this study, ranging from 8700 to 10,693 MPa, underscore the superior stiffness of mixtures with a high steel slag aggregate (SSA) compared to mixtures of natural aggregate (NA), as discussed by Swathi et al. [71]. Prior findings from Swathi et al. [71] indicate that incorporating SSA, due to its high impact resistance and cubic particle shape, substantially enhances stiffness relative to NA, with SSA mixtures exhibiting RM values approximately 1.4 to 1.8 times higher than those of NA mixtures. The notable stiffness difference in our results can probably be attributed not only to the high content of SSA (88.3% to 100%) but also to the use of a stiffer binder (PAC 30/45), both contributing to the observed increase in RM values.

Previous studies, such as those by Xie et al. [72] and Kavussi and Qazizadeh [73], identified significant limitations in using 100% SSA in asphalt mixtures, including increased binder demand and voids in mineral aggregate (VMA) when conventional weight-based gradation methods are applied. These challenges often lead to higher costs and reduced performance under traffic loading, prompting design guidelines to recommend partial replacement of aggregates instead of full substitution [74–76]. Our study addresses this gap by demonstrating that the Bailey method effectively solves these issues, allowing the use of 100% SSA while achieving RM values significantly higher than those reported for NA mixtures. This finding highlights the potential of the Bailey method to improve sustainability and performance in asphalt pavement design by optimizing gradation for high SSA content mixtures.

The Bailey method mixture specifically demonstrated an increase in stiffness in comparison to conventional mixtures, as reflected in these higher RM values. However, it is important to note that a higher stiffness alone does not guarantee better performance in asphalt mixtures. For optimal performance, these mixtures must balance rigidity with flexibility to accommodate deformations and maintain sufficient tensile strength, thus preventing cracking and rupture. In addition, the stiffness of the mixture should align with the structural needs of other layers of the pavement to ensure overall durability and compatibility.

In comparison, Cavalcante et al. [77], using the Superpave methodology, found an RM value of 2241 MPa for HMA containing 57% steel slag, demonstrating a substantial difference from our findings, likely due to the lower content of slag and the different binder used in their studies.

Furthermore, Coelho et al. [78] investigated HMA containing 40% reduction ferronickel slag (RFNS) applied in an experimental parking lot, achieving an RM value of

6323 MPa through the Marshall methodology. Their study confirmed the durability of RFNS as an alternative aggregate, with visual inspections four years later indicating that the pavement remained in excellent condition. This RM value, although lower than those in our study, supports the feasibility of using alternative aggregates in asphalt mixtures.

Finally, Guimarães et al. [23]) evaluated mixtures containing 20% and 17% mining waste, producing RM values of 5137 MPa and 5244 MPa, respectively. Their findings demonstrated that substituting fine aggregate with mining waste in asphalt concrete is viable for local roads without compromising pavement support capacity, although their RM values are also significantly lower than those found in our study with higher slag content.

4.3. Permanent Deformation

Permanent deformation was evaluated using the Flow Number (FN) test. The FN value was obtained by averaging three specimens for each mixture. In Figure 5, the characteristics of the mixtures can be analyzed according to the limit criteria proposed by Nascimento [79] and Bastos [80].

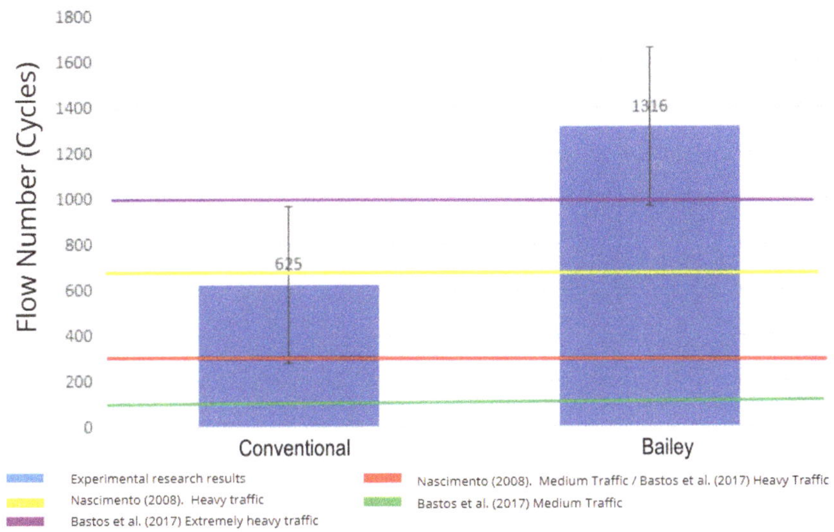

Figure 5. Comparison of FN values obtained for each method compared to the limits proposed by Nascimento [79] and Bastos [80].

According to Medina and Motta [81], the FN is higher the coarser the aggregate gradation and the more consistent (lower penetration) the asphalt binder.

Considering the FN results for the tested mixtures, the conventional mixture showed a considerable value, with a tendency towards medium-to-heavy traffic. However, the FN result for the Bailey mixture, which uses the same steel slag aggregates and the same compaction method, demonstrates the positive influence of the granulometric method on permanent deformation behavior and consequently on wheel track rutting (WTR) defect, resulting in an FN number related to extremely heavy traffic limits, according to Nascimento's research [79].

Additionally, after modeling the deformation curves using the Francken model, the FN results, permanent deformation at FN ($\varepsilon_{p(FN)}$), and FN Index were extracted, as shown in Table 9.

Table 9. Flow number test results for conventional and Bailey methods.

Method	FN (Cycles)	$\varepsilon_{p(FN)}$ (Microstrains)	FN INDEX (Microstrains/Cycles)
Conventional	625	7564.30	12.10
Bailey	1316	8018.10	6.10

It is observed that the Bailey mixture accumulated approximately two times less plastic deformation per cycle than the conventional mixture. This behavior can be attributed to the strength characteristics of the steel slag aggregate combined with the interlocking effect provided by the Bailey gradation. Supporting this finding, Ghuzlan et al. [38] and Golalipou et al. [82] noted that mixtures with finer aggregate arrangements systematically interlocked with coarse aggregates exhibit lower deformation values under loading. This configuration helps distribute stresses more effectively among the coarse aggregates, enabling the mixture to withstand higher load cycles.

Conversely, Oufa and Abdolsamed [83] observed that coarse Bailey mixtures demonstrated superior rutting resistance compared to finer ones, highlighting the influence of gradation design on performance. Despite these differing observations, the results of this study align with findings from Shang et al. [84], Manjunath and Poornachandra [85], and Denneman et al. [86], which collectively emphasize that Bailey gradation mixtures outperform conventional designs such as Superpave and Asphalt Institute gradations. This superiority underscores the effectiveness of the Bailey method in enhancing the structural and functional performance of asphalt mixtures.

4.4. Fatigue Life Analysis

The fatigue life test was conducted in the pavement laboratory of the Military Institute of Engineering, Rio de Janeiro, Brazil, following the standard (DNIT-ME 183 [68]), without the use of LVDT sensors. The test was performed for specimens molded at the determined optimum content for each mixture and subjected to the fatigue test by diametrical compression with the variation in stress level in the range of 15% to 30% of the ITS.

Thus, it was possible to establish the stress difference ($\Delta\sigma$) curves in relation to the number of cycles to failure (Nf). When considering the resilient characteristic of each mixture through the RM, a relationship of resilient strain (ϵ_r) was also established with Nf. The regression coefficients and correlation coefficients are presented in Table 10.

Table 10. Regression parameters of fatigue life curves for the mixtures in this study.

Mixture	Nf = a1 ($\Delta\sigma$) b1			Nf = a1 ($\epsilon\sigma$) b2		
	a1	b1	R^2	a2	b2	R^2
Conventional	35,511	−4.619	0.868	2.00×10^{-14}	−4.059	0.868
Bailey	184,436	−2.682	0.9782	7.00×10^{-8}	−2.682	0.9782

Analyzing Table 10, it can be observed that the values exhibit good determination coefficients R^2, indicating the quality of fit of the models to the observed points. The values obtained from the Bailey method showed less dispersion, i.e., R^2 values closer to 1, while the conventional method values yielded more conservative results.

Júnior [87] explains that the coefficient "b" indicates the slope of the curve, and for stress difference graphs, "b1" indicates susceptibility to stress conditions, meaning higher values indicate greater sensitivity of fatigue behavior of mixtures to variations in stress difference. Thus, according to the data in Table 10, the conventional mixture presents higher values of "b1", therefore being considered more susceptible to variations in stress conditions.

According to Medina and Motta [81], comparing fatigue curves should only be conducted when mixtures exhibit similar stiffness or resilience modulus, since stiffer mixtures "absorb" more stress, resulting in higher tensile stresses as well. Thus, for the same structure (thicknesses and materials of other layers), differences in MR in the pavement will lead to different stress difference values ($\Delta\sigma$).

Given that the RM of the Bailey mixture is approximately 22.9% higher than that of the conventional mixture, it indicates that the mixtures have different moduli. To better understand the fatigue behavior characteristics of these mixtures, a simulation of the design was performed using the Medina software (v 1.5.0) for an arbitrary pavement containing the studied mixtures of HMA with steel slag compared to conventional HMA from the city of Rio de Janeiro/RJ. This simulation can be verified in Section 4.7.

4.5. Influence of the Granulometric Method on Fatigue and Permanent Deformation

With the results of permanent deformation from the FN test and the regression parameters "b1" obtained from the fatigue curves of the mixtures, an analysis of the average effects and interactions among the variable "method" was conducted, guided by the factorial experimental design. This study was parameterized based on the reference variable (conventional method), as observed in Table 11.

Table 11. Comparison of methods for assessing susceptibility to stress state and average flow number effect.

Method	Susceptibility to Stress State	Percentage Over Reference	Average Effect Flow Number	Percentage Over Reference
Conventional	4.059		625	
Bailey	2.682	−34%	1316	110.6%

In relation to the average effects, it can be stated, according to Table 11, that the variation in granulometric composition is a factor that significantly influences the mixture's resistance to permanent deformation and its susceptibility to fatigue life. Overall, the variation from the conventional granulometric method to the Bailey method resulted in a 110.6% increase in the FN.

Regarding its role in fatigue life, it appeared to be more predominant, albeit indicating sensitivity to fatigue through the slope of the curve represented by the parameter "b1". The variation from the conventional granulometric method to the Bailey method made the mixture 34% less susceptible to stress conditions, meaning lower values of "b1" indicate less sensitivity of the mixtureś fatigue behavior to variations in stress difference.

These findings have practical implications for the design and durability of asphalt pavements. The increased FN observed with the Bailey method suggests a higher resistance to permanent deformation, which could lead to improved performance under heavy traffic loads and in regions subject to high temperatures. In contrast, the reduced sensitivity to fatigue in the Bailey method indicates that these mixtures may exhibit better long-term performance in terms of fatigue cracking, which is a critical consideration for the design of pavements in areas subjected to repeated traffic stresses. Therefore, the choice of granulometric method significantly impacts the durability and maintenance needs of asphalt pavements, influencing both the fatigue and deformation behavior of the mixture over time.

4.6. Evaluation of Freeze-Thaw Cycle Damage Resistance

Moisture damage in asphalt mixtures is intrinsically linked to the degradation of asphalt mastic and the interface between mastic and aggregate, with these effects being exacerbated under freeze-thaw cycles. The tensile strength of asphalt mixtures is influenced by various factors, such as aggregate properties, binder characteristics, and gradation; however, the behavior of the aggregate-binder interface is particularly critical [88].

In this study, mixtures produced using the Bailey method and the conventional method exhibited significant differences in the results of the moisture-induced damage test. The

Bailey method showed higher tensile strength in the conditioned state (TSh = 0.55 MPa) compared to the conventional method (TSh = 0.32 MPa), indicating better retention of strength after freeze-thaw cycles. Conversely, mixtures produced with the conventional method presented higher tensile strength in the dry state (TSo = 1.06 MPa vs. 0.88 MPa for the Bailey method), suggesting better initial performance, although this did not translate into improved durability. The TSR was significantly higher for the mixtures produced with the Bailey method (62.5%) compared to the conventional method (30.18%), highlighting greater stability against moisture-induced damage.

These findings align with the work of Guo et al. [89], who demonstrated that the steel slag content significantly influences the long-term water stability of asphalt mixtures. Their results suggest that an optimal steel slag content can enhance resistance to water-induced damage over extended periods. The incorporation of such materials contributes to the improvement of the aggregate-binder interface, supporting the superior performance observed in the mixtures designed using the Bailey method.

Overall, these results emphasize the effectiveness of the Bailey method in optimizing aggregate gradation, promoting better interlocking among particles, and ensuring a more uniform stress distribution within the asphalt matrix. Consequently, mixtures produced using the Bailey method exhibit a greater capacity to mitigate the progressive degradation caused by moisture, making it a more robust alternative for regions with adverse climatic conditions.

4.7. Mechanistic Analysis

As observed in Table 12, the characterization of each asphalt mixture in Medina's database shows a significant influence of gradation composition on the fatigue characteristics of the mixtures. Notably, M2 achieved a fatigue class of 4, indicating superior mechanical performance. In contrast, M1 and M0 exhibited fatigue classes of 1 and 2, respectively.

Table 12. Characteristics added to the Medina software for the analyzed mixtures.

Asphalt Concrete	RJ PAC 30/45 #12.5 mm Sepetiba Reference (M0)	Slag (Conventional Method) (M1)	Slag (Bailey Method) (M2)
	Parameters		
Poisson's Ratio	0.30	0.30	0.30
	Modulus (MPa)		
Constituent Model	Linear Resilient	Linear Resilient	Linear Resilient
Modulus (MPa)	9000	8700	10,693
	Characteristics		
Type of Binder (PAC)	30/45 - PG 70-16	30/45	30/45
Specific Gravity (g/cm^3)	2.4	2.766	2.792
Tensile Strength (MPa)	2.08	1.16	1.09
Asphalt Content (%)	6	8	6.5
Void Volume (%)	5	3.6	4.1
Gradation Size	#12.5 mm	#12.5 mm	#12.5 mm
Los Angeles Abrasion (%)	40	-	-
Standard or Specification	DNIT ES 31 [48]	DNIT ES 31 [48]	DNIT ES 31 [48]
	Fatigue Curve		
Constituent Model	K1 (ϵt^{k2})	K1 (ϵt^{k2})	K1 (ϵt^{k2})
Regression Coefficient (k1)	5.00×10^{-14}	2.00×10^{-14}	7.00×10^{-8}
Regression Coefficient (k2)	−3.992	−4.059	−2.682
Fatigue Class	2	1	4
FFM	0.75	0.69	1.21

In Table 13, a summary of the designed structures for light, medium, and heavy traffic is presented, considering a 95% confidence level for the various mixtures evaluated in this study.

Table 13. Cracked area (CA), rut depth (RD), and thickness values of the pavements under analysis.

Light Traffic (N = 9.96 × 10^5)			**Layer Thickness**			**Total Thickness**
Mixture	**% CA**	**RD (mm)**	**Surface**	**Base**	**Sub-Base**	
M0	12.3	6.3	5 cm	15 cm	15 cm	35 cm
M1	14.2	6.3	5 cm	15 cm	15 cm	35 cm
M2	5.2	6.1	5 cm	15 cm	15 cm	35 cm
Medium Traffic (N = 5 × 10^6)			**Layer Thickness**			**Total Thickness**
Mixture	**% CA**	**RD (mm)**	**Surface**	**Base**	**Sub-base**	
M0	29.1	4.0	10 cm	15 cm	15 cm	40 cm
M1	26.3	3.6	11 cm	15 cm	15 cm	41 cm
M2	15.4	6.8	5 cm	15 cm	15 cm	35 cm
Heavy Traffic (N = 5 × 10^7)			**Layer Thickness**			**Total Thickness**
Mixture	**% CA**	**RD (mm)**	**Surface**	**Base**	**Sub-base**	
M0	28.6	2.4	20.3 cm	20 cm	20 cm	63.0 cm
M1	28.5	2.2	22.1 cm	20 cm	20 cm	62.1 cm
M2	28.9	3.5	12.1 cm	15 cm	15 cm	42.1 cm

It is observed that, for the traffic level of N = 9.96 × 10^5 (Light Traffic) with 95% reliability, the pavements had the same layer thicknesses. However, using the asphalt coating with the Bailey method, the pavement showed the best performance in terms of CA and RD, with 5.2% and 6.1 mm, respectively.

Considering a 95% reliability level and N = 5 × 10^6(Medium Traffic), the conventional mixture would require a 10 cm surface layer, and the slag mixture using the conventional method would need an 11 cm surface layer. Both showed similar performance in terms of the percentage of cracked area and rut depth. Meanwhile, the Bailey mixture would meet the same criteria with a 5 cm surface layer, presenting the lowest percentage of cracked area.

For the reliability level of 95% and N = 5 × 10^7 (Heavy Traffic), the conventional mixture would meet the design criteria with a surface layer thickness of 20.3 cm and a Base + Sub-base of 40 cm; the slag mixture using the Trial method would need a surface layer thickness of 22.1 cm and a Base + Sub-base of 40 cm. Meanwhile, the Bailey mixture would require a 12.1 cm surface layer and a Base + Sub-base of 30 cm. Even with a lower final pavement thickness, the Bailey mixture demonstrated the best performance in terms of pavement CA and RD over the 10-year project analysis period.

In the hypothetical pavement evaluation conducted in this study, the Bailey-graded mixture demonstrated superior performance under varying traffic levels and reliability requirements, achieving lower cracked areas and rut depths with reduced surface layer thicknesses compared to conventional and conventional-graded slag mixtures. Similarly, Swathi et al. [71] highlighted the benefits of SSA in asphalt mixtures, noting that its high impact resistance and cubic shape contribute to reduced rutting potential and improved durability under heavy loads. These findings align with the improved performance observed in the Bailey mixture, which exhibited optimal structural efficiency, supporting its potential as a sustainable alternative for high-traffic pavements. Furthermore, Luan et al. [90] also observed that modifying the aggregate gradation can enhance the use of steel slag in asphalt mixtures. Their study, utilizing a modified Marshall mix design, showed that steel slag mixtures demonstrated superior rutting resistance and low-temperature crack resistance, particularly in stone matrix asphalt (SMA) mixtures. This supports the idea that adjusting gradation, similar to the Bailey method applied in this study, can optimize the use of alternative aggregates like slag, improving both mechanical properties and environmental sustainability of asphalt pavements.

5. Summary and Conclusions

This study aimed to evaluate the feasibility of reusing 100% steel slag aggregates in asphalt concrete, comparing the Bailey and conventional Brazilian gradation methods. Through laboratory testing and a simulation for a hypothetical section in Rio de Janeiro, Brazil, using the Medina software, the following conclusions were drawn:

- The mixture with steel slag produced using the conventional Brazilian method achieved a Resilient Modulus (RM) of 8700 MPa, while the mixture prepared with the Bailey method showed a significantly higher RM of 10,693 MPa. This indicates that the Bailey Method resulted in a stiffer mixture, which is expected to provide improved performance, especially under heavy traffic conditions. These findings highlight the effectiveness of the Bailey Method in enhancing the mechanical properties of steel slag-based asphalt mixtures, making it a promising approach for optimizing pavement design.
- The Bailey mixture under study showed better performance in terms of permanent deformation, with a 110.6% increase in the flow number (FN), making the mixture capable of withstanding extremely heavy traffic, according to the literature. Regarding fatigue life, the Bailey mixture achieved the best result, with a fatigue class of 4, while the conventional mixture obtained a fatigue class of 1.
- Comparing the application of the asphalt mixtures under study in the pavement structure and traffic levels presented, using the Medina software, the Bailey mixture showed the best performance in all analyses conducted. That is, for the same traffic level (light, medium, or heavy), confidence level (95%), and pavement structure from the research, the lowest final asphalt coating layer thicknesses, as well as the lowest overall pavement thicknesses, were obtained.
- In comparison to the control mixture, which contains only natural aggregates, the use of the Bailey method with 100% steel slag resulted in a reduction of up to 35% in the final pavement thickness.
- While this study primarily focused on the technical feasibility of using steel slag in asphalt mixtures, it is worth noting that reducing binder content and reusing steel slag can provide environmental and economic benefits, such as lower material costs and reduced greenhouse gas emissions. These aspects warrant further exploration in future research, including life cycle assessments and cost-benefit analyses.

This demonstrates that steel slag mixtures produced using the Bailey gradation selection method can have advantages in terms of improved fatigue life and permanent deformation, as well as the potential for reducing the binder content in these mixtures. Thus, it is suggested to increase research on gradation selection methods in mixtures with steel aggregates, as well as to study the behavior of these mixtures using the Superpave methodology.

Author Contributions: Conceptualization, W.W.d.S. and L.M.C.; methodology, L.M.C. and A.C.R.G.; software, L.M.C.; formal analysis, W.W.d.S., S.N.M., L.M.C., and A.C.R.G.; investigation, L.M.C.; data curation, L.M.C.; writing—original draft preparation, L.M.C.; writing—review and editing, W.W.d.S., L.M.C., A.C.R.G., and S.N.M.; supervision, A.C.R.G.; project administration, A.C.R.G.; funding acquisition, S.N.M. All authors have read and agreed to the published version of the manuscript.

Funding: This study was financed in part by the Coordination for the Improvement of Higher Education Personnel-Brazil(CAPES)—Finance Code 001.

Data Availability Statement: The original contributions presented in this study are included in the article, further inquiries can be directed to the corresponding authors.

Acknowledgments: The authors would like to thank Ternium Brasil, located in Santa Cruz, Rio de Janeiro.

Conflicts of Interest: The authors declare no conflicts of interest.

References

1. Cunha, D.A.; Pinha, L.C.; Braga, M.J.; Coelho, A.B. Analysis of Market Power in the Crushed Stone Mining Industry of the Metropolitan Region of São Paulo. *Econ. Anal.* **2017**, *35* . [CrossRef]
2. Guimarães, A.C.R.; Filho, J.C.S.; Castro, C.D. Contribution to the use of alternative material in heavy haul railway sub-ballast layer. *Transp. Geotech.* **2021**, *30*, 100524. [CrossRef]
3. Nascimento, P. Use of Granulated Blast Furnace and LD Steel Slag Aggregates in the Production of Cold Asphalt Micro-Surfacing. Master's Thesis, Federal University of Espírito Santo, Vitória, Brazil , 2020.
4. Wang, G.C. The Utilization of Slag in Civil Infrastructure Construction. Woodhead Publishing, 24 June 2016. Available online: https://libproxy.txstate.edu/login?url=https://www.sciencedirect.com/science/book/9780081009949 (accessed on 20 November 2024).
5. Brand, A.S.; Roesler, J.R. Concrete with Steel Furnace Slag and Fractionated Reclaimed Asphalt Pavement. Illinois Center for Transportation. 2014. Available online: https://rosap.ntl.bts.gov/view/dot/27893/dot_27893_DS1.pdf (accessed on 20 November 2024).
6. Delgado, B.; Fonseca, A.; Fortunato, E.; Paixão, A.; Alves, R. Geomechanical assessment of an inert steel slag aggregate as an alternative ballast material for heavy haul rail tracks. *Constr. Build. Mater.* **2021**, *279*, 122438. [CrossRef]
7. Jing, G.; Wang, J.; Wang, H.; Siahkouhi, M. Numerical investigation of the behavior of stone ballast mixed by steel slag in ballasted railway track. *Constr. Build. Mater.* **2020**, *262*, 120015. [CrossRef]
8. Gomes, B.; Guimarães, A.C.R.; do Nascimento, F.A.C.; Tanabe, J. Ballast with siderurgic aggregates: Variation analysis of the shape parameters of particles submitted to triaxial tests through 3D scanner. *Soils Rocks* **2023**, *46*, e2023006122. [CrossRef]
9. Guimarães, A.C.R.; Ávila Costa, K.; de Miranda Reis, M.; Santana, C.S.A.; Castro, C.D. Study of controlled leaching process of steel slag in soxhlet extractor aiming employment in pavements. *Transp. Geotech.* **2021**, *27*, 100485. [CrossRef]
10. Lopes, E.C.; da Silva, T.O.; Pitanga, H.N.; Pedroti, L.G.; Franco de Carvalho, J.M.; Nalon, G.H.; de Lima, G.E.S.; de Araújo, E.N.D. Stabilization of Clayey and Sandy Soils with Slag Fines from Cupola Furnace for Road Construction. *Road Mater. Pavement Des.* **2022**, *24*, 247–266. [CrossRef]
11. Masoudi, S.; Abtahi, S.M.; Goli, A. Evaluation of electric arc furnace steel slag coarse aggregate in warm mix asphalt subjected to long-term aging. *Constr. Build. Mater.* **2017**, *135*, 260–266. [CrossRef]
12. Liu, W.; Li, H.; Zhu, H.; Xu, P. The Interfacial Adhesion Performance and Mechanism of a Modified Asphalt–Steel Slag Aggregate. *Materials* **2020**, *13*, 1180. [CrossRef]
13. Zhong, T.; Zheng, Y.; Chen, Z.; Yao, L.; Zhang, W.; Zhu, Y.; Fu, L. Utilization of steel slag as coarse aggregate and filler in stone mastic asphalt (SMA) mixture: Engineering performance, environmental impact and economic benefits analysis. *J. Clean. Prod.* **2024**, *450*, 141891. [CrossRef]
14. Goli, A. The study of the feasibility of using recycled steel slag aggregate in hot mix asphalt. *Case Stud. Constr. Mater.* **2022**, *16*, e00861. [CrossRef]
15. Alves, R.; Rios, S.; Fortunato, E.; Viana da Fonseca, A.; Guimarães Delgado, B. Mechanical Behaviour of Steel Slag–Rubber Mixtures: Laboratory Assessment. *Sustainability* **2023**, *15*, 1563. [CrossRef]
16. dos Santos, W.W.; Guimarães, A.; Marques, M.E.S. Comprehensive Analysis of Railway Pavement with Box Test: Characterization and Evaluation of Resilient Modulus of Steel Slag Ballast for Sustainable Railway Pavement Viability. In Proceedings of the 5th International Conference on Transportation Geotechnics (ICTG) , Sydney, Australia, 20–22 November 2024; Volume 5. [CrossRef]
17. Guimarães, A.C.R.; Lima, C.D.A.d.; Aragão, F.T.S.; Motta, L.M.G.d.; Santos, J.T.A.d. Mechanical characterization of an alternative laterite gravel used as pavement material. *Soils Rocks* **2024**, *47*, e2024010622. [CrossRef]
18. Silveira, V.L.; Guimarães, A.C.R.; Coelho, L.M.; dos Santos, W.W.; da Silveira, P.H.P.M.; Monteiro, S.N. Recycling Iron Ore Waste through Low-Cost Paving Techniques. *Sustainability* **2024**, *16*, 5570. [CrossRef]
19. Coelho, L.; Guimarães, A.; de Azevedo, A.; Monteiro, S. Sustainable Reclaimed Asphalt Emulsified Granular Mixture for Pavement Base Stabilization: Prediction of Mechanical Behavior Based on Repeated Load Triaxial Tests. *Sustainability* **2024**, *16*, 5335. [CrossRef]
20. Coelho, L.M.; Guimarães, A.C.R. Effect of Compaction Temperature on the Mechanical Behavior of Granular Base Containing RAP. In Proceedings of the 5th International Conference on Transportation Geotechnics (ICTG), Sydney, Australia, 20–22 November 2024; Volume 5; Rujikiatkamjorn, C., Xue, J., Indraratna, B., Eds.; Springer: Singapore, 2025; Lecture Notes in Civil Engineering; Volume 406, pp. 231–238. [CrossRef]
21. dos Santos, W.W.; Coelho, L.M.; Monteiro, S.N.; Marques, M.E.S.; Guimarães, A.C.R. Evaluation of Steel Slag as a Sustainable Alternative Aggregate for Railway Ballast: A Shakedown Theory-Based Approach. *Buildings* **2024**, *14*, 3546. [CrossRef]
22. Kumar, H.; Varma, S. Uma revisão sobre a utilização de escória de aço em asfalto de mistura quente. *Rev. Int. Pesqui. Tecnol. Paviment.* **2021**, *14*, 232–242. [CrossRef]
23. Guimarães, A.C.R.; Arêdes, M.L.A.d.; Castro, C.D.; Coelho, L.M.; Monteiro, S.N. Evaluation of the Mechanical Behavior of Asphaltic Mixtures Utilizing Waste of the Processing of Iron Ore. *Mining* **2024**, *4*, 889–903. [CrossRef]
24. Loures, R.; Guimarães, A.; Silva, B.H.; Castro, C. Cold Mix Containing Steel Slag: A Viable Alternative for Pavement Construction. *Transportes* **2018**, *26*, 54–67. [CrossRef]
25. Pasetto, M.; Baliello, A.; Giacomello, G.; Pasquini, E. Rheological Characterization of Warm-Modified Asphalt Mastics Containing Electric Arc Furnace Steel Slags. *Adv. Mater. Sci. Eng.* **2016**, *2016*, 9535940. [CrossRef]

26. Plati, C.; Tsakoumaki, M.; Loizos, A. Quality Assurance of Steel Slag Asphalt Mixtures for Sustainable Pavement Surface Courses. *Recycling* **2024**, *9*, 91. [CrossRef]
27. Li, J.; hui Zhang, J.; Qian, G.; long Zheng, J.; Zhang, Y. Three-Dimensional Simulation of Aggregate and Asphalt Mixture Using Parameterized Shape and Size Gradation. *J. Mater. Civ. Eng.* **2019**, *31*, 3 [CrossRef]
28. Mansour, T.N.; Putman, B. Influence of Aggregate Gradation on the Performance Properties of Porous Asphalt Mixtures. *J. Mater. Civ. Eng.* **2013**, *25*, 281–288. [CrossRef]
29. Hassan, H.; Wu, K.; Huang, W.; Chen, S.; Zhang, Q.; Xie, J.; Cai, X. Study on the influence of aggregate strength and shape on the performance of asphalt mixture. *Constr. Build. Mater.* **2021**, *294*, 123599. [CrossRef]
30. Wang, F.; Xiao, Y.; Cui, P.; Ma, T.; Kuang, D. Effect of aggregate morphologies and compaction methods on the skeleton structures in asphalt mixtures. *Constr. Build. Mater.* **2020**, *263*, 120220. [CrossRef]
31. Liang, H.; Wang, D.; Shi, L.; Liang, X.; Tang, C. Use of digital images for fracture performance evaluation of asphalt mixtures. *Constr. Build. Mater.* **2020**, *253*, 119152. [CrossRef]
32. Sulaiman, E.D.; abed, J.M.; Al-Saffar, Z.H. Improving the Gradation of Aggregate used in Asphalt Mixtures Using Bailey Method. *NTU J. Eng. Technol.* **2023**, *2*. [CrossRef]
33. Rahi, A.; Albayati, A. The Suitability of Bailey Method for Design of Local Asphalt Concrete Mixture. *Civ. Eng. J.* **2021**, *7*, 827–839. [CrossRef]
34. Hafeez, I.; Kamal, M.A.; Ishaq, M.A.; Ahmad, N.; Khanzada, S. A Laboratory-Based Research Study to Investigate the Aggregate Packing Characteristics and Its Influence on Asphaltic Mixtures Performance. *Arab. J. Sci. Eng.* **2015**, *40*, 3119–3134. [CrossRef]
35. Komba, J.J.; Maina, J.W.; Horak, E.; Steyn, W.J. Relationship between Aggregate Packing Characteristics and Compactability of Hot-Mix Asphalt Mixes. In Proceedings of the Airfield and Highway Pavements 2019: Testing and Characterization of Pavement Materials, Chicago, IL, USA, 21–24 July 2019 ; pp. 183–194.
36. Blaauw, S.A.; Maina, J.W.; Horak, E. Towards a Mix Design Model for the Prediction of Permeability of Hot-Mix Asphalt. *Constr. Build. Mater.* **2019**, *221*, 637–642. [CrossRef]
37. Horak, E.; Maina, J.; Myburgh, P.; Sebaaly, H. Monitoring Permeability Potential of Hot Mix Asphalt via Binary Aggregate Packing Principles Correlated with Bailey Ratios and Porosity Principles. *J. South Afr. Inst. Civ. Eng.* **2019**, *61*, 32–44. [CrossRef]
38. Ghuzlan, K.A.; Al-Mistarehi, B.W.; Al-Momani, A.S. Rutting Performance of Asphalt Mixtures with Gradations Designed Using Bailey and Conventional Superpave Methods. *Constr. Build. Mater.* **2020**, *261*, 119941. [CrossRef]
39. Imaninasab, R.; Loria-Salazar, L.; Carter, A. Impact of Aggregate Structure Restoration on Rutting Resistance of Asphalt Mixtures with Very High Percentages of RAP. *Road Mater. Pavement Des.* **2023**, *25* (Suppl. S1), 28–44. [CrossRef]
40. Alsheyab, M.A.; Al-sugaier, A.O.; Rahman, M.L.; Emami Ahari, H. Performance evaluation of Bailey method used in asphalt mixtures containing natural river sands. *Int. J. Transp. Sci. Technol.* **2023**. [CrossRef]
41. Joumblat, R.; Kassem, H.; Elkordi, A.; Khatib, J. 17—Use of alternative recycled fillers in bituminous mixtures: A review. In *Advance Upcycling of By-Products in Binder and Binder-Based Materials*; Kırgız, M.S., Ed.; Woodhead Publishing Series in Civil and Structural Engineering; Woodhead Publishing: Cambridge, UK, 2024 ; pp. 335–356. [CrossRef]
42. Hu, Y.; Sreeram, A.; Airey, G.D.; Li, B.; Si, W.; Wang, H. Comparative analysis of time sweep testing evaluation methods for the fatigue characterisation of aged bitumen. *Constr. Build. Mater.* **2024**, *432*, 136698. [CrossRef]
43. Hu, Y.; Ryan, J.; Sreeram, A.; Allanson, M.; Pasandín, A.R.; Zhou, L.; Singh, B.; Wang, H.; Airey, G.D. Optimising the dosage of bio-rejuvenators in asphalt recycling: A rejuvenation index based approach. *Constr. Build. Mater.* **2024**, *433*, 136761. [CrossRef]
44. Hu, Y.; Allanson, M.; Sreeram, A.; Ryan, J.; Wang, H.; Zhou, L.; Airey, G.D. Characterisation of bitumen through multiple ageing-rejuvenation cycles. *Int. J. Pavement Eng.* **2024**, *25*, 2365350. [CrossRef]
45. Vavrik, W.R.; Huber, G.; Pine, W.E. Bailey Method for Gradation Selection in HMA Mixture Design. *Transp. Res. Board: Transp. Res. Circ.* **2002**, *E-C044*. http://onlinepubs.trb.org/onlinepubs/circulars/ec044.pdf.
46. *AASHTO T-19*; Standard Method of Test for Bulk Density (Unit Weight) and Voids in Aggregate. American Association of State Highway and Transportation Officials (AASHTO): Washington, DC, USA, 2009 .
47. Bernucci, L.B.; Motta, L.M.G.; Soares, J.B.; Ceratti, J.A.P. *Asphalt Pavement: Basic Training for Engineers*; S.l: Rio de Janeiro, Brazil, 2010.
48. *DNIT ES 31*; Flexible Pavements—Asphalt Concrete. National Department of Transport Infrastructure (DNIT): Brasília, Brazil, 2006 .
49. *DNER—National Department of Highways, ME 083*; Aggregate—Gradation Analysis.; DNER—National Department of Highways, ME: Rio de Janeiro, Brazil, 1998.
50. *DNER—National Department of Highways, ME 084*; Fine Aggregate—Determination of Real Density. DNER—National Department of Highways, ME: Rio de Janeiro, Brazil, 1995.
51. *DNER—National Department of Highways, ME 081*; Aggregates—Determination of Absorption and Density of Coarse Aggregate. DNER—National Department of Highways, ME: Rio de Janeiro, Brazil, 1997.
52. *DNER—National Department of Highways, ME 054*; Sand Equivalent. DNER—National Department of Highways, ME: Rio de Janeiro, Brazil, 1997.
53. *DNER—National Department of Highways, ME 078*; Coarse Aggregate—Adhesion to Bituminous Binder. DNER—National Department of Highways, ME: Rio de Janeiro, Brazil, 1994.

54. *DNER—National Department of Highways, EM 262;* Steel Slag for Road Pavements. DNER—National Department of Highways, ME: Rio de Janeiro, Brazil, 1994.
55. *DNIT—National Department of Transport Infrastructure 113;* Road Pavement–Artificial Aggregate–Evaluation of the Expansion Potential of Steel Slag—Test Method. DNIT—National Department of Transport Infrastructure: Brasília, Brazil, 2009.
56. *DNER—National Department of Highways, ME 399;* Aggregates—Determination of Impact Loss in the Treton Apparatus. DNER—National Department of Highways, ME: Rio de Janeiro, Brazil, 1999.
57. *DNER—National Department of Highways, ME 035;* Aggregates—Determination of Los Angeles Abrasion. DNER—National Department of Highways, ME: Rio de Janeiro, Brazil, 1998.
58. *DNER—National Department of Highways ME 148;* Bituminous Material—Determination of Flash and Fire Points. DNER—National Department of Highways ME: Rio de Janeiro, Brazil, 1994.
59. *DNIT—National Department of Transport Infrastructure 155;* Asphaltic Material—Determination of Penetration—Test Method. DNIT—National Department of Transport Infrastructure: Brasília, Brazil, 2010.
60. *DNIT Standard 131/2010—ME;* Asphaltic Materials-Determination of Softening Point-Ring and Ball Method—Test Method. DNIT—National Department of Infrastructure and Transportation: Brasília, Brazil, 2010 .
61. *NBR 15184;* Bituminous Materials—Determination of Viscosity at Elevatedn Temperatures Using a Rotational Viscometer. ABNT: São Paulo, Brazil, 2005.
62. *ASTM 2872;* Standard Test Method for Effect of Heat and Air on a Moving Film of Asphalt Binder (Rolling Thin-Film Oven Test). American Society for Testing and Materials (ASTM): West Conshohocken, PA, USA, 2012.
63. *DNIT Standard 178/2018—PRO;* Asphalt Pavement-Preparation of Test Specimens for Mechanical Tests Using Superpave Gyratory Compactor or Marshall Compactor—Procedure; DNIT—National Department of Infrastructure and Transportation: Rio de Janeiro, Brazil, 2018 .
64. Coelho, L.M. Influence of Gradation on the Mechanical Behavior of Asphalt Concrete Containing Steel Aggregate. Master's Thesis, Military Institute of Engineering, Rio de Janeiro, Brazil , 2021.
65. *DNIT Standard 136/2018—ME;* Asphalt Pavement-Asphalt Mixtures-Determination of Tensile Strength by Diametrical Compression—Test Method. DNIT—National Department of Infrastructure and Transportation: Rio de Janeiro, Brazil, 2018.
66. *DNIT Standard 135/2018—ME;* Asphalt Pavement-Asphalt Mixtures-Determination of Resilience Modulus—Test Method. DNIT—National Department of Infrastructure and Transportation: Rio de Janeiro, Brazil, 2018 .
67. *DNIT Standard 184/2018–ME;* Pavement-Asphalt Mixtures-Repeated Load Uniaxial Test for Determination of Permanent Deformation Resistance—Test Method. DNIT—National Department of Infrastructure and Transportation: Rio de Janeiro, Brazil, 2018 .
68. *DNIT Standard 183/2018—ME;* Asphalt Pavement-Fatigue Test by Controlled Stress Diametrical Compression—Test Method. DNIT—National Department of Infrastructure and Transportation: Rio de Janeiro, Brazil, 2018 .
69. Franco, F. Mechanistic-Empirical Design Method of Asphalt Pavements—SISPAV. Ph.D. Thesis, Federal University of Rio de Janeiro, Rio de Janeiro, Brazil, 2007.
70. DNIT—Departamento Nacional de Infraestrutura de Transportes. MeDiNa: Software. Available online: https://www.gov.br/dnit/pt-br/assuntos/planejamento-e-pesquisa/ipr/medina/solicitacao-de-download-do-medina (accessed on 31 January 2021).
71. Swathi, M.; Andiyappan, T.; Guduru, G.; Amarnatha Reddy, M.; Kuna, K.K. Design of asphalt mixes with steel slag aggregates using the Bailey method of gradation selection. *Constr. Build. Mater.* **2021**, *279*, 122426. [CrossRef]
72. Xie, J.; Wu, S.; Lin, J.; Cai, J.; Chen, Z.; Wei, W. Recycling of basic oxygen furnace slag in asphalt mixture: Material characterization and moisture damage investigation. *Constr. Build. Mater.* **2012**, *36*, 467–474. [CrossRef]
73. Kavussi, A.; Qazizadeh, M.J. Fatigue characterization of asphalt mixes containing electric arc furnace (EAF) steel slag subjected to long term aging. *Constr. Build. Mater.* **2014**, *72*, 158–166. [CrossRef]
74. Kandhal, P.S.; Hoffman, G.L. Evaluation of Steel Slag Fine Aggregate in Hot-Mix Asphalt Mixtures. *Transp. Res. Rec.* **1997**, *1583*, 28–36. [CrossRef]
75. Ahmedzade, P.; Sengoz, B. Evaluation of steel slag coarse aggregate in hot mix asphalt concrete. *J. Hazard. Mater.* **2009**, *165*, 300–305. [CrossRef]
76. FHWA. User Guidelines for Byproducts and Secondary Use Materials in Pavement Construction. *Technical Report FHWA-RD-97-14;* Federal Highway Administration: Georgetown Pike, McLean, VA, USA, 2008 .
77. Cavalcante, E.; Bessa, I.; Castelo Branco, V.; Soares, J.B. Dosagem de misturas asfálticas segundo a metodologia Superpave utilizando escória de aciaria como agregado. In Proceedings of the IV Simpósio Internacional de Avaliação de Pavimentos e Projetos de Reforço, Fortaleza, CE, Brazil, 7–9 October 2009.
78. Coelho, L.; Guimarães, A.; Alves Moreira, C.; dos Santos, G.; Monteiro, S.; da Silveira, P. Feasibility of Using Ferronickel Slag as a Sustainable Alternative Aggregate in Hot Mix Asphalt. *Sustainability* **2024**, *16*, 8642. [CrossRef]
79. Nascimento, L.A.H. New Approach to the Design of Dense Asphalt Mixtures Using the Gyratory Compactor with Focus on Permanent Deformation. Master's Thesis, COPPE/UFRJ , Rio de Janeiro, Brazil, 2008.
80. Bastos, J.B.S. Considerations on Permanent Deformation of Asphalt Pavements in Mechanistic-Empirical Design. Ph.D. Thesis, Federal University of Ceará, Fortaleza, Brazil, 2017.
81. Medina, J.; Motta, L.M.G. *Pavement Mechanics*, 3rd ed.; Interciência: Rio de Janeiro, Brazil, 2015 .

82. Golalipour, A.; Jamshidi, E.; Niazi, Y.; Afsharikia, Z.; Khadem, M. Effect of aggregate gradation on rutting of asphalt pavements. *Procedia-Soc. Behav. Sci.* **2012**, *53*, 440–449. [CrossRef]
83. Oufa, M.S.; Abdolsamedb, A.A. Controlling rutting performance of hot mix asphalt. *Int. J. Sci. Eng. Res* **2016**, *6*, 872–882.
84. Shang, G.T.; Takahashi, O.; Maekawa, R. Recommended Combination of the Bailey Parameters in Superpave Gradation Design for Japanese Airfield Pavements. *Int. J. Pavement Res. Technol.* **2013**, *6*, 704–713. [CrossRef]
85. Manjunath, K.; Poornachandra Dev, N. Design of Hot Mix Asphalt Using Bailey Method of Gradation. *Int. J. Res. Eng. Technol.* **2014**, *3*, 386–393. [CrossRef]
86. Denneman, E.; Verhaeghe, B.M.; Sadzik, E. Aggregate packing characteristics of good and poor performing asphalt mixes. In Proceedings of the 26th Southern African Transport Conference, Pretoria, South Africa, 9–12 July 2007.
87. Júnior, T.D.C. Study of the Mechanical Properties of Conventional and Modified Asphalt Mixtures. Master's Thesis, Federal University of Rio Grande do Sul (UFRGS), Porto Alegre, Brazil, 2008.
88. Chen, Z.; Gong, Z.; Jiao, Y.; Wang, Y.; Shi, K.; Wu, J. Moisture stability improvement of asphalt mixture considering the surface characteristics of steel slag coarse aggregate. *Constr. Build. Mater.* **2020**, *251*, 118987. [CrossRef]
89. Guo, Y.; Wu, H.; Shen, A.; Yang, X.; Cui, T. Estudo da estabilidade hídrica a longo prazo de misturas asfálticas contendo agregado de escória de aço. *J. Adhes. Sci. Technol.* **2019**, *34*, 877–902. [CrossRef]
90. Luan, Y.; Zhang, W.; Zhang, A.W.; Zhao, Y.; Pan, Z.; Niu, Z.; Zeng, K.; Chen, X.; Mohammad, L.N. Mechanical Property Evaluation for Steel Slag in Asphalt Mixture with Different Skeleton Structures Using Modified Marshall Mix Design Methodology. *J. Mater. Civ. Eng.* **2022**, *34*, 04014014. [CrossRef]

buildings

MDPI

Article

From Waste to Roads: Improving Pavement Performance and Achieving Sustainability with Recycled Steel Slag and Low-Density Polyethylene

Syed Amir Mehmood [1], Muhammad Imran Khan [2], Sarfraz Ahmed [1], Rania Al-Nawasir [3] and Rafiq M. Choudhry [2,*]

1 Department of Transportation Engineering, MCE College, National University of Sciences & Technology (NUST), Islamabad 44000, Pakistan; samehmood.pg19mce@student.nust.edu.pk (S.A.M.); sarfraz@mce.nust.edu.pk (S.A.)

2 Civil Engineering Department, College of Engineering, Imam Mohammad Ibn Saud Islamic University (IMSIU), Riyadh 11623, Saudi Arabia; mekhan@imamu.edu.sa

3 Roads and Transport Engineering Department, College of Engineering, University of Al-Qadisiyah, Ad Diwaniyah 58002, Iraq; raniaihsanali@qu.edu.iq

* Correspondence: rchoudhry@imamu.edu.sa

Abstract: The use of waste, recycled, and modified materials is increasingly popular in roadway construction for sustainability and pavement longevity. This research examines the combination of steel slag (SS) and low-density polyethylene (LDPE), commonly used in plastic bags and steel manufacturing by-products, to mitigate environmental pollution. LDPE was tested as a binder modifier in two bitumen grades, 60–70 and 80–100, at concentrations of 3%, 5%, and 7% by weight. SS was used as a replacement for coarse aggregate. The physical properties of both modified and unmodified bitumen grades and SS were analyzed before creating and testing hot-mix asphalt (HMA) samples. The dynamic modulus of these samples was measured at temperatures of 4.4 °C, 21.1 °C, 37.8 °C, and 54.4 °C with frequencies of 0.1 Hz, 0.5 Hz, 1 Hz, 5 Hz, 10 Hz, and 25 Hz. Master curves were developed, and the dynamic modulus data underwent design of experiment (DOE) and computational intelligence (CI) analyses. Using KENPAVE, a mechanistic–empirical tool, the analysis assessed the design life and enhancements in damage ratio for each modifier and grade. The results showed that adding LDPE increases the softening point and penetration grade but decreases ductility due to increased bitumen stiffness, leading to premature fatigue failure at higher LDPE levels. Both 3% LDPE and 3% SS-modified LDPE improved Marshall Stability and dynamic modulus across all temperature and frequency ranges. Specifically, 3% LDPE enhanced stability by 13–16% and 3% SS-LDPE by 30–32%. The KENPAVE results for 3% LDPE showed a design life improvement of 19–25% and a damage ratio reduction of 15–18%. In comparison, 3% SS-LDPE demonstrated a design life improvement of 50–60% and a damage ratio reduction of 25–35%. Overall, this study concludes that 3% LDPE- and 3% SS-LDPE-modified HMA in both bitumen grades 60–70 and 80–100 provide optimal results for improving pavement performance.

Keywords: waste utilization; low-density polyethylene; steel slag; sustainability; dynamic modulus; computational intelligence

1. Introduction

A country's transportation infrastructure is crucial to its growth and a mark of an industrialized country and developed nation. Most roads are typically made of hot-mix asphalt (HMA), which is a combination of bitumen and aggregates [1,2]. Many authors have researched the stability of bitumen and the replacement of aggregates to improve the bonding between them, either to provide better stability or to improve ride quality [3–5]. The need for aggregate replacements increases the utilization of the waste produced during different construction and demolition projects, which makes up for the shortcoming of aggregates [6]. Similarly, the need for the stability of bitumen increases to maintain a pavement's load-carrying capability, prevent moisture damage, and improve the bond between aggregates and bitumen [7,8].

Shortage of materials is a rising issue in construction work; thus, using sustainable materials is now a developing area of research [9]. The current situation requires an economic and attractive road system with low economic costs. This method has become important for the conservation and management of our resources. Pavement repair has come a long way in the last 25 years. It is a professional and environmentally friendly process to modify and improve existing asphalt [10]. Many researchers have used different industrial wastes such as recycled construction waste, wastes from the steel industry in the form of slag, over-burnt brick waste, reclaimed asphalt waste, treated soil, etc., which are recommend materials for achieving sustainability [11–17]. With the steel industry's aggressive growth, the use of steel slag has expanded [18]. Asphalt concrete can be successfully mixed with steel slag in place of aggregate and can improve its stability and thermal efficiency [19]. The use of steel slag is restricted to substitution either as coarse aggregates or as a fine aggregate only, because the entire replacement increases angularity, which can cause an increase in the void ratio and bulking [20]. Flexible pavement suffers from premature failure due to friction and fatigue cracks, which causes the road materials to fail before they reach the end of their intended lifespan. This is mainly due to the dramatic increase in road traffic that has occurred in the last twenty years. Shear flow, a form of rutting, is a typical failure of road materials that is mostly caused by vehicle stress, high temperatures, and heavy ESALs [21]. Three main causes of road material disintegration have been identified, namely temperature and surrounding conditions, heavy vehicle use, and negligence or quality compromise in construction.

The temperature of bitumen greatly influences the hardness and stiffness properties of bituminous mix, which is solely due to the viscoelastic nature of bitumen, which changes with the thermal gradient [22]. Lower winter temperatures stiffen the bitumen, which causes fatigue cracking, while higher summer temperatures increase the viscosity of bitumen, which causes rutting. As a result, higher temperatures, as seen in Pakistan, significantly degrade the strength of the asphalt material. In addition, if a sustained load is maintained on the asphalt for a longer period, damage is greatly increased, and shear deformation and plastic flow occur due to excessive shear and overloaded truck stresses [23].

Asphalt modification and recycling meet all the social goals of providing efficient and safe roads while significantly reducing energy (oil) consumption and environmental impact compared to road improvements [24]. Thus, it has a lasting impact and allows supplying agencies to increase their available funds. However, they must be used properly, as not all highways are ideal candidates for asphalt recycling [25]. This will provide a reliable and safe driving surface for travelers.

Bitumen is a viscoelastic material that is obtained from crude oil and is mainly composed of hydrocarbons; very few materials can replace it. One of the problems of bitumen is its temperature susceptibility, specifically its propensity to become brittle at lower temperatures and softened at higher temperatures, which may be modified by some additives and polymers [26]. Additives such as polyethylene, styrene, crumb rubber, ultra-high-molecular-weight polyethylene, and many others have been mixed in asphalt concrete to improve certain engineering characteristics of asphaltic mix [27–30]. In comparison to conventional asphaltic pavements, the deterioration of pavement materials can be reduced by partially adding waste plastic to bitumen [31].

The physical behavior and mechanical characteristics of flexible pavements after partially replacing waste plastic with low- and high-density polyethylene reveal that the waste both reduces the density and increases the stability [32]. Zhang et al. [33] studied the influence of polyethylene wax and selected three types of additives with different molecular weights. The results indicated that all types of additives improved the asphalt mixture's ability to resist moisture and maintain stability at high temperatures. Recycling waste materials in HMA is the best choice for constructing pavements while keeping other necessary volumetric properties since several motorway restoration projects have been planned with HIR and CIR technologies (hot-in-place and cold-in-place recycling, respectively), which have already been deployed and tested in a few locations [34].

Incorporating bitumen additives into waste recycled coarse aggregates (steel slag) can reduce the overall project cost while also increasing the dynamic stability of the pavements. The steel manufacturing industry produces steel slag, a waste product with high mechanical properties and excellent surface morphology. In addition to its high mechanical properties, steel slag requires proper treatment due to its high-water absorption and volume expansion. Researchers have applied and compared various treatments, including polyvinyl alcohol (PVA), silane coupling agent (SCA), cement paste (CP), epoxy acrylic organic-silicone resin (EAOR), and low-density polyethylene (LDPE), to the surface of steel slag [35,36]. The treatment results showed that treating steel slag made the material much more stable, improved Marshall and flow values, and made it more resistant to freeze–thaw. It also made volume expansion problems a lot less severe [37].

The increasing demand for sustainable construction practices has prompted the investigation of innovative solutions for addressing waste management issues. Steel slag, a by-product of steel production, is plentiful but frequently underused, often ending up in landfills. Similarly, LDPE is commonly utilized in packaging and is considerably contributing to plastic pollution owing to its low recycling rate. The integration of these materials in road construction provides a dual advantage: minimizing waste accumulation and optimizing pavement design life.

To improve pavement performance and durability, many researchers have explored recycled waste materials as an alternative to conventional coarse aggregates, such as rubber from tire waste [38], reclaimed asphalt pavement (RAP) [39], recycled crushed concrete (RCC) [40], and ceramic tile waste (CTW) from construction and demolition operations [41]. Despite the economic and environmental benefits of using these recycled wastes in bitumen pavement, they are also subject to limitations, for example, the use of RCA [42], while the use of rubber waste reduces the mechanical properties due to its elastic nature and thus reduces the stiffness of bitumen, leading to increased deformation under heavy loads [43]. Additionally, despite the widespread application of RAP and its use as a highly expanded aggregate in various paving applications as an alternative to aggregates, it is also subject to many challenges, including processing requirements, significant variation in material quality depending on the source, age, and condition of the reclaimed materials, mix design, workability, compaction, and separation risks. Also, a high RAP content may weaken resistance to rutting under heavy loads if not designed properly [44,45]. Similarly, the use of CTW produced from residential building waste as an alternative to aggregate also has problems, including the tendency of its particles to crack early due to its brittle nature and high porosity, which makes the asphalt mixture susceptible to damage resulting from moisture sensitivity. Its excessive use also weakens its workability during mixing and compaction due to its angular and coarse texture, and it also requires more energy for the crushing and treatment process [46,47]. Compared to these recycled materials, steel slag shows high durability compared to natural aggregate due to its high strength and density, better resistance to slip due to its rough texture, and high stability under harsh weather conditions due to its strong and stable composition, which allows it to be chemically stable, being less susceptible to alkali and carbonate reactions. These advantages make steel slag a promising alternative to natural aggregate in the production of HMA [48–50]. This study addresses several critical gaps in the field of asphalt pavement research and introduces a novel, effective, and sustainable approach for transport infrastructure design by combining LDPE as a binder modifier and steel slag as a coarse aggregate replacement. Firstly, while LDPE and steel slag have been studied individually for their use in asphalt mixtures, little attention has been paid to their combined effects on pavement performance. Secondly, prior research lacks comprehensive evaluations of the dynamic modulus over a wide range of temperatures and frequencies, which is critical for understanding material behavior under diverse traffic and climate conditions. Thirdly, this work contributes to sustainability by utilizing industrial waste materials and offering practical solutions for reducing environmental impacts. Lastly, the integration of computational intelligence and advanced modeling techniques, such as multi-expression programming, fills a methodological gap in optimizing material design and performance predictions.

2. Scope and Objectives of the Study

This study seeks to address environmental sustainability by utilizing industrial waste while improving the mechanical and thermal properties of asphalt pavements to mitigate premature failures. The primary aim of this study is to evaluate the combined effect of low-density polyethylene (LDPE) with varying percentages (3%, 5%, and 7%) as a binder modifier in two different grades of bitumen (PEN 60/70 and 80/100) and steel slag (SS) as a coarse aggregate replacement in enhancing the performance of hot-mix asphalt (HMA) to provide practical solutions for durable and sustainable road construction. The LDPE-modified binders were evaluated for their physical properties to determine their performance characteristics. These binders were then used to prepare Marshall samples, both with virgin aggregates and with SS as a coarse aggregate replacement. To assess the effectiveness of these modifications, the dynamic modulus was measured at temperatures of 4.4 °C, 21.1 °C, 37.8 °C, and 54.4 °C and frequencies of 0.1 Hz, 0.5 Hz, 1 Hz, 5 Hz, 10 Hz, and 25 Hz. The dynamic modulus results were analyzed using the NCHRP09-29_Mastersolver2-2 Excel spreadsheet. Furthermore, prediction equations for the modified materials were derived from a Python 3.12-based dataset using multi-expression programming (MEPX), an artificial intelligence (AI) technique. Statistical analyses were conducted using a two-level factorial design in Minitab to investigate the influence of LDPE, SS, and their interactions on asphalt performance.

3. Materials and Methods

3.1. Material Selection

The materials used in this study include fine aggregate, coarse aggregate, and bitumen with two different grades (grades 60–70 and 80–100). Steel slag was obtained from a local factory (see Figure 1) as the by-product produced from impurities during steel making in the blast furnace. Steel slag is taken in its raw form, then crushed into coarse aggregates using a crushing machine, and cleaned with a brush to remove dust and suspended metals, similar to the process applied to coarse aggregate. LDPE was used as a modifier and was obtained from local industry in the form of pellets, as shown in Figure 2. The physicochemical properties of steel slag and LDPE are displayed in Table 1 and Table 2, respectively.

Figure 1. Steel aggregate from blast furnace slag.

Figure 2. Low-density polyethylene (LDPE) pellets and mixing in binder.

Table 1. Physicochemical properties of steel slag (SS).

Chemical Properties	
Property	**Value**
SiO_2 (%)	15.04
Al_2O_3 (%)	4.12
Fe_2O_3 (%)	22.55
CaO (%)	41.53
MgO (%)	6.17
K_2O (%)	0.05
Na_2O (%)	0.14
SO_3 (%)	0.08
LOI (%)	0.25
Physical properties	
Color	Light to dark brown
Shape and Surface texture	Highly angular and rough
Specific gravity (gm/cm^3)	3.25
Bulk density (Kg/m^3)	1911.11
Moisture content (%)	0.54

Table 2. Physicochemical properties of LDPE.

Property	Value
Color	white
Specific gravity (gm/cm^3)	0.921
Tensile strength (MN/m^2)	10.11
Flexural modulus (GN/m^2)	0.203
Melting point (°C)	113.20
Thermal degradation temperature (°C)	406
Chemical form	$(-CH_2-CH_2-)_n$

The physical properties of aggregates/steel slag and bitumen grades 60–70 and 80–100 are given in Table 3 and Table 4, respectively. All the volumetric properties of control asphalt mix samples are given in Table 5. All the results of the tests performed were within the limits of specifications. The aggregate gradation used in preparing HMA samples is shown in Table 6. In this study, bitumen was modified with 3, 5, and 7% LDPE by weight of bitumen. Moreover, the coarse aggregate in HMA was also replaced by 3% steel slag (SS).

Table 3. Test results performed on natural aggregates and steel slag.

Description of Test	Test Standard		Result		Limits
			Natural Aggregates	Steel Slag	
Abrasion Value	ASTM C-131 [51]		26	25	45 (max)
Impact Value	BS 812-112 [52]		15	18	30 (max)
Crushing Strength	BS 812-110 [53]		27	26	30 (max)
Specific Gravity	Coarse	ASTM C-127 [54]	2.73	3.25	-
	Fine	ASTM C-128 [55]	2.66	-	-
	Filler	AASHTO T-133 [56]	2.63	-	-
Water Absorbance	Coarse	ASTM C-127 [54]	0.53	0.83	3% (max)
	Fine	ASTM C-128 [55]	0.78	-	3% (max)

Table 4. Test results of bitumen physical properties (grade 60–70 and 80–100).

Test Description	Test Standard	Bitumen Grade 60–70		Bitumen Grade 80–100	
		Result	Specification	Result	Specification
Penetration, 25 °C, 0.1 mm	ASTM D-5 [57]	65	60/70	96	80/100
Softening Point, °C	ASTM D-36 [58]	47	(46–57) °C	45	(43–54) °C
Specific Gravity, 25 °C, cm	ASTM D-70 [59]	0.99	0.99 (min)	1.01	0.99 (min)
Ductility, g/cm^3	ASTM D-113 [60]	>100	100 cm (min)	>100	100 cm (min)

Table 5. Marshall results at OBC for HMA (grade 60–70 and 80–100).

Descriptions	Measured Value (60–70)	Measured Value (80–100)	Standards
Optimum Asphalt Contents (%)	4.0	4.0	4% voids—air
Unit Weight (g/cm^3)	2.290	2.275	n/a
Voids in Mineral Aggregate (%)	13.25	13.75	13 (Minimum)
Voids filled with Asphalt (%)	70.5	72	65 to 75
Stability (kN)	11.38	11.07	8.0 (Minimum)
Flow(mm)	3.50	3.00	2.0 to 3.5

Table 6. NHA specifications for gyratory sample preparations.

Sieve Size (mm)	NHA-B Specification Range (% Passing)	Selection (% Passing)
19	100	100
12.5	75–90	82.5
9.5	60–80	70
4.75	40–60	50
2.38	20–40	30
1.18	5–15	10
0.075	3–8	5.5

3.2. Testing Methods

Initial testing focused on characterizing the aggregates and bitumen to establish baseline properties. This phase also included the calculation of the optimum binder content (OBC) required to achieve 4% air voids. Standardized tests were conducted to determine the physical and mechanical properties of aggregates and bitumen, ensuring compliance with relevant ASTM and BS standards (shown in Table 4).

After basic tests, the consistency and performance of LDPE-modified bitumen for two grades, 60–70 and 80–100, were evaluated. Low-density polyethylene (LDPE) was incorporated into the bitumen at modification levels of 3%, 5%, and 7% by total binder weight. The blending process was carried out at 140–155 °C with a mixing rate of 500 rpm for 30 min to ensure uniform dispersion. The resulting modified bitumen was tested for consistency and other critical performance parameters.

The Marshal Mix design was conducted to determine the volumetric properties of both control and modified samples for the selected bitumen grades. The aggregate gradations and the optimum bitumen content data are presented in Tables 5 and 6. The modification included the addition of LDPE at 3%, 5%, and 7%, as well as the partial replacement of coarse aggregate with steel slag (particle sizes ranging from 19 mm to 4.75 mm). The impact of these modifications on the volumetric properties was thoroughly evaluated, providing insights into the effects of LDPE and steel slag on the mix design.

Dynamic modulus testing was performed to assess the mechanical behavior of the specimens under varying temperature and frequency conditions. Specimens were cored from gyratory compacted samples with dimensions of 100 mm in diameter and 150 mm in height. The gyratory compaction process, carried out at a mixing temperature of 160–170 °C, involved 125 gyrations to simulate high-traffic conditions equivalent to ESALs $\geq$ 30 million [61]. The prepared specimens were tested at four temperature levels (4.4 °C, 21.1 °C, 37.8 °C, and 54.4 °C) and six frequency levels (25, 10, 5, 1, 0.5, and 0.1 Hz) [62]. The incorporation of LDPE and steel slag in the specified percentages was examined to determine their influence on the dynamic modulus.

4. Results and Discussion

4.1. Aggregate Testing

In this portion, steel slag and natural aggregates were tested for their physical properties to check and compare their abrasion, crushing strength, impact, specific gravity, and water absorption. The test result in Figure 3 shows that steel slag performs better in abrasion and crushing strength by 1%, while in the case of the impact, its value is reduced up to 3%, and the water absorption of steel slag is slightly more than natural aggregates by 0.3% with a specific gravity of 3.25. All the aggregate testing results were within the limits of standard specifications.

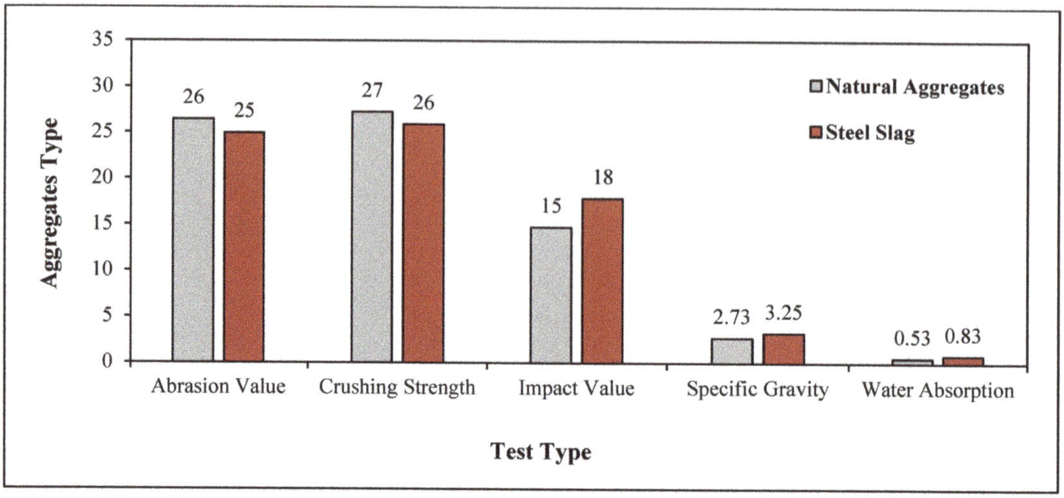

Figure 3. Physical test results on natural aggregates and steel slag.

4.2. Bitumen Consistency Testing

This portion includes the test results on modified samples. The LDPE-modified samples were tested for penetration, softening point, ductility, and specific gravity. The bitumen was heated at a temperature of 100 to 100 °C, and then LDPE pellets were mixed in at a rate of 500 rpm for 30 min at 140 to 155 °C. After mixing, the required modified mixture was kept in an oven for 1 h at 110 °C to retain its morphology. LDPE in two bitumen grades (60–70 and 80–100) was taken into the mix at 3%, 5%, and 7% by weight of bitumen. The test results show that adding more LDPE to a sample makes the penetration grade and softening point values better in both bitumen grades. However, as the modification rate goes up, the ductility value goes down, as shown in Figures 4 and 5. The 3% LDPE modifier mixed with bitumen in both grades had a lower ductility value but a higher softening point and penetration value. The softening point went up by 9 °C in grade 60–70 and by 8 °C in grade 80–100. The penetration value went down to 52 in grade 60–70 and to 61 in grade 80–100. The ductility value went down by 3 cm in grade 60–70 and by 4 cm in grade 80–100, as shown in Table 7. The modification resulted in a reduction in the penetration value and an increase in the softening point value, indicating an increase in the stiffness and stability of the bitumen. However, the ductility decreased significantly in grades 5 and 7%, indicating a large surface energy value, which may not be desirable in certain situations, particularly in lower temperatures and slow movement conditions. All the bitumen test results were within the limits, except ductility in cases of 5% and 7% in both grades.

Table 7. Effect on bitumen consistency by adding LDPE.

Description of Test	Bitumen Grade 60–70				Bitumen Grade 80–100			
	0%	3%	5%	7%	0%	3%	5%	7%
Penetration Value	65	52	38	32	96	61	49	33
Softening Point	47	56	62.25	68.4	45	53	57.35	67.1
Ductility Value	104	101	41	31	106	102	47	36
Specific Gravity	0.989	0.968	0.974	0.961	1.011	1.000	0.977	0.969

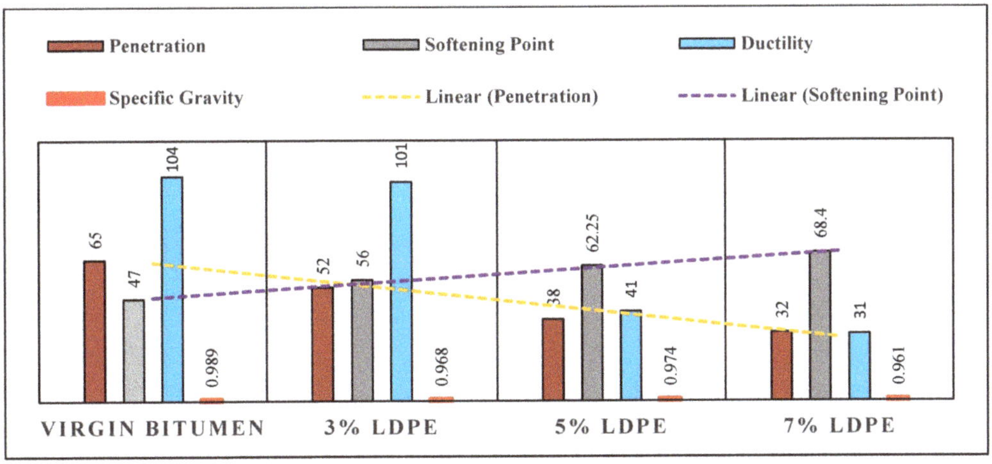

Figure 4. Effect on bitumen (grade 60/70) consistency by adding LDPE.

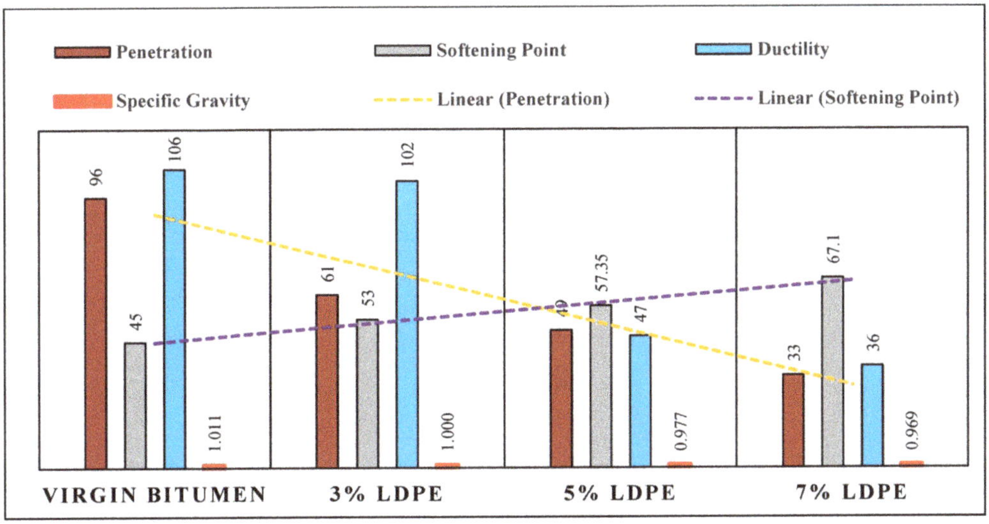

Figure 5. Effect on bitumen (grade 80/100) consistency by adding LDPE.

4.3. Marshall Stability and Volumetric Properties

The volumetric properties of two grades of bitumen, 60/70 and 80/100, were determined using an incremental percentage of bitumen starting from 3% to 5% with a 0.5 increment and a coaction rate of 75 blows on each side. An OBC of 4% was achieved after the Marshall Test on a virgin binder for both grades; this OBC was adopted in LDPE- and SS-LDPE modified HMA. The test results and volumetric properties are given in Table 8. In cases of stability, 3% LDPE and 3% LDPE with SS show the highest values, with up to a 15% increase in the case of 3% LDPE and up to a 32% increase in the case of 3% SS-LDPE in both the bitumen grades, as shown in Figure 6. The flow values for all the modifications were within the limits of ASTM standard specifications in both the bitumen grades, as shown in Figure 7. The voids filled with asphalt (VFA) and voids in mineral aggregates (VMA) for 3% LDPE and 3% LDPE with SS have almost the same value or less deviation from the control sample in both the bitumen grade conditions and within the limits of standard specifications, as shown in Figures 8 and 9. The greater value of stability is the indication of a more stable and compact mix that withstands higher loads with the best performance. The value of the Marshall quotient for 3% SS-LDPE in bitumen grade 60–70 is 436, and in the case of bitumen grade 80–100, this value is 425, which is the highest response among all the modifications being considered and is an indication of stiffer blends. For the performance of HMA pavements, VFA and VMA are of greater use and are indirectly an indication of a good mix with greater durability values and control of bleeding in the HMA. The control sample was selected according to optimum binder content, and the 3% LDPE and 3% SS-LDPE values of VFA and VMA closer to it show a durable mix.

Table 8. Marshall results of LDPE- and SS-LDPE-modified samples.

Samples No.	Description	Air Voids (%)	VFA (%)	Unit Weight of Mix (g/cc)	Stability (KN)	Flow (mm)	VMA (%)
				Bitumen Grade 60–70			
1	Control Sample	3.963	70.469	2.280	11.37	3.5	13.420
2	3% LDPE Modifier	3.952	70.970	2.242	13.14	3	13.614
3	5% LDPE Modifier	3.837	71.332	2.213	10.84	3.25	13.385
4	7% LDPE Modifier	3.795	71.732	2.215	11.04	3.25	13.427
5	3% LDPE Modifier with SS	4.143	70.199	2.236	14.96	3.5	13.903
				Bitumen Grade 80–100			
1	Control Sample	4.040	70.067	2.274	11.08	3	13.496
2	3% LDPE Modifier	3.910	70.693	2.238	12.54	3.5	13.340
3	5% LDPE Modifier	3.430	73.279	2.216	11.12	3.25	12.835
4	7% LDPE Modifier	3.192	74.656	2.207	11.24	3.5	12.593
5	3% LDPE Modifier with SS	4.263	69.349	2.245	14.60	3.5	13.907

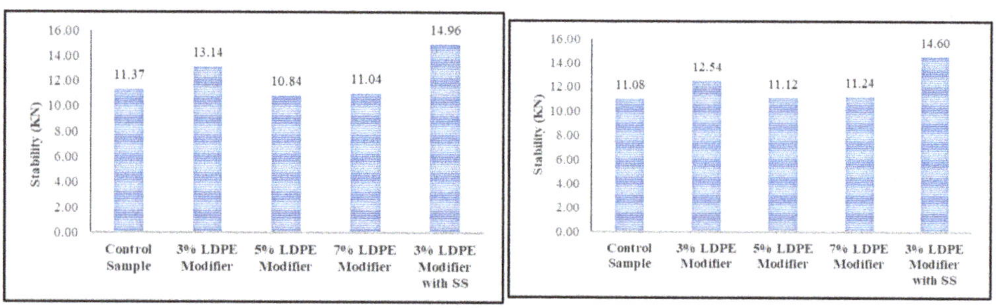

(a) (b)

Figure 6. Marshall Stability of modified mixes with LDPE and SS (**a**) for grade 60/70 and (**b**) for grade 80/100.

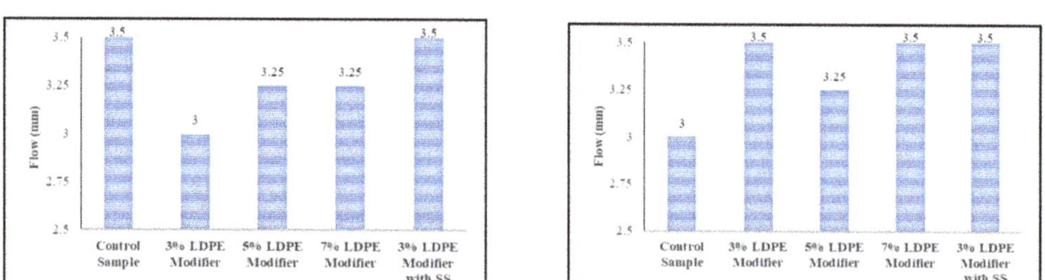

(a) (b)

Figure 7. Marshall flow of modified mixes with LDPE and SS (**a**) for grade 60/70 and (**b**) for grade 80/100.

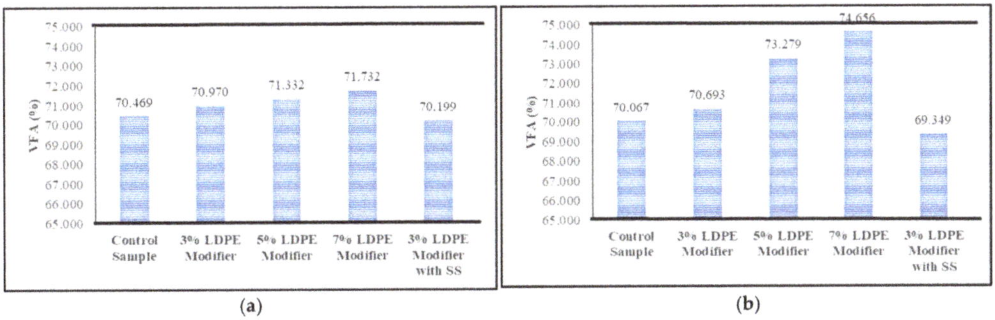

Figure 8. VFA of modified mixes with LDPE and SS (**a**) for grade 60/70 and (**b**) for grade 80/100.

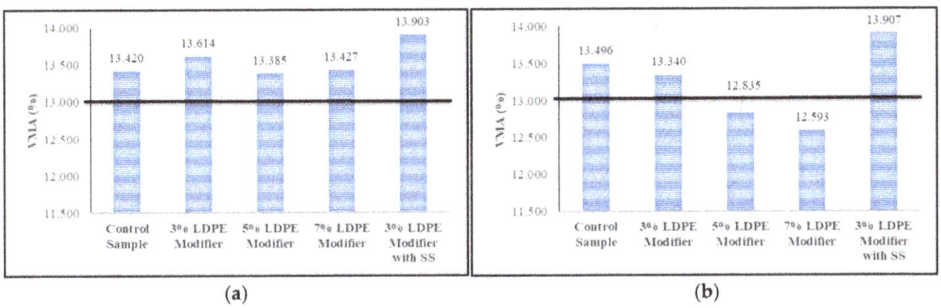

Figure 9. VMA of modified mixes with LDPE and SS (**a**) for grade 60/70 and (**b**) for grade 80/100.

4.4. Dynamic Modulus Using SPT

A dynamic modulus test was performed on the gyratory compacted specimens that were cored to 100 mm diameter and trimmed to 150 mm height under four temperature periods, i.e., 4.4 °C, 21.1 °C, 37.8 °C, and 54.4 °C, and under six frequency levels, i.e., 25 Hz, 10 Hz, 5 Hz, 1 Hz, 0.5 Hz, and 0.1 Hz. A total of thirty samples of modified HMA specimens were tested, including three samples each of LDPE-modified grades 60/70 and 80/100 bitumen with LDPE percentages of 0%, 3%, 5%, and 7%. Furthermore, 3% LDPE-modified bitumen grades 60/70 and 80/100 were used for SS-modified HMA samples. The dynamic modulus values for samples modified with LDPE are shown graphically in Figures 10 and 11 at four different test temperatures and in Figures 11 and 12 at six different load frequencies for grade 60–70 and 80–100 modified samples.

The SPT dynamic modulus (DM) (|E*|) of test samples is directly proportional to frequency and inversely proportional to test temperature, i.e., the dynamic modulus (DM) increases with increasing frequency and decreases with increasing temperature. Based on the dynamic modulus (DM) results, it is hypothesized that it is temperature-sensitive; in colder places, HMA pavements perform better with more stiffness as opposed to regions with higher temperatures, where there is more flexibility in HMA. The dynamic modulus increases with the addition of LDPE and SS, as shown in Figures 10 and 11 for isochronal and isothermal curves, respectively, in grade 60–70 samples. HMA grade 60–70 modified samples perform better in all cases as compared to the control sample, and samples containing 7% LDPE and 3% SS-LDPE show the highest dynamic modulus at both lower and higher temperatures. Figures 12 and 13 for isothermal and isochronal curves, respectively, demonstrate how the dynamic modulus increases with the addition of LDPE and SS in grade 80–100 samples.

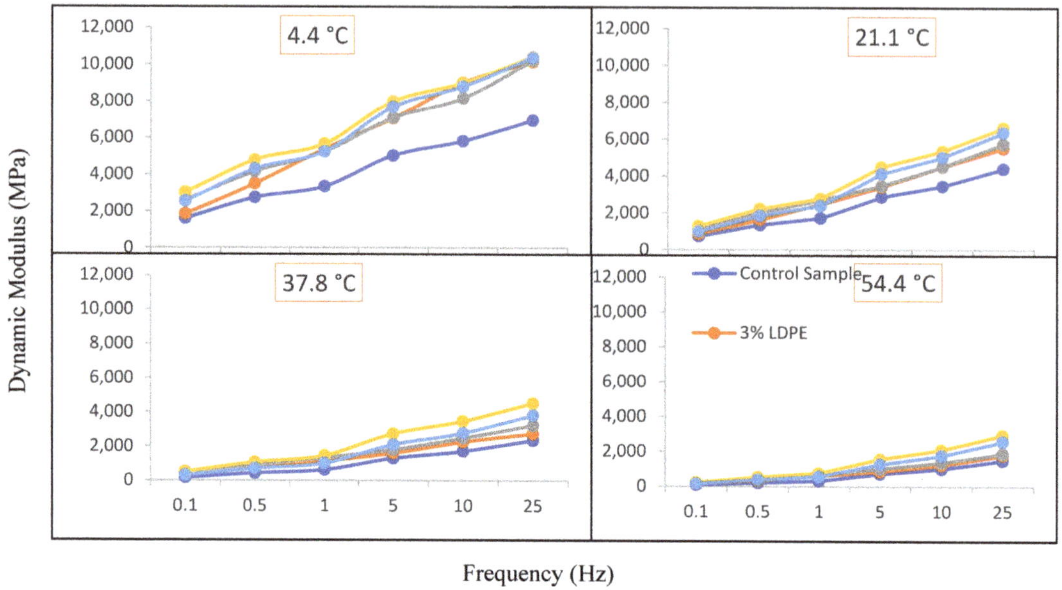

Figure 10. Dynamic modulus of LDPE- and SS-LDPE-modified grade 60/70 bitumen (isochronal curves).

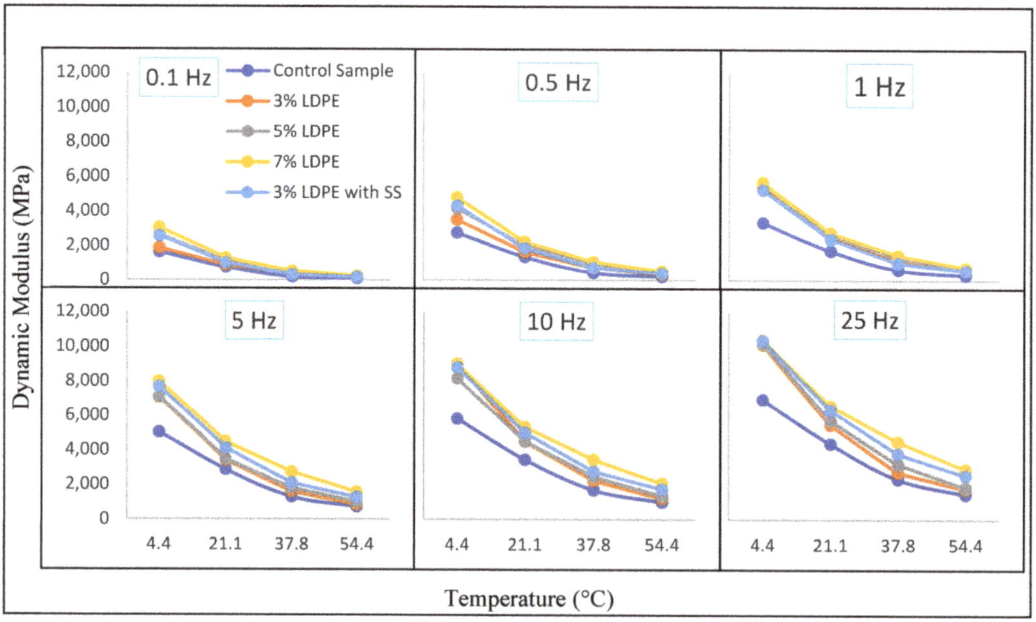

Figure 11. Dynamic modulus of LDPE- and SS-LDPE-modified grade 80/100 bitumen (isochronal curves).

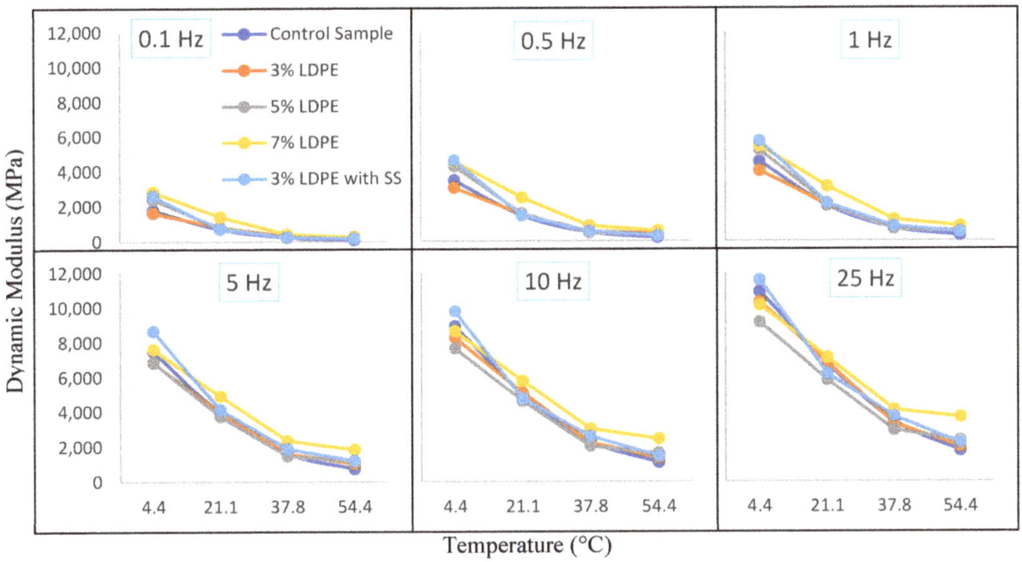

Figure 12. Dynamic modulus of LDPE- and SS-LDPE-modified grade 80/100 bitumen (isothermal curves).

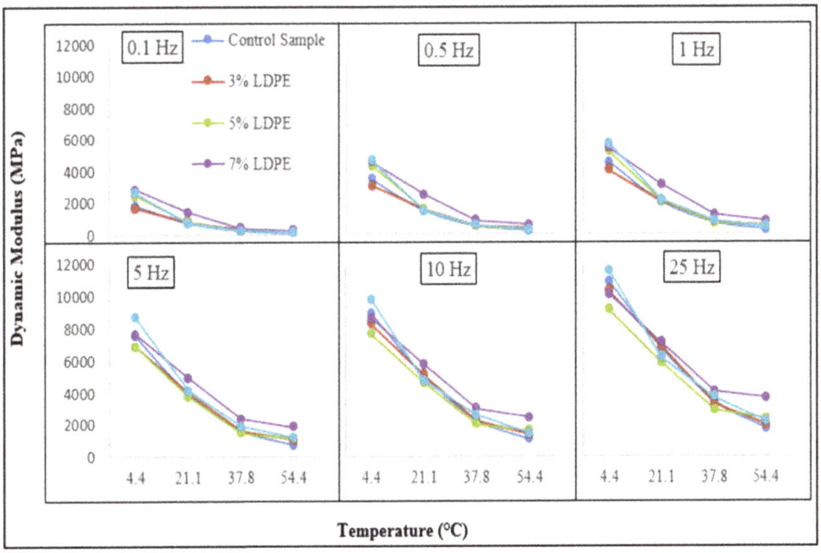

Figure 13. Dynamic modulus of LDPE- and SS-LDPE-modified grade 60/70 bitumen (isothermal curves).

In comparison to previous LDPE-modified samples, HMA grade 80–100 modified samples with 7% LDPE and 3% SS-LDPE exhibited the highest dynamic modulus in both lower and higher temperature situations; however, a slight increase in the dynamic modulus values of grade 80–100 control samples at 4 °C was observed as compared to grade 60–70 control samples, which showed less impact than modified grade 80–100 samples at lower temperatures. This is due to the higher penetration value of grade 80–100 samples as they perform better in low-temperature conditions. LDPE increases the stiffness of bitumen, and the SS improves the stability as steel slag is denser than Babuzai slag. The best response from the isothermal and isochronal curves was again taken at 3% SS-LDPE because the 7% LDPE in both the bitumen grades shows an abnormal result, with the highest dynamic modulus values being in higher temperature conditions, the average response being in low-temperature conditions, and the lowest response at being low frequency levels, which can further be seen from the master curves in Figure 14 for grade 60–70 and in Figure 15 for grade 80–100 HMA.

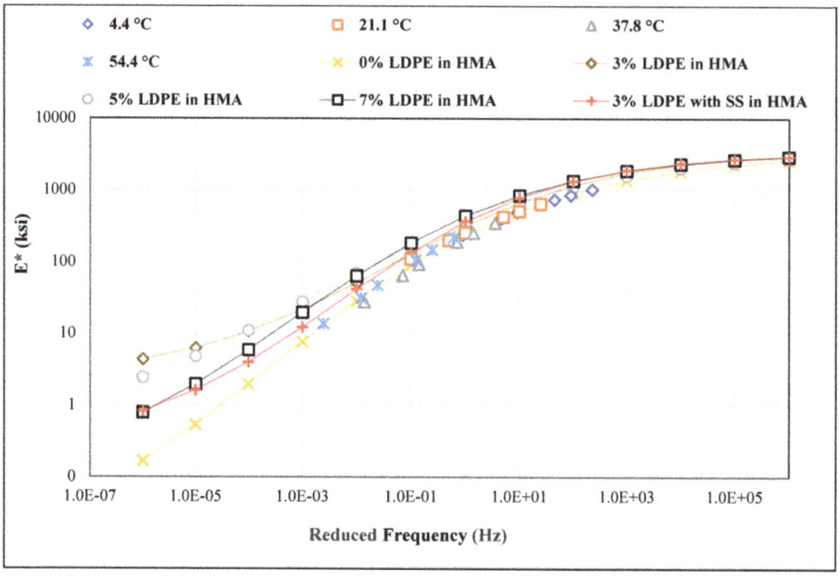

Figure 14. Dynamic modulus |E*| master curves of LDPE- and SS-LDPE-modified grade 60/70 bitumen.

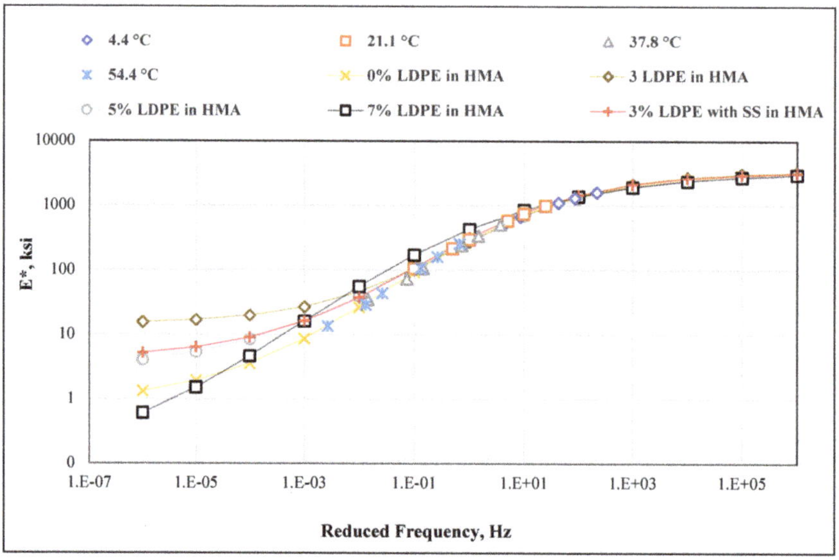

Figure 15. Dynamic modulus |E*| master curves of LDPE- and SS-LDPE-modified grade 80/100 bitumen.

4.4.1. Master Curve Development from Dynamic Modulus Dataset

The temperature and time superposition rule at 21.1 °C was adopted for the dynamic modulus (DM) dataset to develop a master curve. These datasets have been obtained from different combinations of LDPE modification results that help in the design process of asphaltic pavement. While testing at −10 °C was not practicable with the AMPT apparatus in Pakistani environmental conditions, the master curve was designed from the procedure of NCHRP-Proj: 09-36. The NCHRP data-solver MS Excel add-in tool depicts the best-fit line which restricts SSE for creating master curves. Figures 14 and 15 show master curves, which display the (DM) dataset responses with frequencies at levels of high–low for LDPE-modified HMA grades 60/70 and 80/100, respectively, where, accordingly, low and high frequencies correspond to high and low temperatures.

Master curves show that LDPE-modified mix samples have greater stiffness values at both low and high temperatures as compared to regular mix samples because of the mix's similar performance at low temperatures. The 7% LDPE- and 3% SS-LDPE modified mixtures have the best results at high and low temperatures, according to Figures 14 and 15 of the master curve. Also, a high frequency, like 25 Hz, depicts a high-speed movement of traffic, whereas a low frequency, like 0.1 Hz, shows a low-speed movement of traffic. Based on the master curve frequency distribution, at higher frequencies, LDPE-modified samples perform better, but when the frequency gradually reduces to 0.1 Hz and as the LDPE content increases from 3% to 7%, its dynamic modulus shows a weaker response. The findings corroborate the idea that LDPE-modified samples of 7% LDPE and 3% LDPE-SS have optimum (DM) values for high-speed cars at low as well as high temperatures, whereas 3% LDPE-modified samples have ideal (DM) values for low-speed cars at higher temperature ranges. The shift factor, reduced frequency, and inputs for the ME Pavement design guide are all included in the master solver worksheet so that each gradation can be used as a guide. The master curve findings demonstrate an increase in stiffness at both high and low temperatures for the LDPE- and SS-LDPE-modified HMA mixtures, particularly for the optimal 3% LDPE and 3% SS-LDPE combinations. This increase in stiffness indicates improved resistance to deformation under repetitive loading, which directly correlates to better fatigue resistance. Stiffer mixtures reduce the strain energy absorbed under loading cycles, thus slowing crack initiation and propagation in the pavement. R^2 of LDPE-modified and SS-LDPE-modified material for two bitumen grades is mentioned in Table 9.

Table 9. R2 of LDPE- and SS-LDPE-modified HMA.

Description of a Modifier	R^2	Se/Sy
	Bitumen Grade 60/70	
0% LDPE	0.9522	0.15
3% LDPE	0.9209	0.20
5% LDPE	0.9540	0.15
7% LDPE	0.9802	0.10
3% LDPE with SS	0.9538	0.15
	Bitumen Grade 80/100	
0% LDPE	0.9899	0.07
3% LDPE	0.9190	0.20
5% LDPE	0.7460	0.36
7% LDPE	0.8490	0.27
3% LDPE with SS	0.9028	0.22

4.4.2. Experimental Two-Level Factorial Design for SPT Dynamic Modulus

Experimentation probabilistic design using Minitab 21.1 software was performed on the data obtained by conducting dynamic modulus testing on LDPE-modified grade 60–70 and 80–100 bitumen. In this study, a two-level factorial mathematical design is selected, which means high and low levels of modifier, temperature, and frequency were taken as input variables and the output or response variable was dynamic modulus, as shown in Table 10.

Table 10. Two-level factors for factorial design (SS-LDPE-modified HMA).

Factors	Measured Units	Low-Level Factors		High-Level Factors	
		Grade 60/70	Grade 80/100	Grade 60/70	Grade 80/100
Temperature	Centigrade	4.4	4.4	54.4	54.4
SS-LDPE	%	0	0	3	3
Frequency	Hertz	0.1	0.1	25.0	25.0

Tables 11 and 12 show the main and interaction effects of SS-LDPE-modified grades 60–70 and 80–100 HMA, respectively. The main effect is the individual average high and average low responses of factors within themselves, while the interaction effect is the average high and low response of each factor to another. Temperature is represented by A, frequency by B, and modifier by C, as indicated in the tables.

Table 11. Effect estimates for dynamic modulus—grade 60–70 SS-LDPE-modified HMA.

	Main			Interactions			
	A: Temp	B: Freq	C: LDPE	AB	BC	AC	ABC
Avg. High	1077	5355.75	3919	2059.25	3670.25	2832	3048.25
Avg. Low	5387.5	1108.75	2545.5	4405.25	2794.25	3632.5	3416.25
Effect	−4310.5	4247	1373.5	−2346	876	−800.5	−368
Coefficients	−2155.25	2123.5	686.75	−1173	438	−400.25	−184
p-Value	0.000	0.000	0.000	0.000	0.000	0.000	0.003

Table 12. Effect estimates for dynamic modulus—grade 80–100 SS-LDPE-modified HMA.

	Main			Interactions			
	A: Temp	B: Freq	C: LDPE	AB	BC	AC	ABC
Avg. High	1052.75	6608.5	4137	2089	3916.25	3768.5	3950.75
Avg. Low	6722.5	1166.75	3638.25	5686.25	3859	4006.75	3824.5
Effect	−5669.75	5441.75	498.75	−3597.25	57.25	−238.25	126.25
Coefficients	−2834.87	2720.875	249.375	−1798.62	28.625	−119.12	63.125
p-Value	0.000	0.000	0.000	0.000	0.040	0.593	0.250

In Tables 11 and 12, the main and interaction effects of frequency, modifier, and temperature on the dynamic modulus (DM) of SS-LDPE-modified grades 60–70 and 80–100 hot-mix asphalt are presented with R2 of 99.67% and 99.79%, respectively. Positive and negative signs of the effect indicate the nature of the relationship; a positive sign shows a direct relationship, while a negative sign is an indication of an inverse relationship. From this observation, it can be concluded that the dynamic modulus value rises with the rise in frequency and modifier and falls with the temperature rise. The design of this experiment was performed with a 95% confidence level. The p-value demonstrates the significance of each factor in this research work, apart from AC (two-way interaction) and ABC (three-way interaction), of the grade 80–100 SS-LDPE modifier. Therefore, the design of the experiment (DOE) excluded these factors for further processing.

Analysis of Variance (ANOVA) results are displayed in Table 13 for grade 60–70 SS-LDPE-modified HMA and in Table 14 for grade 80–100 SS-LDPE-modified HMA.

Table 13. ANOVA for dynamic modulus—grade 60–70 SS-LDPE-modified HMA.

Source Description	DF	Adjusted SS	Adj MS	F-Value	*p*-Value
Given Model	9	274,123,233	30,458,137	472.4	0
Linear	3	231,009,162	77,003,054	1194.29	0
Temperature	1	111,473,841	111,473,841	1728.92	0
Frequency	1	108,213,560	108,213,560	1678.35	0
SS-LDPE	1	11,321,761	11,321,761	175.6	0
2-Way Interactions	3	41,458,467	13,819,489	214.34	0
Temperature*Frequency	1	33,012,913	33,012,913	512.02	0
Temperature*SS-LDPE	1	3,844,801	3,844,801	59.63	0
Frequency*SS-LDPE	1	4,600,753	4,600,753	71.36	0
3-Way Interactions	1	813,280	813,280	12.61	0.003
Temperature*Frequency*SS-LDPE	1	813,280	813,280	12.61	0.003
Error	14	902,664	64,476		
Total	23	275,025,896			

Table 14. ANOVA for dynamic modulus—grade 80–100 SS-LDPE-modified HMA.

Source Description	DF	Adjusted-SS	Adj MS	F-Value	*p*-Value
Given Model	7	450,194,120	64,313,446	984.24	0
Linear	3	372,044,758	124,014,919	1897.9	0
Temperature	1	192,876,390	192,876,390	2951.75	0
Frequency	1	177,675,858	177,675,858	2719.12	0
SS-LDPE	1	1,492,509	1,492,509	22.84	0
2-Way Interactions	2	77,982,777	38,991,389	596.72	0
Temperature*Frequency	1	77,641,245	77,641,245	1188.21	0
Temperature*SS-LDPE	1	341,532	341,532	5.23	0.036
Error	16	1,045,490	65,343		
Total	23	451,239,610			

The *p*-values in the ANOVA tables indicate statistical significance, with values < 0.05 confirming that the effects of temperature, frequency, and SS-LDPE, as well as their significant interactions, are not due to random variation. The F-values quantify the relative influence of these factors, with higher values indicating stronger effects. For grade 60–70 HMA (Table 13), all main factors and interactions, including the three-way interaction, are significant, as shown by their *p*-values and high F-values. For grade 80–100 HMA (Table 14), the significant main factors and two-way interactions, particularly temperature and frequency, dominate the response, while the interaction term temperature*SS-LDPE is relatively less impactful but still significant ($p = 0.036$). These results confirm the robustness of the experimental design and validate the influence of the studied factors on the dynamic modulus. From Table 13, the degree of freedom (DOF) for the grade 60–70 model is 9:3 DOF for one-way interaction, 3 DOF for two-way interaction, and 1 DOF for three-way interaction. In this case, all the parameters are significant in the sense of F-value and *p*-value. From Table 14, the DOF for the 80–100 model is 7:3 DOF for one-way interaction, 2 DOF for two-way interaction, and no DOF for three-way interaction, as it was neglected previously from the design after the interaction effect's insignificance.

The normal plot of standardized effects for grade 60–70 SS-LDPE-modified HMA is shown in Figure 16, and that for grade 80–100 SS-LDPE-modified HMA is shown in Figure 17. In both cases, the significance level was taken as 0.05 with three parameters, namely temperature, frequency, and modifier, along with main and interaction effects. All the parameters in the grade 60–70 and grade 80–100 models shown in red are significant, while the insignificant ones were removed from the grade 80–100 model after a higher *p*-value of the interaction effect. From the normalized graphs, it can be concluded that the bitumen modifier has a significant effect on the design model.

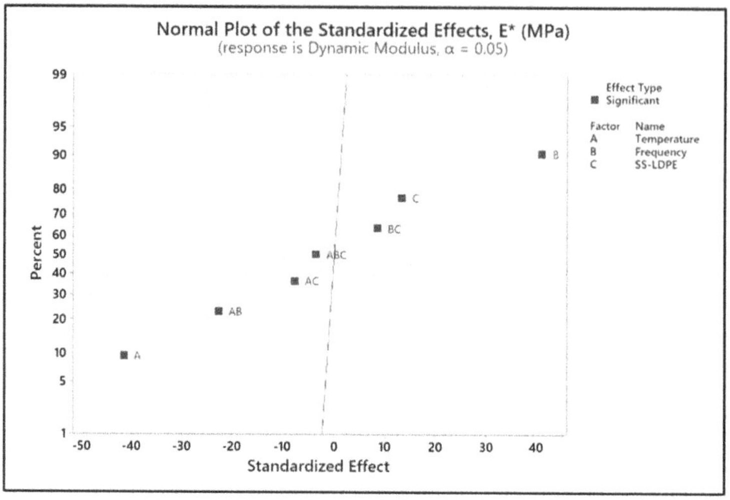

Figure 16. Normal plot for DM (for grade 60–70 and SS–LDPE-modified HMA).

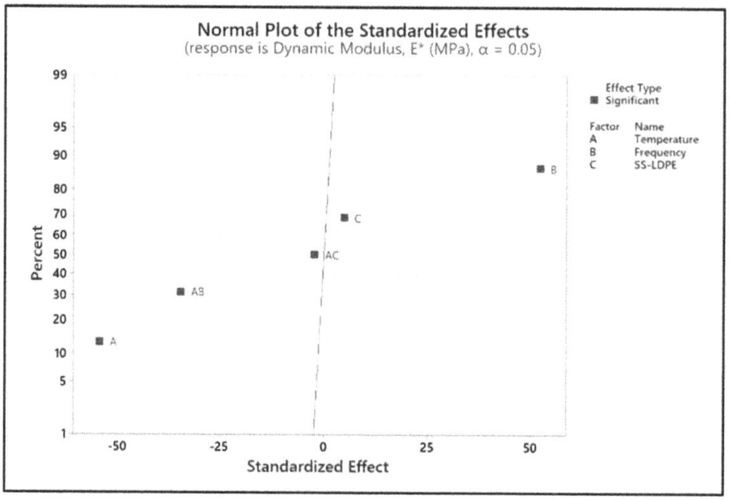

Figure 17. Normal plot for DM (for grade 80–100 and SS–LDPE-modified HMA).

The main effect plots were based on the high and low levels of each parameter, namely temperature, frequency, and modifier, as shown in Figure 18 for |E*| DM of grade 60–70 SS-LDPE-modified HMA and in Figure 19 for DM of grade 80–100 SS-LDPE-modified HMA. The steepness of the lines drawn shows the correlation of the response variable with the considered parameters. From the main effect plots, the SS-LDPE modifier is responsible for bringing about a positive change in HMA behavior in both cases. The SS-LDPE modifier has a strong impact on HMA in grade 60–70 as compared to grade 80–100.

The interaction effect plots of each parameter with another parameter in the form of a two-way interaction and a three-way interaction were computed for their high and low levels for each factor, as shown in Figures 20 and 21 for the dynamic modulus of grade 60–70 and 80–100 SS-LDPE-modified HMA, respectively. In interaction plots, parallel lines indicate the insignificance of the considered factors, while non-parallel lines show the interaction between those factors. It also has a considerable effect because of the steepness of the lines. From the plots, we can observe some non-parallel lines in grade 60–70 SS-LDPE-modified HMA; it can be inferred that there is an interaction effect between the variables with the response dynamic modulus, while in grade 80–100 SS-LDPE-modified HMA, the BC plot has parallel lines, which means there is no interaction effect between them.

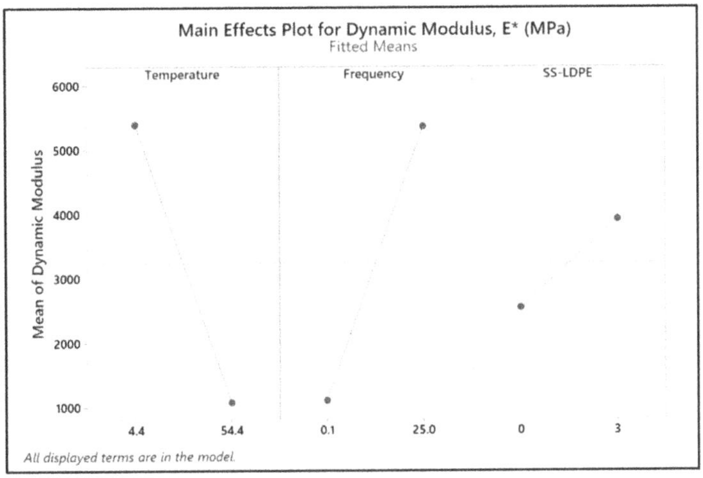

Figure 18. Main effect plot for DM (for grade 60–70 and SS-LDPE-modified HMA).

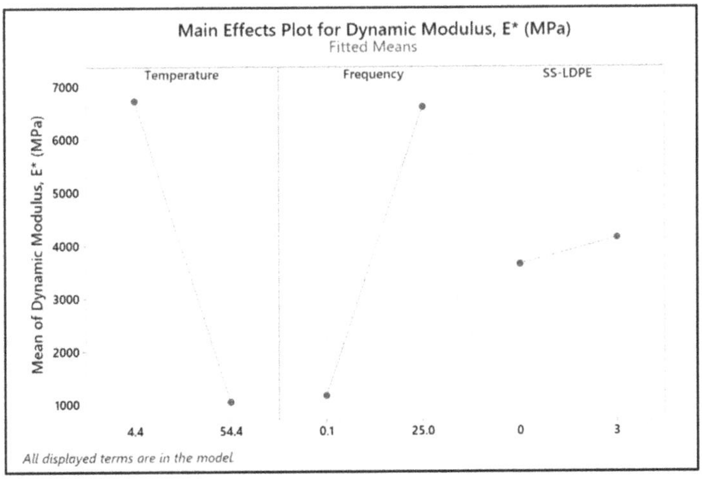

Figure 19. Main effect plot for DM (for grade 80–100 and SS-LDPE-modified HMA).

Pareto charts for the dynamic modulus of grade 60–70 SS-LDPE-modified HMA are shown in Figure 22, and those for the dynamic modulus of grade 80–100 SS-LDPE-modified HMA are shown in Figure 23. Pareto charts show the mean response of the interaction and the importance of the components included in the design, together with the influence of relative importance. A level of significance of 0.05 was taken, which is indicated by the red dotted line in the charts. In the grade 60–70 SS-LDPE-modified HMA, all the factors in either a one-way interaction, two-way interaction, or three-way interaction are significant, while in the grade 80–100 SS-LDPE-modified sample, two-way interaction BC and three-way interaction ABC were non-significant, so they were removed from the model observations.

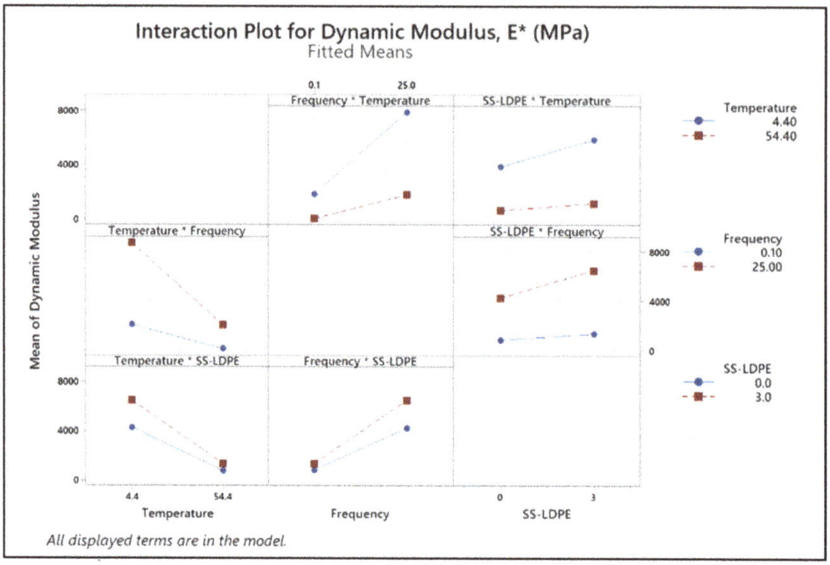

Figure 20. Interaction effect plot for DM (for grade 60–70 and SS-LDPE-modified HMA).

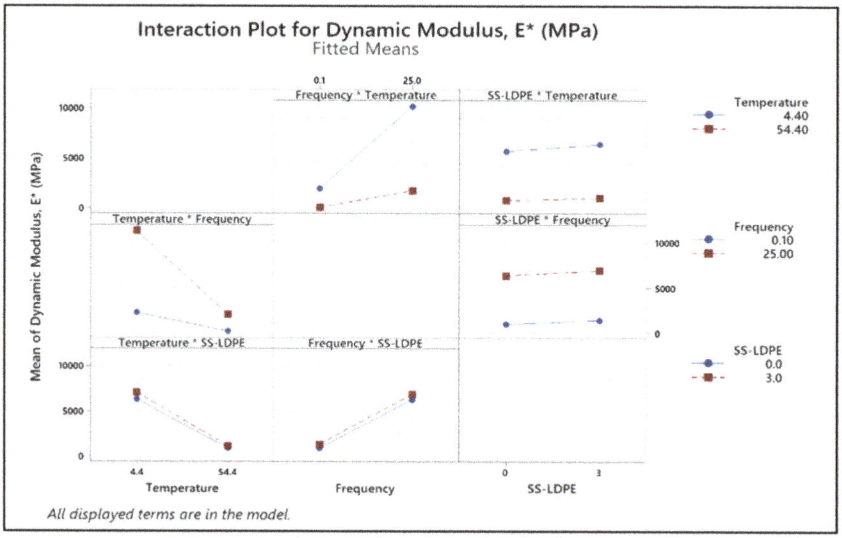

Figure 21. Interaction effect plot for DM (for grade 80–100 and SS-LDPE-modified HMA).

The cube plots for the dynamic modulus of grade 60–70 SS-LDPE-modified HMA are shown in Figure 24, and those for grade 80–100 SS-LDPE-modified HMA are shown in Figure 25. Cube plots clearly show that the highest response dynamic modulus values may be obtained at the lowest temperature and highest frequency and modifier when employed. The dynamic modulus values rise at lower temperatures because the material stiffens at that temperature, and they fall as the temperature rises because the bitumen softens with high strains, which has a detrimental effect on the dynamic modulus. On the other hand, lower frequency levels, which are indications of slow-moving vehicles, have negative impacts on the response factor. Hence, it is evident from the cube plot that the parameters, namely temperature, frequency, and modifier, are all significant factors affecting the response dynamic modulus.

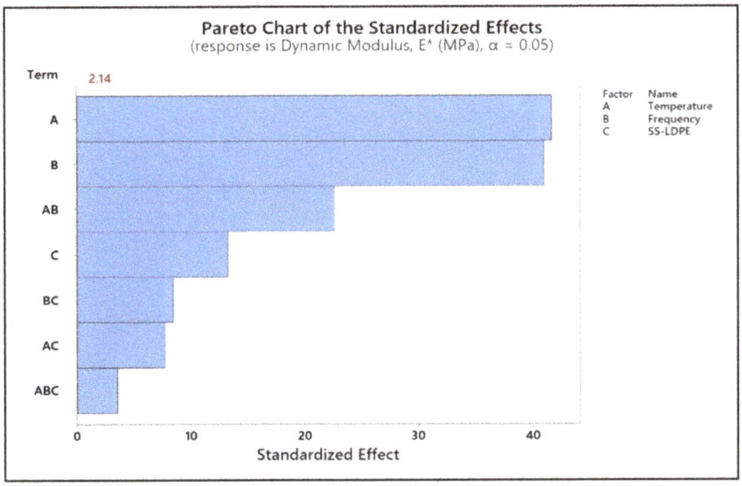

Figure 22. Pareto chart for DM (for—grade 60–70 and SS-LDPE-modified HMA).

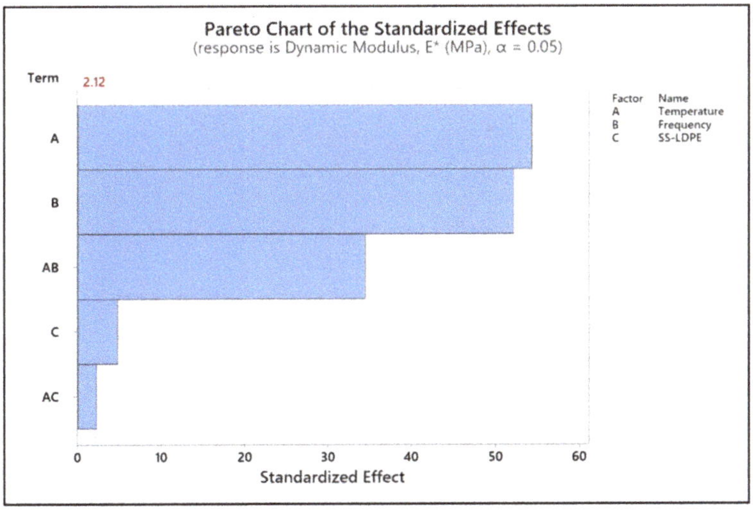

Figure 23. Pareto chart for DM (for—grade 80–100 and SS-LDPE-modified HMA).

The contour plots for the dynamic modulus of grade 60–70 SS-LDPE-modified HMA in terms of low levels of modifier are shown in Figure 26a, and those for high levels are shown in Figure 26b; similarly, the dynamic modulus of grade 80–100 SS-LDPE-modified HMA in terms of low levels of modifier is shown in Figure 26c, and that for high levels is shown in Figure 26d. The dynamic modulus has an inverse relationship with test temperature and a direct relationship with frequency and modifier. Response coefficients can be visualized as a topographical map since they are represented by contour plot lines. These line curves show how temperature and frequency interact at high and low levels. By drawing a line perpendicular to the vertical axis at each temperature point and maintaining a constant frequency, it is possible to determine the dynamic response of the model. At any frequency, keeping the temperature constant can also give us the response dynamic modulus; the dynamic modulus was anticipated at every frequency level.

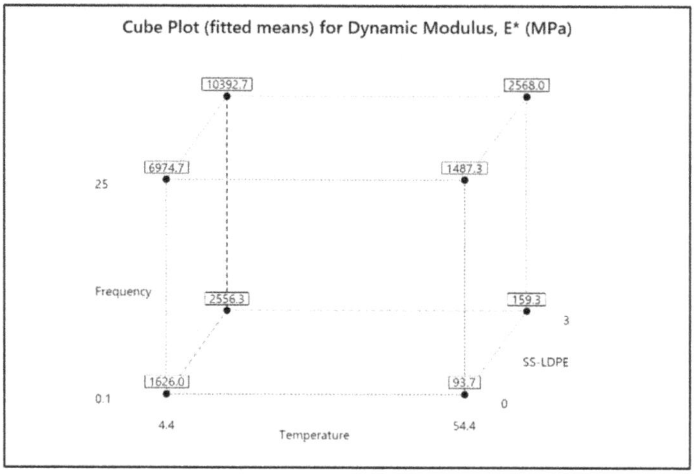

Figure 24. Cube plot for DM (for—grade 60–70 and SS-LDPE-modified HMA).

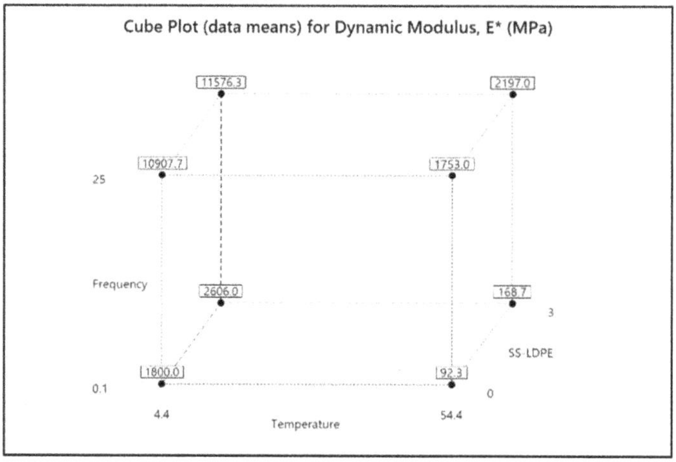

Figure 25. Cube plot for DM (for—grade 80–100 and SS-LDPE-modified HMA).

Figure 27a displays the surface plot for the dynamic modulus of grade 60–70 SS-LDPE-modified HMA at a low level of modifier, while Figure 27b displays it at a high level. Similarly, Figure 27c displays the surface plot for the dynamic modulus of grade 80–100 SS-LDPE-modified HMA at a low level of modifier, while Figure 27d displays it at a high level. The surface plot provides a 3D view of the design, combining the response variable, dynamic modulus, with other factors like temperature and frequency. From the plots, at lower temperature and higher frequency levels, the response gives us the highest dynamic modulus, and at higher temperature and lower frequency levels, the response gives us the lowest dynamic modulus. These plots are used to obtain any response value by choosing random values of the given components by simply drawing a line straight from frequency and temperature to the axis of the |E*| (DM), which are not specified in the dataset of lab-evaluated test values.

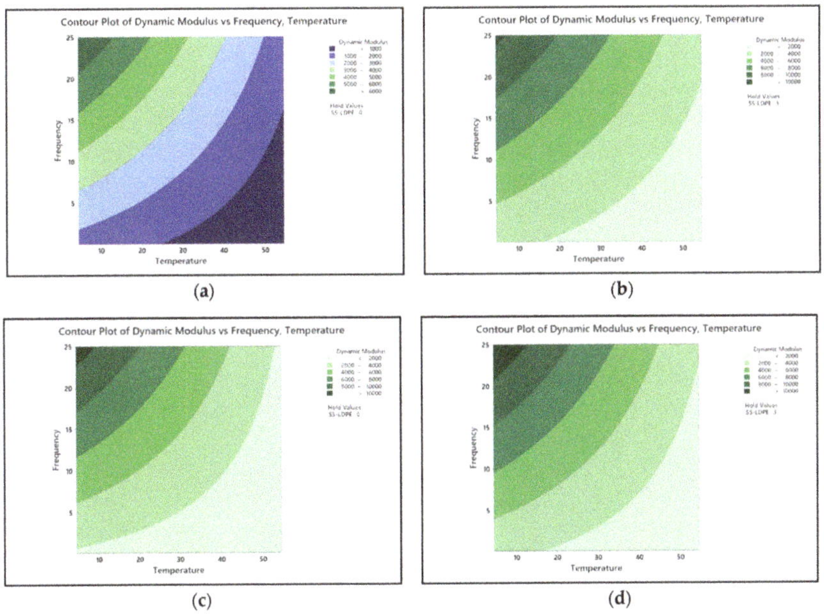

Figure 26. Contour plot of DM of SS-LDPE-modified HMA. (**a**) Contour plot of DM at low level—grade 60–70 SS-LDPE-modified HMA. (**b**) Contour plot of DM at high level—grade 60–70 SS-LDPE-modified HMA. (**c**) Contour plot of DM at low level—grade 80–100 SS-LDPE-modified HMA. (**d**) Contour plot of DM at high level—grade 80–100 SS-LDPE-modified HMA.

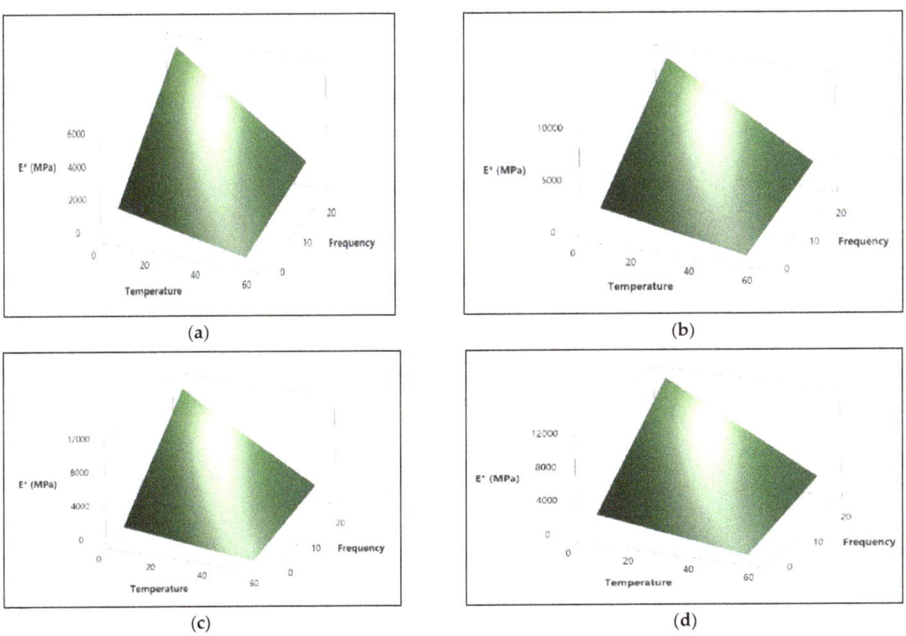

Figure 27. Surface plot of DM for SS-LDPE-modified HMA. (**a**) Surface plot of DM at low level—grade 60–70 SS-LDPE-modified HMA. (**b**) Surface plot of DM at high level—grade 60–70 SS-LDPE-modified HMA. (**c**) Surface plot of DM at low level—grade 80–100 SS-LDPE-modified HMA. (**d**) Surface plot of DM at high low level—grade 80–100 SS-LDPE-modified HMA.

4.5. Evolutionary Algorithm for Dynamic Modulus Using CI

Multi-expression programming was adopted in this study to examine genetic programming using a Python dataset to encode multiple parameters into the same chromosome in computational language, leading to an evolutionary computation algorithm. Performance is a measure of how the pavement responds to several factors, such as loading rate, pavement temperature, atmospheric condition, etc. Pavements need to perform as expected during the design phase to guarantee their long-term viability. The test temperature, loading frequency, and modifier content (such as low-density polyethylene and steel slag) all influence the dynamic modulus, a performance measure.

$$Dynamic\ Modulus = f\,(temperature, frequency,\ modifier) \quad (1)$$

The data from Equation (1) were utilized to create a model that can estimate the material's dynamic modulus for values other than the test dataset. Based on the trial-and-error process, different models were produced, namely regression models using Minitab and MS Excel, Cobb–Douglas models, Witczak models, and Hirsch models. However, there was a drawback in each model, as it overpredicted the dynamic modulus values when model validation was performed on each. As a result, multi-expression programming was utilized, using computational regression intelligence models that best represented the data, with a training best error below 100 for accuracy and targeted output, as shown in Figure 28a,b for grade 60–70 LDPE and 80–100 LDPE-modified HMA, respectively.

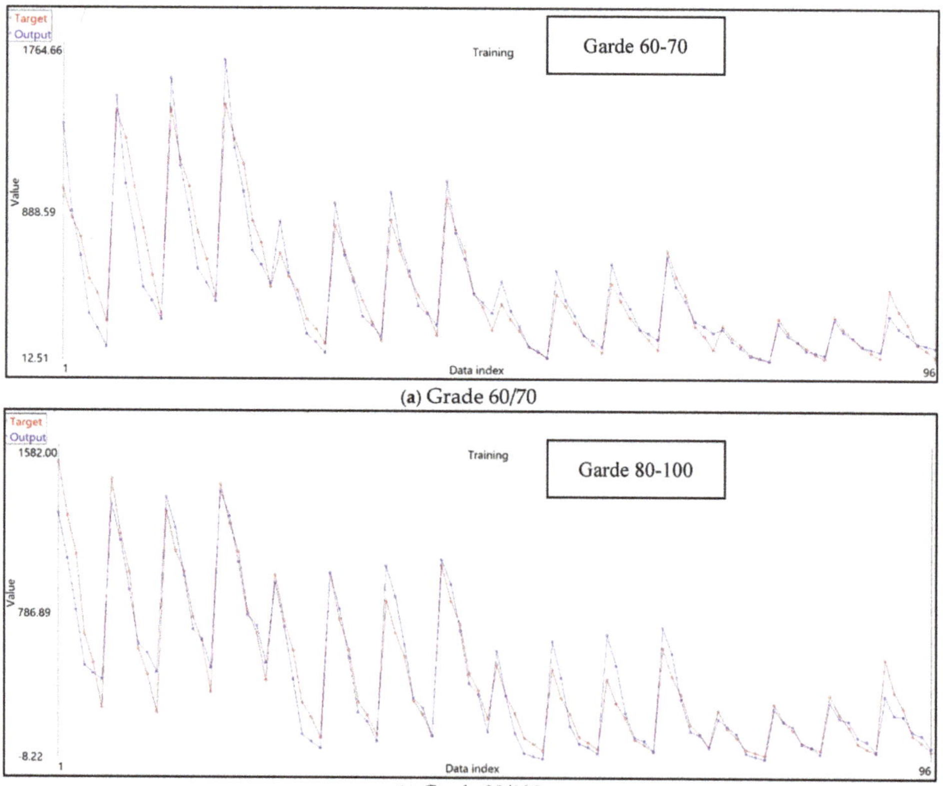

(**a**) Grade 60/70

(**b**) Grade 80/100

Figure 28. (**a**,**b**) MEP training best error for DM for LDPE-modified HMA.

Grade 60–70 bitumen has a relatively lower viscosity compared to grade 80–100, which may have led to inconsistent interaction between the LDPE modifier and the bitumen, especially at lower concentrations. This could have resulted in variations in the mixture's rheological behavior and contributed to the observed error. Moreover, LDPE's modifying effect is dependent on its interaction with the base bitumen. For grade 60–70 bitumen, which has a different chemical composition compared to grade 80–100, the modification may not have been as effective in improving the performance characteristics, leading to higher deviation from experimental data. Furthermore, the computational intelligence models may have shown increased sensitivity to changes in the properties of LDPE-modified HMA at the lower bitumen grade, as the interaction between the variables could have been more complex or nonlinear for grade 60–70.

For generating the model using MEP, different parameters were set, such as the number of subpopulations (10), subpopulation size (1000), code length (30), and number of generations (2000). The obtained result was encoded in Python, and it was decoded to obtain the models for grade 60–70 with a training error of 72.44 and those for grade 80–100 with a training error of 68.18.

The final version of the grade 60–70 LDPE-modified HMA model can be written as Equation (2):

$$|E^*| = 2 \times \left[132\sqrt{2F} + 35M - 16\sqrt{2TF} - 4M\sqrt{T}\right] \tag{2}$$

The final version of the grade 80–100 LDPE-modified HMA model can be written as Equation (3):

$$|E^*| = \left|24F + 12M - 11T - \frac{M}{F} + 258\right| - \left[2F^2 + FM - 62F - 31M + 11T + 2\frac{M}{F} - 246\right] \tag{3}$$

where $|E^*|$ = Projected Absolute Dynamic Modulus (ksi); M = Added Modifier (LDPE), % (0 to 7); F = loading frequency of SPT, Hz (0.1 to 25); and T = test temperature, °C (4.4 to 54.4).

Multi-Expression Programming Model Validation

Model validation was performed on LDPE-modified HMA models produced using the MEP computing algorithm. In our study, the mean absolute percentage error (MAPE) technique was adopted for checking model validation. MAPE is the difference between the lab-tested (actual) dynamic modulus and the calculated or projected (predicted) dynamic modules, divided mathematically by the actual dynamic modulus, then extracted as absolute values of all the differences, taking the mean of the overall sum, and finally converted into a percentage. Equation (4) for yielding the mean absolute percentage error is given below.

$$MAPE = \frac{100\%}{n} \sum_{i=1}^{n} \left| \frac{A_t - P_t}{A_t} \right| \tag{4}$$

where A_t = Actual Dynamic Modulus and P_t = Predicted Dynamic Modulus.

The MAPE values for the grade 60–70 LDPE-modified HMA and the grade 80–100 LDPE-modified HMA models are 23.42% and 5.15%, respectively. The MEP models are valid through MAPE, but in the case of grade 60–70, the model shows some weak response but is still acceptable; in the case of grade 80–100, the model represents a strong relationship with the data. The model's prediction shows a weak response when there is temperature involved in the affecting factors, which is true for this case study as well. Table 15 shows the MAPE score and accuracy of the model.

Table 15. Interpretation of MAPE score.

MAPE Score (%)	Model Accuracy (%)	Interpretation of Score
<10	>90	Excellent Model
10–20	80–90	Good Model
20–50	50–80	Relatively Good Model
>50	<50	Poor Model

Figure 29 shows a grade 60–70 LDPE-modified HMA validation plot, and Figure 30 demonstrates a grade 80–100 LDPE-modified HMA validation plot. The plots with 45-degree lines show the validation of the dataset; the closer the dataset points to the 45-degree line, the more accurately the model represents the data, so Figures 29 and 30 indicate that the models are reliable and valid. However, the developed MEP model for the grade 60–70 LDPE-modified HMA shows a slight deviation from its 45-degree line, with dotted lines denoting the trend line below the solid 45-degree line.

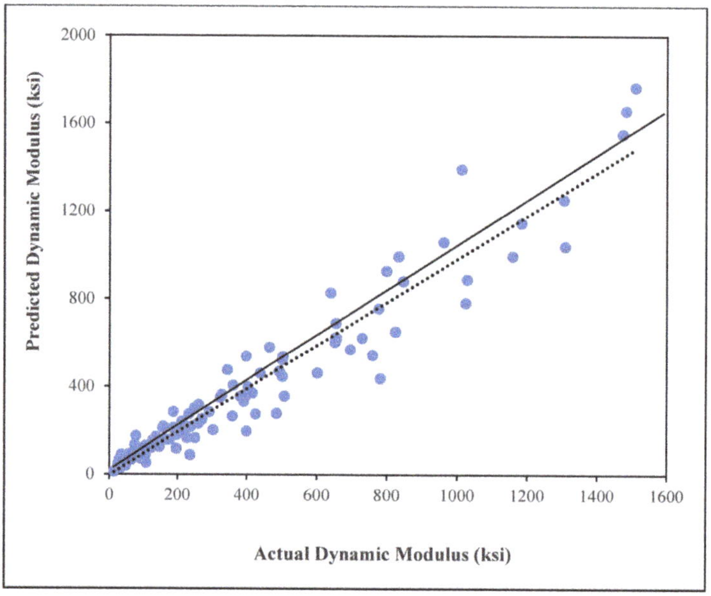

Figure 29. Validation plot—grade 60–70 LDPE-modified HMA.

4.6. KENPAVE Damage Analysis

An analysis of hot-mix asphalt (HMA) pavements using the mechanistic–empirical method with KENPAVE 1.0 software was conducted in this study. The primary objective was to evaluate the effect of various modifiers on the design life of flexible pavements. Dynamic modulus values were incorporated, accounting for seasonal variations in modulus at a constant frequency.

Two bitumen grades, 60–70 and 80–100, were analyzed with modifiers, including LDPE at 3%, 5%, and 7%, and 3%-SS for each grade. Dynamic modulus values for these grades, along with the modifiers, were evaluated across three temperature ranges (4.4 °C, 21.1 °C, and 54.4 °C) at a constant frequency of 5 Hz during KENLAYER analysis. The loading configuration included single-axle dual-tire trucks with an allowable load repetition of 50,000 cycles, a contact pressure of 100 psi, a dual spacing of 13 inches, and a contact radius of 4.2 inches.

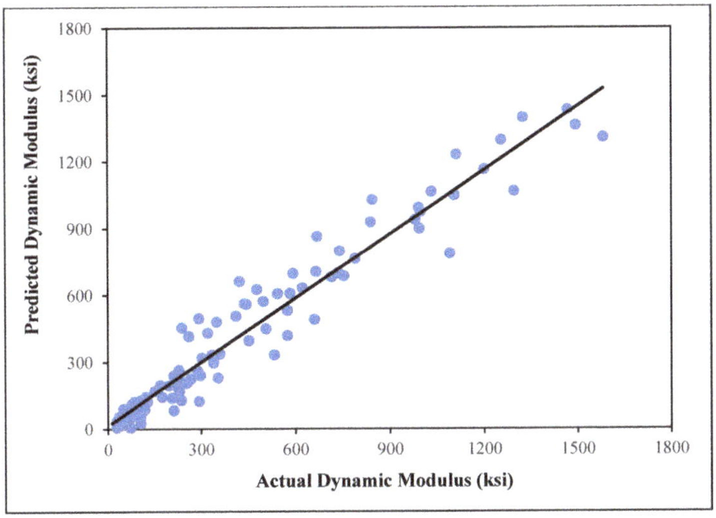

Figure 30. Validation plot—grade 80–100 LDPE-modified HMA.

For the damage analysis, two critical points were examined, i.e., at the bottom of the HMA layer and at the top of the subgrade. Table 16 presents the results of the damage analysis, design life, and percentage improvement for bitumen grades 60–70 and 80–100 due to LDPE and SS modifications.

Table 16. KENPAVE analysis of HMA samples.

Period of Year	Description	Percent of Modifier in Bitumen Grade 60–70					Percent of Modifier in Bitumen Grade 80–100				
		0	3	5	7	3-SS	0	3	5	7	3-SS
Number 1 (Temp: 4.4 °C; Freq: 5 Hz)	Dynamic Modulus (ksi)	729	1023	1027	1157	1113	1090	993	996	1105	1257
	Allowable Load at Asphalt Bottom (N_f)	1,498,000	2,154,000	2,163,000	2,487,000	2,375,000	2,318,000	2,082,000	2,089,000	2,355,000	2,750,000
	Damage Ratio	0.03337	0.02322	0.02311	0.0201	0.02105	0.02157	0.02402	0.02393	0.02123	0.01818
	Allowable Load Repetition at Subgrade Top (N_d)	645,200	1,119,000	1,127,000	1,387,000	1,296,000	1,249,000	1,064,000	1,069,000	1,279,000	1,611,000
	Damage Ratio	0.07749	0.04467	0.04438	0.03604	0.03859	0.04003	0.047	0.04676	0.03908	0.03104
Number 2 (Temp: 21.1 °C; Freq: 5 Hz)	Dynamic Modulus (ksi)	417	494	504	651	598	574	575	543	714	601
	Allowable Load at Asphalt Bottom (N_f)	931,000	1,057,000	1,074,000	1,343,000	1,242,000	1,198,000	1,200,000	1,142,000	1,468,000	1,249,000
	Damage Ratio	0.0537	0.04731	0.04655	0.03723	0.04024	0.04173	0.04167	0.04377	0.03406	0.04003
	Allowable Load Repetition at Subgrade Top (N_d)	295,300	368,400	378,500	543,900	480,500	453,200	454,300	419,200	625,000	484,600
	Damage Ratio	0.1693	0.1357	0.1321	0.09193	0.1041	0.1103	0.1101	0.1193	0.08	0.1032
Number 3 (Temp: 54.4 °C; Freq: 5 Hz)	Dynamic Modulus (ksi)	105	123	146	227	184	107	151	169	267	175
	Allowable Load at Asphalt Bottom (N_f)	593,200	583,300	586,100	656,700	611,400	591,400	588,200	599,200	708,000	603,600
	Damage Ratio	0.08429	0.08572	0.08531	0.07613	0.08177	0.08455	0.085	0.08344	0.07062	0.08284
	Allowable Load Repetition at Subgrade Top (N_d)	74,330	84,560	97,870	148,200	120,700	75,460	100,800	111,600	175,500	115,000
	Damage Ratio	0.6727	0.5913	0.5109	0.3374	0.4142	0.6626	0.496	0.4482	0.2848	0.4347
Summary	Sum of Damage Ratio (Fatigue)	0.17136	0.15625	0.15497	0.13346	0.14306	0.14785	0.15069	0.15114	0.12591	0.14105
	Sum of Damage Ratio (Deformation)	0.91949	0.77167	0.68738	0.46537	0.55689	0.81293	0.6531	0.61426	0.40388	0.56894
	Summative Damage (F.C + P.D)	1.09085	0.92792	0.84235	0.59883	0.69995	0.96078	0.80379	0.7654	0.52979	0.70999
	Percentage Reduction in Damage Ratio	0	14.94	22.78	45.10	35.83	0	16.34	20.34	44.86	26.10
	Design Life (Years)	1.09	1.3	1.45	2.15	1.8	1.23	1.53	1.63	2.48	1.76
	Percentage Improvement	0	19.27	33.03	97.25	65.14	0	24.39	32.52	101.63	43.09

4.6.1. Reduction in Damage Ratios

The allowable load repetition at the bottom of the HMA (Nf) and the top of the subgrade (Nd) for each season at a single vehicle speed was used to compute the damage ratios. The damage ratios of modifiers with control samples were taken for both grades 60–70 and 80–100 HMA. The result in Table 16 shows that in the case of all modifiers, damage ratios were significantly reduced as compared to the control sample.

4.6.2. Percentage Improvement in Design Life

The design life of pavement in years was calculated from the highest damage ratio among the amounts of fatigue cracking at the bottom of the HMA and the permanent deformation at the top of the subgrade. From Table 16, the design life of a pavement can be significantly improved by adding LDPE and SS to the HMA.

5. Conclusions

This study evaluates the combined effects of low-density polyethylene (LDPE) at varying percentages (3%, 5%, and 7%) as a binder modifier and steel slag (SS) as a coarse aggregate replacement to enhance hot-mix asphalt (HMA) performance for durable and sustainable road construction. The following conclusions are drawn:

- LDPE and steel slag (SS) modifications significantly improved the performance of hot-mix asphalt (HMA) for both bitumen grades (60–70 and 80–100), with notable upgrades in physical and mechanical properties and Performance Grade (PG).
- SS exhibited superior abrasion resistance and crushing strength compared to conventional aggregates.
- LDPE-modified bitumen improved penetration and softening point values, enhancing high-temperature performance and upgrading the PG.
- The 3% LDPE yielded optimal results by balancing stiffness, stability, and ductility.
- Specimens with 3% LDPE showed improvements in Marshall Stability by 15.57% for grade 60–70 and 13.18% for grade 80–100, with minimal reductions in air voids and unit weight.
- Dynamic modulus values were higher across all frequencies and temperatures, reducing fatigue cracking and permanent deformation by 14.94% to 16.34%.
- Combining SS with 3% LDPE improved Marshall Stability by over 31%, reduced damage ratio by up to 35.83%, and significantly enhanced the design life of HMA.
- The design life of grade 60–70 HMA increased by 65.14%, while grade 80–100 saw a 43.09% improvement.
- Damage ratio reductions were observed as 35.83% for grade 60–70 and 26.10% for grade 80–100.
- Dynamic modulus improvements were observed across all temperature ranges, especially at extremely low and high temperatures.
- These findings confirm the potential of SS and LDPE modifications for durable and sustainable flexible pavement solutions.

6. Recommendations

This study recommends employing 3% LDPE as a bitumen modifier in both 60–70 and 80–100 grades. This target percentage of the LDPE modifier improved overall performance while slightly lowering some parameters compared to the control sample. Similarly, the same LDPE modifier percentage has good stability values, with the best dynamic modulus at a lower frequency and higher temperature ranges and ultimately increases the damage resistance of asphaltic pavements according to results from the KENPAVE software. Ultimately, the adoption of LDPE as a binder modifier and steel slag (SS) as a substitute for coarse aggregates in hot-mix asphalt enhanced its performance across all frequency levels and temperature ranges, particularly at the highest and lowest temperatures. This study strongly suggests using this approach for roads in Pakistan with higher temperature zones, aiming to harness the steel industry's waste, reduce plastic waste accumulation, and prolong the pavement's design life.

7. Limitations

The limitations of utilizing recycled steel slag (SS) and LDPE for pavement construction can be summarized in the following points:

- Waste Material Compatibility: The physical and chemical properties of LDPE and SS vary depending on the origin and nature of the materials and may be subject to volume expansion resulting from moisture and carbonation, causing volume instability that affects their bonding with the bitumen binder and may therefore cause cracks or deformation in the pavement.
- Cost Implications: Recycling materials may appear cost-effective, but the expenses of processing, shipping, transportation, and quality control might exceed the advantages in certain circumstances.
- Environmental Issues: SS may contain trace heavy metals, presenting possible leaching hazards. LDPE, being a plastic substance, may deteriorate into microplastics under certain circumstances, posing environmental concerns.
- Lab-Scale Testing: The findings are based on laboratory tests, and field validation under real-world conditions is needed for broader adoption.

8. Future Scope of Work

Future research on the integration of recycled steel slag (SS) and low-density polyethylene (LDPE) in hot-mix asphalt (HMA) for sustainable pavement solutions should focus on optimizing the blend composition for enhanced mechanical performance and long-term durability. Advanced experimental studies can investigate the microstructural interactions between SS and LDPE, aiming to improve stability and bonding. Life cycle assessment (LCA) and cost–benefit analysis should be conducted to establish the economic and environmental feasibility of scaling up this approach for widespread implementation. Furthermore, field-scale trials and performance monitoring under varying traffic loads and climatic conditions would validate their practical applicability. Exploring the use of other industrial by-products alongside SS and LDPE in HMA could further enhance sustainability and performance metrics. The following points are suggested for further studies:

- More research and cooperation with bitumen plants and road agencies are required to conserve industrial resources by replacing aggregates with different waste materials to achieve sustainability and lower costs.
- Further testing on HMA mixes, such as four-point beam fatigue, semi-circular bending, fuel leakage, Hamburg wheel truck, skid resistance, abrasion, and other features, should be carried out.
- Computational intelligence based on advanced computational methods and predictive models should be used to optimize pavement design further.

Author Contributions: Conceptualization, S.A.M. and M.I.K.; methodology, S.A.M. and S.A.; software, S.A.M.; validation, S.A.M., M.I.K. and S.A.; formal analysis, S.A.M. and R.A.-N.; investigation, S.A.M. and M.I.K.; resources, S.A. and R.M.C.; data curation, S.A.M.; writing—original draft preparation, S.A.M.; writing—review and editing, M.I.K., S.A., R.M.C. and R.A.-N.; visualization, S.A.M.; supervision, S.A. and M.I.K.; project administration, S.A. and R.M.C.; funding acquisition, R.M.C. All authors have read and agreed to the published version of the manuscript.

Funding: This research received no external funding.

Data Availability Statement: The original contributions presented in this study are included in the article. Further inquiries can be directed at the corresponding author.

Acknowledgments: The authors acknowledge the support of the National University of Sciences & Technology (NUST), Islamabad, Pakistan, and Imam Mohammad Ibn Saud Islamic University (IMSIU), Riyadh, Saudi Arabia, for completing this study.

Conflicts of Interest: The authors declare no conflicts of interest.

References

1. Yu, H.; Zhu, Z.; Zhang, Z.; Yu, J.; Oeser, M.; Wang, D. Recycling waste packaging tape into bituminous mixtures towards enhanced mechanical properties and environmental benefits. *J. Clean. Prod.* **2019**, *229*, 22–31. [CrossRef]
2. Al-Nawasir, R.I.; Al-Humeidawi, B.H. Qualitative Evaluation for Asphalt Binder Modified with SBS Polymer. *Tikrit J. Eng. Sci.* **2023**, *30*, 88–101. [CrossRef]
3. Heydari, S.; Hajimohammadi, A.; Haji Seyed Javadi, N.; Khalili, N. The use of plastic waste in asphalt: A critical review on asphalt mix design and Marshall properties. *Constr. Build. Mater.* **2021**, *309*, 125185. [CrossRef]
4. Yadykova, A.Y.; Ilyin, S.O. Nanocellulose-stabilized bitumen emulsions as a base for preparation of nanocomposite asphalt binders. *Carbohydr. Polym.* **2023**, *313*, 120896. [CrossRef] [PubMed]
5. Xue, H.; Zhu, H.; Guo, M.; Shao, S.; Zhang, S.; Zhang, Y. Modeling for predicting triaxial mechanical properties of recycled aggregate concrete considering the recycled aggregate replacement. *Constr. Build. Mater.* **2023**, *368*, 130447. [CrossRef]
6. Zhao, H.; Wang, S.; Wang, R.; Shen, L.; Wang, Q. Utilization of raw coal gangue as coarse aggregates in pavement concrete. *Constr. Build. Mater.* **2023**, *378*, 131062. [CrossRef]
7. Aldagari, S.; Hung, A.M.; Shariati, S.; Faisal Kabir, S.K.; Ranka, M.; Bird, R.C.; Fini, E.H. Enhanced sustainability at the bitumen-aggregate interface using organosilane coating technology. *Constr. Build. Mater.* **2022**, *359*, 129500. [CrossRef]
8. Al-Nawasir, R.; Al-Humeidawi, B.; Khan, M.I.; Khahro, S.H.; Memon, Z.A. Effect of glass waste powder and date palm seed ash based sustainable cementitious grouts on the performance of semi-flexible pavement. *Case Stud. Constr. Mater.* **2024**, *21*, e03453. [CrossRef]
9. Quan, X.; Wang, S.; Li, J.; Luo, J.; Liu, K.; Xu, J.; Zhao, N.; Liu, Y. Utilization of molybdenum tailings as fine aggregate in recycled aggregate concrete. *J. Clean. Prod.* **2022**, *372*, 133649. [CrossRef]
10. Caputo, P.; Algieri, V.; Maiuolo, L.; De Nino, A.; Sicilia, E.; Ponte, F.; Calandra, P.; Oliviero Rossi, C. Waste additives as biopolymers for the modification of bitumen: Mechanical performance and structural analysis characterization. *Colloids Surf. A Physicochem. Eng. Asp.* **2023**, *663*, 131079. [CrossRef]
11. Asi, I.M.; Qasrawi, H.Y.; Shalabi, F.I. Use of steel slag aggregate in asphalt concrete mixes. *Can. J. Civ. Eng.* **2007**, *34*, 902–911. [CrossRef]
12. Arisha, A.M.; Gabr, A.R.; El-Badawy, S.M.; Shwally, S.A. Performance Evaluation of Construction and Demolition Waste Materials for Pavement Construction in Egypt. *J. Mater. Civ. Eng.* **2018**, *30*, 04017270. [CrossRef]
13. Khorshidi, M.; Goli, A.; Orešković, M.; Khayambashi, K.; Ameri, M. Performance Evaluation of Asphalt Mixtures Containing Different Proportions of Alternative Materials. *Sustainability* **2023**, *15*, 13314. [CrossRef]
14. Kazmee, H.; Tutumluer, E.; Beshears, S. Using Accelerated Pavement Testing to Evaluate Reclaimed Asphalt Pavement Materials for Pavement Unbound Granular Layers. *J. Mater. Civ. Eng.* **2017**, *29*, 04016205. [CrossRef]
15. Arulrajah, A.; Piratheepan, J.; Aatheesan, T.; Bo, M.W. Geotechnical Properties of Recycled Crushed Brick in Pavement Applications. *J. Mater. Civ. Eng.* **2011**, *23*, 1444–1452. [CrossRef]
16. Vo, H.V.; Park, D.-W. Lightweight Treated Soil As a Potential Sustainable Pavement Material. *J. Perform. Constr. Facil.* **2016**, *30*, C4014009. [CrossRef]
17. Suresh, M.; Pal, M. Effect of centrifuged latex on moisture damage of hot-mix asphalt with brick aggregate. *Proc. Inst. Civ. Eng. Transp.* **2021**, *174*, 99–109. [CrossRef]
18. Shen, D.-H.; Wu, C.-M.; Du, J.-C. Laboratory investigation of basic oxygen furnace slag for substitution of aggregate in porous asphalt mixture. *Constr. Build. Mater.* **2009**, *23*, 453–461. [CrossRef]
19. Chen, Z.; Wu, S.; Xiao, Y.; Zeng, W.; Yi, M.; Wan, J. Effect of hydration and silicone resin on Basic Oxygen Furnace slag and its asphalt mixture. *J. Clean. Prod.* **2016**, *112*, 392–400. [CrossRef]
20. Paul, D.; Suresh, M.; Pal, M. Utilization of fly ash and glass powder as fillers in steel slag asphalt mixtures. *Case Stud. Constr. Mater.* **2021**, *15*, e00672. [CrossRef]
21. Zhang, C.; Tan, Y.; Gao, Y.; Fu, Y.; Li, J.; Li, S.; Zhou, X. Resilience assessment of asphalt pavement rutting under climate change. *Transp. Res. Part D Transp. Environ.* **2022**, *109*, 103395. [CrossRef]

22. Liu, S.; Qi, X.; Shan, L. Effect of molecular structure on low-temperature properties of bitumen based on molecular dynamics. *Constr. Build. Mater.* **2022**, *319*, 126029. [CrossRef]
23. Alhaji, M.M.; Alhassan, M.; Adejumo, T.W.; Abdulkadir, H. Road pavement collapse from overloaded trucks due to traffic diversion: A case study of Minna-Kateregi-Bida Road, Nigeria. *Eng. Fail. Anal.* **2022**, *131*, 105829. [CrossRef]
24. Zhao, W.; Yang, Q. Life cycle assessment and multi-index performance evaluation of semi-flexible pavement after composite modification by using fly ash, rubber particles, warm mixing asphalt and recycled asphalt pavement. *Constr. Build. Mater.* **2023**, *364*, 129945. [CrossRef]
25. Hasan, A.; Hasan, U.; Whyte, A.; Al Jassmi, H. Lifecycle Analysis of Recycled Asphalt Pavements: Case Study Scenario Analyses of an Urban Highway Section. *Civil Eng* **2022**, *3*, 242–262. [CrossRef]
26. Yang, J.; Liu, L.; Zhang, G.; Ding, Q.; Sun, X. The Preparation of Ground Blast Furnace Slag-Steel Slag Pavement Concrete Using Different Activators and Its Performance Investigation. *Buildings* **2023**, *13*, 1590. [CrossRef]
27. Kakar, M.R.; Mikhailenko, P.; Piao, Z.; Bueno, M.; Poulikakos, L. Analysis of waste polyethylene (PE) and its by-products in asphalt binder. *Constr. Build. Mater.* **2021**, *280*, 122492. [CrossRef]
28. Khan, M.I.; Sutanto, M.H.; Sunarjono, S.; Room, S.; Yusoff, N.I.M. Effect of Crumb Rubber, Epolene (EE-2), and Date Palm Ash as Modifiers on the Performance of Binders and Mixtures: A Sustainable Approach. *Sustainability* **2019**, *11*, 6484. [CrossRef]
29. Khan, N.; Sutanto, M.H.; Mohammed, B.S.; Yousafzai, A.K.; Khan, M.I.; Memon, A.M.; Al-Nawasir, R. Thermal Performance Evaluation of PCM-Impregnated Aggregates in Hot Mix Asphalt: Mitigating Urban Heat Island Effects. *J. Adv. Res. Fluid Mech. Therm. Sci.* **2024**, *124*, 1–12. [CrossRef]
30. Hossein Hamedi, G.; Ghalandari Shamami, K.; Mazhari Pakenari, M. Effect of ultra-high-molecular-weight polyethylene on the performance characteristics of hot mix asphalt. *Constr. Build. Mater.* **2020**, *258*, 119729. [CrossRef]
31. Biswas, A.; Goel, A.; Potnis, S. Performance comparison of waste plastic modified versus conventional bituminous roads in Pune city: A case study. *Case Stud. Constr. Mater.* **2020**, *13*, e00411. [CrossRef]
32. Ullah, S.; Raheel, M.; Khan, R.; Tariq Khan, M. Characterization of physical & mechanical properties of asphalt concrete containing low- & high-density polyethylene waste as aggregates. *Constr. Build. Mater.* **2021**, *301*, 124127. [CrossRef]
33. Zhang, Y.; Song, Q.; Lv, Q.; Wang, H. Influence of different polyethylene wax additives on the performance of modified asphalt binders and mixtures. *Constr. Build. Mater.* **2021**, *302*, 124115. [CrossRef]
34. Zhong, H.; Huang, W.; Yan, C.; Zhang, Y.; Lv, Q.; Sun, L.; Liu, L. Investigating binder aging during hot in-place recycling (HIR) of asphalt pavement. *Constr. Build. Mater.* **2021**, *276*, 122188. [CrossRef]
35. Ding, X.-H.; Luo, B.; Zhou, H.-T.; Chen, Y.-H. Generalized solutions for advection–dispersion transport equations subject to time-and space-dependent internal and boundary sources. *Comput. Geotech.* **2025**, *178*, 106944. [CrossRef]
36. Su, Y.; Luo, B.; Luo, Z.; Xu, F.; Huang, H.; Long, Z.; Shen, C. Mechanical characteristics and solidification mechanism of slag/fly ash-based geopolymer and cement solidified organic clay: A comparative study. *J. Build. Eng.* **2023**, *71*, 106459. [CrossRef]
37. Gan, Y.; Li, C.; Ke, W.; Deng, Q.; Yu, T. Study on pavement performance of steel slag asphalt mixture based on surface treatment. *Case Stud. Constr. Mater.* **2022**, *16*, e01131. [CrossRef]
38. Rashed, A.M.; Al-Hadidy, A. Comparative performance of DG mixes and SMA mixes with waste crumb rubber as aggregate replacement. *Case Stud. Constr. Mater.* **2023**, *19*, e02615. [CrossRef]
39. AL-Ghurabi, S.B.; Al-Humeidawi, B.H. Evaluation of performance of hot mix asphalt contained various sizes of Reclaim asphalt pavement and polymer modified Bitumen. *Mater. Today Proc.* **2023**, *80*, 2905–2910. [CrossRef]
40. Kang, A.; Jiang, Y.; Kou, C.; Xiao, P.; Chen, J. Performance evaluation of hot mix asphalt containing recycled concrete aggregates treated by physical and chemical methods. *Constr. Build. Mater.* **2024**, *432*, 136649. [CrossRef]
41. Silvestre, R.; Medel, E.; García, A.; Navas, J. Using ceramic wastes from tile industry as a partial substitute of natural aggregates in hot mix asphalt binder courses. *Constr. Build. Mater.* **2013**, *45*, 115–122. [CrossRef]
42. Liu, Z.; Feng, T.; Zhu, X.; Gao, J.; Hu, K.; Guo, M.; Gu, F.; Li, F. Bird's-eye view of recycled solid wastes in road engineering. *J. Road Eng.* **2024**, *4*, 93–150. [CrossRef]
43. Ma, F.; Dai, J.; Fu, Z.; Liu, J.; Dong, W.; Huang, Z. A New type of crumb rubber asphalt mixture: A dry process design and performance evaluation. *Appl. Sci.* **2020**, *10*, 372. [CrossRef]
44. Luo, L.; Yang, S.-H.; Oeser, M.; Liu, P. Moisture damage mechanism of asphalt mixtures containing reclaimed asphalt pavement binder: A novel molecular dynamics study. *J. Clean. Prod.* **2024**, *475*, 143711. [CrossRef]
45. Rout, M.D.; Biswas, S.; Shubham, K.; Sinha, A.K. A systematic review on performance of reclaimed asphalt pavement (RAP) as sustainable material in rigid pavement construction: Current status to future perspective. *J. Build. Eng.* **2023**, *76*, 107253. [CrossRef]
46. Ali, R.I.; Al-Humeidawi, B.H. A scientometric study and a bibliometric review of the literature on the design and construction of semi-flexible pavement. *Al-Qadisiyah J. Eng. Sci.* **2023**, *16*, 82–91. [CrossRef]
47. Al-Nawasir, R.; Al-Humeidawi, B.; Shubbar, A. Influence of Sustainable Grout Material on the Moisture Damage of Semi-flexible Pavement. *Period. Polytech. Civ. Eng.* **2024**, *68*, 961–973. [CrossRef]

48. Loureiro, C.D.; Moura, C.F.; Rodrigues, M.; Martinho, F.C.; Silva, H.M.; Oliveira, J.R. Steel slag and recycled concrete aggregates: Replacing quarries to supply sustainable materials for the asphalt paving industry. *Sustainability* **2022**, *14*, 5022. [CrossRef]
49. Zalnezhad, M.; Hesami, E. Effect of steel slag aggregate and bitumen emulsion types on the performance of microsurfacing mixture. *J. Traffic Transp. Eng. (Engl. Ed.)* **2020**, *7*, 215–226. [CrossRef]
50. Abendeh, R.; Alhorani, R.; Ahmad, H.; Baker, M.B. Effect of steel slag as fine and coarse aggregate on pore structure and freeze-thaw resistance of high-strength concrete. *Jordan J. Civ. Eng.* **2021**, *15*, 4.
51. *ASTM C-131;* Standard Test Method for Resistance to Degradation of Small-Size Coarse Aggregate by Abrasion and Impact in the Los Angeles Machine. ASTM International West: Conshohocken, PA, USA, 2010.
52. *BS 812-112;* Testing Aggregates. Method for Determination of Aggregate Impact Value. British Standards Institution: London, UK, 1990.
53. *BS 812-110;* Methods for Determination of Aggregate Crushing Value. British Standards Institution: London, UK, 1990.
54. *ASTM C-127;* Standard Test Method for Relative Density (Specific Gravity) and Absorption of Coarse Aggregate. ASTM International West: Conshohocken, PA, USA, 2024.
55. *ASTM C-128;* Standard Test Method for Relative Density (Specific Gravity) and Absorption of Fine Aggregate. ASTM International West: Conshohocken, PA, USA, 2017.
56. *AASHTO T-133;* Standard Method of Test for Density of Hydraulic Cement. Highway Transportation Officials: Washington, DC, USA, 2019.
57. *ASTM D-5;* Standard Test Method for Penetration of Bituminous Materials. ASTM International West: Conshohocken, PA, USA, 2019.
58. *ASTM D-36;* Standard Test Method for Softening Point of Bitumen (Ring-and-Ball Apparatus). ASTM International West: Conshohocken, PA, USA, 2014.
59. *ASTM D-70;* Standard Test Method for Density of Semi-Solid Asphalt Binder (Pycnometer Method). ASTM International West: Conshohocken, PA, USA, 2021.
60. *ASTM D-113;* Standard Test Method for Ductility of Asphalt Materials. ASTM International West: Conshohocken, PA, USA, 2023.
61. *ASTM D-15;* Standard Test Method for Preparation and Determination of the Relative Density of Asphalt Mix Specimens by Means of the Superpave Gyratory Compactor. ASTM International West: Conshohocken, PA, USA, 2015.
62. *AASHTO TP 62-07;* Standard Method of Test for Determining Dynamic Modulus of Hot Mix Asphalt (HMA). Highway Transportation Officials: Washington, DC, USA, 2007.

MDPI AG
Grosspeteranlage 5
4052 Basel
Switzerland
Tel.: +41 61 683 77 34

Buildings Editorial Office
E-mail: buildings@mdpi.com
www.mdpi.com/journal/buildings

www.ingramcontent.com/pod-product-compliance
Lightning Source LLC
Chambersburg PA
CBHW040816250926
51850CB00007BA/1183
* 9 7 8 3 7 2 5 8 7 9 4 5 8 *